D1825181

1 MONTH OF
FREE
READING

at

www.ForgottenBooks.com

By purchasing this book you are eligible for one month membership to ForgottenBooks.com, giving you unlimited access to our entire collection of over 1,000,000 titles via our web site and mobile apps.

To claim your free month visit:

www.forgottenbooks.com/free907439

* Offer is valid for 45 days from date of purchase. Terms and conditions apply.

ISBN 978-0-265-90439-8
PIBN 10907439

This book is a reproduction of an important historical work. Forgotten Books uses
state-of-the-art technology to digitally reconstruct the work, preserving the original format
whilst repairing imperfections present in the aged copy. In rare cases, an imperfection in
the original, such as a blemish or missing page, may be replicated in our edition. We do,
however, repair the vast majority of imperfections successfully; any imperfections that
remain are intentionally left to preserve the state of such historical works.

Forgotten Books is a registered trademark of FB &c Ltd.
Copyright © 2018 FB &c Ltd.
FB &c Ltd, Dalton House, 60 Windsor Avenue, London, SW19 2RR.
Company number 08720141. Registered in England and Wales.

For support please visit www.forgottenbooks.com

THE

OURNAL

OF

AL RESE

Continuation of the

OF THE BOSTON SOCIETY OF MEDICAL SCI

THE

JOURNAL

OF

MEDICAL RESEARCH

EDITED BY

HAROLD C. ERNST, M.D.

VOLUME XXV.

(New Series, Vol. XX.)

SEPTEMBER, 1911, TO FEBRUARY, 1912.

Numbers 127, 128, 129

BOSTON
MASSACHUSETTS
U.S.A.

UNIV. OF
CALIFORNIA

TRB1
J6
v. 25
~~BIOLOGY~~
~~LIBRARY~~
Bios

NUMBER 127 . . . pp. 1–261
NUMBER 128 . . . pp. 263–408
NUMBER 129 . . . pp. 409–514

INDEX OF SUBJECTS.

A.

PAGE

Albumin. — The absorption of . . . without digestion - - 399
Amebæ. — Pure cultures of . . . parasitic in mammals - - 263
Appendicitis. — Fatty compounds as a factor in the etiology
of . . . - - - - - - - - - - - - 359
Arteritis. — A study of primary intimal . . . of syphilitic
origin - - - - - - - - - - - - - 85

B.

Bacillus typhosus. — An investigation on the permeability of slow
sand filters to . . . - - - - - - - - - 101
Bacillus typhosus. — The isolation of . . . from butter - - 231
Bacilli. — The rapid isolation of typhoid, paratyphoid and
dysentery . . - - - - - - - - - - 95
Bacteria. — Certain fundamental principles relating to the activity
of . . . in the intestinal tract. Their relation to thera-
peutics - - - - - - - - - - - - - 117
Breath. — Organic matter in the expired . . . - - - - 35
Butter. — The isolation of bacillus typhosus from . . - - 231

C.

Calcification. — Studies on . . . and ossification. IV. - - 373
Cancer. — On the relative local influence of coexisting tuberculous
inflammation and . . . in the lung - - - - - 503
Carcinoma. — . . . involving the entire kidney - - - 239
Cerebellar abscess. — A case of . . . with isolation of Micro-
coccus cereus albus - - - - - - - - - 393

D.

Digestion. — The absorption of albumin without . . . - - 399
Dog. — The metabolism of the hypophysectomized . . . - - 409
Dysentery. — The rapid isolation of typhoid, paratyphoid and
. . . bacilli - - - - - - - - - - 95

E.

Enzyme. — Note on a peptid-splitting . . . in woman's milk - 235

516846

PAGE

F.

Fatty compounds. — . . . . as a factor in the etiology of appendicitis - - - - - - - - - - - - - - 359
Functional activity. — The identity in dog and man of the sequence of changes produced by . . . in the Purkinje cell of the cerebellum - - - - - - - - - - - 285

G.

Ground squirrels. — Tuberculosis among . . . (Citellus Beecheyi, Richardson) - - - - - - - - - 189

H.

"Hormone." — The value of the . . . theory of the causation of new growth - - - - - - - - - - 259
Human tuberculosis. — The relative importance of the bovine and human types of tubercle bacilli in the different forms of . . . 313

I.

Intestinal tract. — Certain fundamental principles relating to the activity of bacteria in the . . . Their relation to therapeutics - - - - - - - - - - - - - 117

M.

Metabolism. — The . . . of the hypophysectomized dog - - 409
Micrococcus cereus albus. — A case of cerebellar abscess with isolation of . . . - - - - - - - - - 393
Milk. — Note on a peptid-splitting enzyme in woman's . . . - 235

N.

New growth. — The value of the "Hormone" theory of the causation of . . . - - - - - - - - - 259

O.

Organic matter. — . . . in the expired breath - - - 35
Ossification. — Studies on calcification and . . . IV. - - 373

P.

Paratyphoid. — The rapid isolation of typhoid . . . , and dysentery bacilli - - - - - - - - - - 95
Purkinje cell. — The identity in dog and man of the sequence of changes produced by functional activity in the . . . of the cerebellum - - - - - - - - - - - 285

PAGE

R.

Rats. — Notes on twenty-two spontaneous tumors in wild . . .
(M. Norvegicus) - - - - - - - - - - - 205
Reaction curve. — The . . . in glycerin broth as an aid in
differentiating the bovine from the human type of tubercle
bacillus - - - - - - - - - - - - - 335
Renal epithelium. — A note on the regeneration of . . . in the
intact cat kidney - - - - - - - - - - - 369

S.

Sand filters. — An investigation on the permeability of slow
. . . to bacillus typhosus - - - - - - - - 101
Sugar. — The isolation of typhoid bacilli from urine and feces with
the description of a new double . . . tube medium - - 217
Syphilis. — Precipitation tests for . . . - - - - - 199

T.

Thrombo-angitis obliterans. — A study of a case of . . . - 247
Tubercle bacillus. — The reaction curve in glycerin broth as an
aid in differentiating the bovine from the human type
of . . . - - - - - - - - - - - - 335
Tubercle bacilli. — The relative importance of the bovine and
human types of . . . in the different forms of human tuber-
culosis - - - - - - - - - - - - - 313
Tuberculosis. — . . . among ground squirrels (Citellus Bee-
cheyi, Richardson) - - - - - - - - - - 189
Tuberculosis. — The vaccination of cattle against . . . - - 1
Tuberculous inflammation. — On the relative local influence of
coexisting . . . and cancer in the lung - - - - 503
Tumors. — Notes on twenty-two spontaneous . . . in wild rats
(M. Norvegicus) - - - - - - - - - - 205
Typhoid. — The rapid isolation of . . . , paratyphoid and
dysentery bacilli - - - - - - - - - - 95
Typhoid bacilli. — The isolation of . . . from urine and feces
with the description of a new double sugar tube medium - - 217

V.

Vaccination. — The . . . of cattle against tuberculosis. II. - 1

INDEX OF NAMES.

PAGE

A.

Amoss, Harold L. 35
Anthony, Bertha Van Houten 359

B.

Beasley, Edward B. 101
Benedict, Francis G. 409
Bergey, D. H. 231
Brooks, Harlow 247

C.

Chapin, Charles W. 189

D.

Day, Alexander A. 95
Dolley, David H. 285

G.

Grant, P. A. 399
Grund, M. 335
Gurd, Fraser B. 85

H.

Henry, Gladys R. 373
Holmes, Harriet F. 373
Homans, John 409

K.

Karsner, Howard T. 393
Kendall, Arthur I. 95, 117
Krumwiede, Charles, Jr. 313

L.

Levin, I. 259

M.

MacNider, Wm. de B. 369
McCoy, George W. 189
Milne, Lindsay S. 239

PAGE

O.

Oertel, Horst 503

P.

Park, Wm. H. 313

R.

Rosenau, Milton J. 35
Russell, F. F. 217

S.

Sittenfield, M. J. 259
Smith, Theobald 1
Strong, Lawrence W. 199

V.

Van Alstyne, V. N. 399

W.

Wade, H. W. 85
Warfield, Louis M. 235
Wells, H. Gideon 373
Wherry, Wm. B. 205
Williams, Anna W. 263
Woolley, Paul G. 205

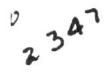

THE
Journal of Medical Research.

(New Series, Volume XX.)

Vol. XXV., No. 1.   SEPTEMBER. 1911.   Whole No. 127.

THE VACCINATION OF CATTLE AGAINST TUBERCULOSIS. — II.*

THE PATHOGENIC EFFECT OF CERTAIN CULTURES OF THE HUMAN TYPE ON CALVES.

Theobald Smith, M.D.

(From the Department of Comparative Pathology of the Medical School and the Bussey Institution of Harvard University.)

Among the various questions which have arisen during the investigations made to test the practical value of Behring's bovo-vaccination with cultures of human tubercle bacilli is the one which concerns the effect of the vaccination upon the calf. The present article contains some data bearing upon this question. Three different strains of human tubercle bacilli, denominated respectively X., XII., and XXIV., were used as vaccines. These were chosen at random from a larger collection of cultures. These cultures are described in earlier publications by the writer and the essentials are re-stated in the following table:

TABLE I.

Culture No.	Date of Isolation.	Source.
X¹..	February, 1902.	Lung of adult (advanced pulmonary tuberculosis).
XII¹..	April, 1902.	Mesenteric lymph nodes of child.
XXIV²..	June, 1906.	Lung of adult (pulmonary tuberculosis).

* Investigations carried on under a grant from the Massachusetts Society for Promoting Agriculture. Received for publication June 8, 1911.

It will be seen from the table that these strains had been under cultivation for a number of years on glycerine-agar media before they were used as vaccines in this series. The same strains had been used among others in the first vaccination experiment.[3]

The calves were kept on the Bussey farm. Owing to inadequate means for the proper isolation of animals only a small number could be safely handled together. At no time during these tests were any bovine tubercle bacilli used and there was but one cow on the farm, which did not, however, come in contact with the calves, so that it may be safely assumed that the latter remained unexposed to bovine tuberculosis over the period included in this report.

Each calf soon after arrival was tested with tuberculin. No reacting animals were kept. No other criteria were used in selecting calves. They were purchased in the open market and represented various grades of black and white, red, and red and white stock. No bull calves were used. The tuberculin used was prepared and tested in this laboratory. As far as possible the same lot was used for the successive tests. That made from the human type of bacilli is denominated human tuberculin, that made from cultures of both human and bovine types mixed together during the concentration of the fluid is called mixed tuberculin. The tuberculin was always diluted with nine volumes of water before injection.

The calves were at first fed on boiled milk and subsequently on hay and some grain. No unboiled milk was fed at any time. They were kept in a roomy stable and allowed to run in a small pasture the greater part of the day in summer and on pleasant days in winter.

The preparation of the cultures or vaccines for injection is, briefly, as follows: Either five per cent glycerine-agar or glycerine bouillon cultures were used. When the former were used, the growth was transferred directly to weighed porcelain crucibles for weighing. When the fluid cultures were used the flakes from the membrane were first laid on sterile filter paper to remove excess of bouillon and then transferred to the crucibles. After determining the weight of the bacillar mass it was transferred to a boiled agate mortar and ground during the gradual addition of sterile salt solution. The final dilution usually contained one centigram of moist bacilli to one cubic centimeter. This was placed in sterile bottles containing some glass beads to break up any clumps which might form and to maintain a uniform suspension during the injection of several animals. The suspension was injected a few hours after preparation unless otherwise stated. Usually the suspension was finely granular or flocculent. A homogeneous suspension without flocculi just visible to the unaided eye was not obtainable with the method employed. The injection was made in all cases into the right jugular vein. In no case did a local swelling appear at the site of inoculation.

DETAILED STATEMENT OF VACCINATION OF CALVES WITH THREE STRAINS OF HUMAN TYPE.

Before discussing the results of the experiment as a whole I shall give the main details of the experimental record of each animal by itself.

The first lot to be treated included Nos. 209–211, which were vaccinated with culture Human X.

No. 209. — Red and white calf weighing about 130 lbs. when received May 18, 1909:

May 24. 9 p.m., injected .3 cubic centimeter human tuberculin V.

May 25. 6 a.m., 101.8; 8 a.m., 101.9; 10 a.m., 102; 12 noon, 102; 2 p.m., 102.4; 4 p.m., 102.8; 6 p.m., 102.3.

May 31. Glycerine-agar culture Human X., thirty-four days old, used. Four cubic centimeters of a suspension, equivalent to four centigrams moist bacilli, injected into right jugular vein. The temperature began to rise June 20 and reached its highest point, 105° F., June 25. It then fell to normal within three days. A slight rise occurred between July 10 and 17. As a rule a high evening temperature was recorded during heat waves when the calves had been out in the pasture through the day.

June	2–5.	Diarrhea.	September	3.	Weight, 226 lbs.
July	2.	Weight, 163 lbs.	"	10.	"	235 "
"	9.	"	170 "	"	17.	"	248 "
"	16.	"	178 "	October	1.	"	253 "
"	23.	"	186 "	"	8.	"	264 "

July   30. Weight, 190 lbs.   October   15. Weight, 270 lbs.
August 6.   "   198 "   22.   "   275 "
  "   13.   "   207 "   November 20.   "   295 "
  "   20.   "   215 "   December 18.   "   327 "
  "   27.   "   218 "

Jan. 28, 1910. 9 p.m., .5 cubic centimeter mixed tuberculin IX. injected.

Jan. 29, 1910. 7 a.m., 104 F.; 9 a.m., 104.6; 11 a.m., 104.1; 1 p.m., 101.8; 3 p.m., 102.5; 5 p.m., 101.9.

April 1, 1910. 9 p.m., .5 cubic centimeter tuberculin IX. injected.

April 2,   "   6 a.m., 102; 8 a.m., 102.9; 10 a.m., 102.7; 12 a.m., 102.1; 2 p.m., 102; 4 p.m., 101.6.

June 5, 1910. 6 a.m., 101.6; 5 p.m., 101.
  " 6, "   "   "   101; 5 p.m., 101.2.
  " 7, "   "   "   101; "   "   101.4.
  " 7, "   9 p.m., .5 cubic centimeter mixed tuberculin IX.

June 8,   "   6 a.m., 101.8; 8 a.m., 102; 10 a.m., 101.8; 12 noon, 102; 2 p.m., 101.2; 4 p.m., 101.4; 6 p.m., 102.2.

June 9, 1910. 6 a.m., 101.6; 5 p.m., 101.6.

Transferred to large infected herd A.

No. 210.—Black calf received with No. 209 May 18, 1909. Weight about 145 lbs.:

May 24. 9 p.m., injected .3 cubic centimeter human tuberculin V.

May 25. 6 a.m., 102.4; 8 a.m., 102.2; 10 a.m., 101.8; noon, 102.1; 2 p.m., 101.7; 4 p.m., 102 3; 6 p.m., 102 4.

May 31. 6 p.m., injected into right jugular vein six cubic centimeters of the same suspension used on No. 209. Dose, six centigrams moist bacilli. In this case the temperature began to rise June 8, reaching a maximum of 106° F. June 11. It remained high until June 27, fluctuating between 104° and 105° F. Slight evening elevations occurred during heat waves in the summer.

June 2-10. Diarrhea.   September 3. Weight, 196 lbs.
July   2. Weight, 145 lbs.   "   10.   "   204 "
  "   9.   "   162 "   "   17.   "   209 "
  "   16.   "   161 "   October   1. .   "   217 "
  "   30.   "   168 "   "   8.   "   226 "
August 6.   "   175 "   "   15.   "   233 "
  "   13   "   178 "   November 20.   "   256 "
  "   20.   "   187 "   December 18.   "   272 "
  "   27.   "   191 "

December 19. On right cheek a general swelling appears, raising cheek one-half to one inch above normal level. No constitutional disturbance noticeable.

December 30. Abscess opens spontaneously. The discharging pus examined microscopically and injected into two guinea-pigs. These were killed respectively thirty-eight and eighty-five days later. Both normal.

Jan. 28, 1910. 9 p.m., injected .5 cubic centimeter mixed tuberculin IX.

January 29. 7 a.m., 105.3; 9 a.m., 104.2; 11 a.m., 104.1; 1 p.m., 104.6; 3 p.m., 104.9; 5 p.m., 103.1.

April 1. 9.30 p.m., injected .5 cubic centimeter mixed tuberculin IX.

April 2. 6 a.m., 103 9; 8 a.m., 105.1; 10 a.m., 104; 12 noon, 104.4; 2 p.m , 103 7; 4 p.m., 103.8.

June 5. 6 a.m., 102.4; 5 p.m., 101.4.

"   6.   6 "   101.4; 5 "   102.

"   7.   6 "   101.6; 5 "   101.7.

"   7.   9 "   injected .5 cubic centimeter mixed tuberculin IX.

June 8. 6 "   104; 8 a.m., 104; 10 a.m., 103.5; 12 noon, 103; 2 p.m., 102.3; 4 p.m., 102.6; 6 a.m., 102.5.

June 9. 6 a.m., 102; 5 p.m., 102.4.

July 12. 9 p.m., injected .5 cubic centimeter mixed tuberculin IX.

July 13. 6 a.m., 103.2; 8 a.m., 103.2; 10 a.m., 103; 12 noon, 103; 2 p.m., 102.8; 4 p.m., 102.7; 6 p.m., 102.4.

July 14. 6 a.m., 101.6.

July 26. Weight, 475 lbs. Killed by blow on head and severing arteries of neck.

Very careful examination of thoracic and abdominal viscera as well as the brain showed no abnormalities beyond a slight localized thickening of pleura and a small (two to three millimeters) subpleural spot, suggesting collapse.   This portion of the lungs, as well as another piece, and a portion of the large dorsal mediastinal lymph node were torn up in salt solution and the turbid suspension injected into the abdomen of six guineapigs.   These were chloroformed at intervals from three to five months after inoculation and all found normal.

TABLE II.
Inoculation of guinea-pigs with tissue from heifer No. 210.

Number of Guinea-pig.	Tissue Injected and Amount.	Result.
429......	2 cc. turbid suspension lung tissue (b).	Chloroformed in 90 days; no lesion.
427......	2 cc. turbid suspension lung tissue (b).	Chloroformed in 110 days; no lesion.
433......	2 cc. turbid suspension lung tissue (a).	Chloroformed in 119 days; no lesion.
430......	2 cc. turbid suspension lung tissue (a).	Chloroformed in 172 days; no lesion.
431......	4 cc. milky suspension caudal mediastinal lymph node.	Chloroformed in 116 days; no lesion.
432......	4 cc. milky suspension caudal mediastinal lymph node.	Chloroformed in 149 days; no lesion.

No. 211.— Red calf, received May 18, 1909, with Nos. 209 and 210. Weight about 120 lbs.:

May 24. 9 p.m., injected .5 cubic centimeter human tuberculin V.

May 25. 6 a.m., 102.4; 8 a.m., 101.9; 10 a.m., 102.2; 12 noon, 101.9; 2 p.m., 101.7; 4 p.m., 102.1; 6 p.m., 102.1.

May 31. Injected into right jugular vein two cubic centimeters of the same suspension used on Nos. 209 and 210. Dose, two centigrams moist bacilli. In this animal the temperature rose abruptly June 10, reaching a maximum of 106.4° F. June 15. It did not return to a normal level until June 30, fluctuating between 103° and 105° F. During the summer, the morning and evening oscillations were quite marked (see Figure, page 8).

July	2.	Weight, 137 lbs.		September 10.	Weight, 203 lbs.		
"	9.	"	143 "	"	17.	"	207 "
"	16.	"	149 "	"	24.	"	212 "
"	23.	"	154 "	October	1.	"	222 "
"	30.	"	159 "	"	8.	"	227 "
August	6.	"	167 "	"	15.	"	232 "
"	13.	"	175 "	"	22.	"	240 "
"	20.	"	181 "	November 20.		"	267 "
"	27.	"	187 "	December 18.		"	303 "
September 3.		"	190 "				

Jan. 28, 1910. 9 p.m., injected .5 cubic centimeter mixed tuberculin IX.

Jan. 29, 1910. 7 a.m., 105; 9 a.m., 104.8; 11 a.m., 103 8; 1 p.m., 104.4; 3 p.m., 104.2; 5 p.m., 104.5.

April 1, 1910. 9 p.m., injected .5 cubic centimeter mixed tuberculin IX.

April 2, 1910. 6 a.m., 102.7; 8 a.m., 103.2; 10 a.m., 102; 12 noon, 101.8; 2 p.m., 101.5; 4 p.m., 102.2.

June 5, 1910. 6 a.m., 102.1; 5 p.m., 101.5.

"   6,   "   6   "   101.2; 5   "   101.6.

"   7,   "   6   "   101.8; 5   "   102.

June 7, 1910. 9 p.m., injected .5 cubic centimeter mixed tuberculin IX.

June 8, 1910. 6 a.m., 102.4; 8 a.m., 104; 10 a.m., 102.6; 12 noon, 102.4; 2 p.m., 102; 4 p.m., 102.3; 6 p.m., 102.6.

June 9, 1910. 6 a.m., 102.2; 5 p.m., 101.6.

From the foregoing records it will be noted that in the three calves a rise of temperature occurred: in Nos. 210 and 211 about ten days, in No. 209 about twenty days after the injection. It was a severe reaction in Nos. 210 and 211 and lasted nearly three weeks. In No. 209 it was relatively slight. No. 210 received the largest dose and manifested the severest reaction. Its weight lagged also slightly behind that of the others. The tuberculin reaction persisted longest in this case, and even as late as the fourteenth month after injection there was a slight reaction. Owing to this the animal was killed and guinea-pigs inoculated with tissues to determine whether any injected bacilli or their progeny had survived, but none were found. Nor were signs of the vaccination detected anywhere in the organs. The two remaining animals at this time were in fine condition and placed in a large herd where tuberculosis was prevalent.

The next lot of three calves were vaccinated with another strain of human tubercle bacilli, No. XXIV. The details of the experiment are as follows:

No. 212. — Black calf with white feet. Received Aug. 31, 1909. Weight, 99 lbs. (see Figure, page 8).

September 13. 9 p.m., weight, 110 lbs. Received .4 cubic centimeter human tuberculin V.

September 14. 6 a.m., 101.2; 8 a.m., 102; 10 a.m., 101.7; 12 noon, 101.6; 2 p m., 101 7; 4 p.m., 101.6; 6 p.m., 102.

October 16. 1 p.m., weight, 132 lbs. Received into right jugular vein twenty milligrams tubercle bacilli of culture, Human XXIV., grown since September 30 on glycerine agar. Growth rich at this time. The suspension in salt solution so prepared that each cubic centimeter contained ten milligrams of moist bacilli.

November 12. Weight, 138 lbs.

"   14–17. Diarrhea.

November 16. Respirations about 100 per minute. Occasional slight cough.

November 20. Respirations 140 to 150. Frequent coughing. Weight, 124 lbs.

November 25. 4 p.m. Calf lying down. Respiration very rapid, shallow, accompanied by a slight grunt (see page 8 for temperature curve).

November 26. Dies at 8.10 a.m.

Thorax: Almost entire lungs involved in disease. As a rule ordinary pneumonia involves primarily the cephalic half of the lungs and extends more rarely to contiguous areas of the large caudal lobes. In this case, the greater portion of the caudal lobes is firm, somewhat nodular to the touch. The smaller cephalic lobes less firm. Portions float in water. The entire lungs are congested. The pleural surface of the smaller lobes is more or less mottled as if the infundibula were filled with a lighter material than the parenchyma. Occasional sub-pleural yellowish points one to two millimeters in diameter are seen over the entire lung. In one bronchus a consistent ball of mucus six to eight millimeters in diameter.

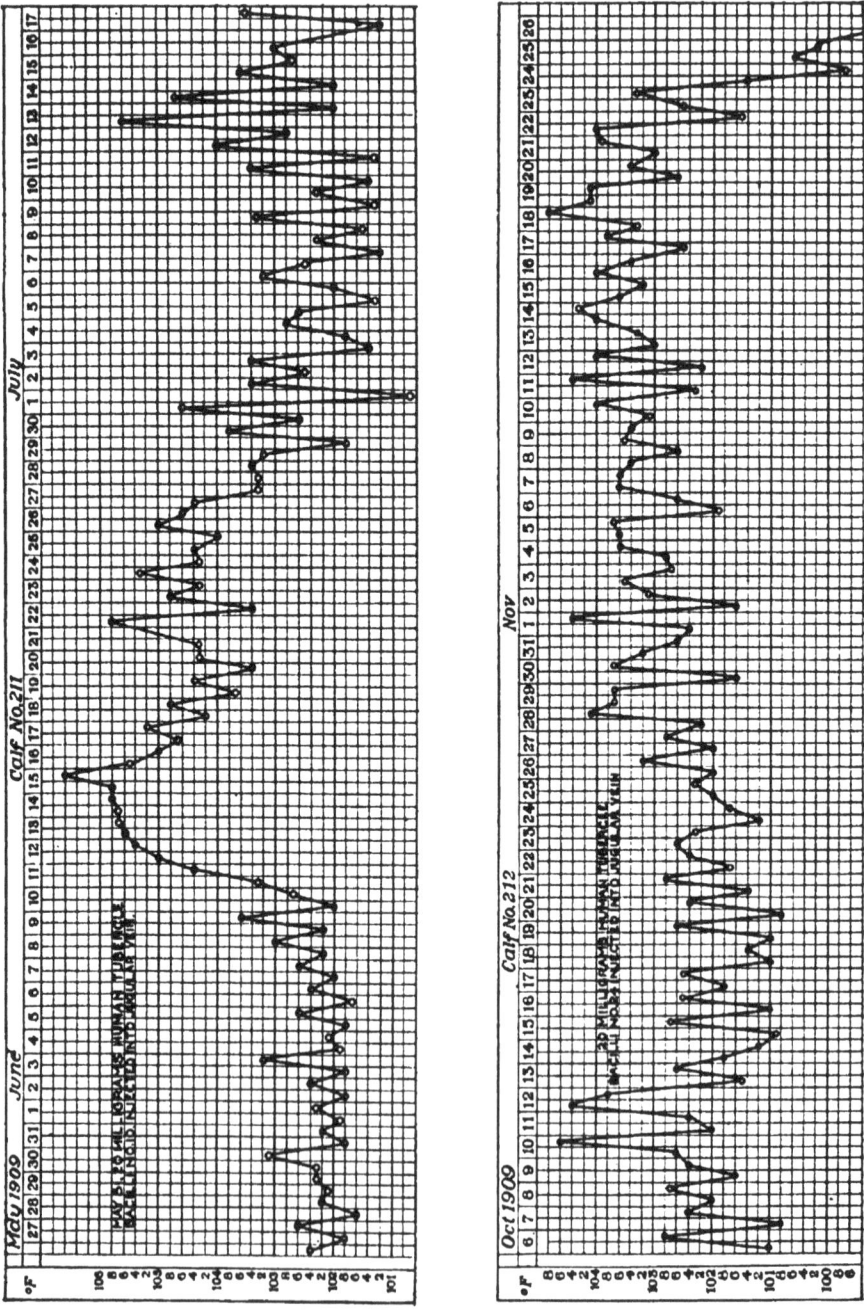

All dorsal mediastinal and bronchial lymph nodes are enlarged to twice
the normal size; opaque, grayish-yellow on section. In abdomen there
is marked gelatinous edema along duodenum, bile-duct, and around

pancreas. Liver, pale with yellowish tint. Bile ropy, thick, like syrup and slightly clouded. Spleen, small; kidneys congested. The remaining organs show nothing abnormal.

Histological examination : Sections of lung tissue were made from six different regions, but in all the same changes were found. The alveolar structure is largely obliterated by changes in the alveolar walls, leading to desquamation of cells, migration of endothelial leucocytes into the lumen and proliferation of cells composing the parenchyma. The alveolar walls are thus greatly broadened and encroach on alveolar spaces. Polynuclear leucocytes and lymphocytes are scarce; the chief type of cell being endothelial. The capillaries of the alveolar walls are extremely distended. Mitoses not infrequent. In addition to these changes there is more or less exudation of a very compact, delicately fibrillar, fibrin-like substance into many alveoli which stains deeply red with eosin.

Roundish masses of cells resembling the epithelioid cells of tubercles are found scattered through the tissues. These are as a rule without giant cells and are surrounded with a thin connective-tissue capsule. These tubercles are about .3 to .5 millimeter in diameter, some of the largest with necrotic centers.

In the thoracic lymph nodes there are many isolated and confluent giant cells and some larger (.5 millimeter) foci of epithelioid cells with central necrosis. These various lesions are, as usual, restricted to the follicles. Similar foci though far fewer in number are found in the lymph nodes of the mesentery. In the liver there is extensive vacuolation of the cells centrally and extending nearly to periphery of lobule (fatty degeneration). In some of these areas there is present amorphous eosin-staining material representing necrotic liver cells. Some invasion by polynuclear leucocytes of these necrotic areas.

In sections stained with carbol fuchsin and decolorized with sulphuric acid very many bacilli are found not only within the tubercles, but throughout the lung tissue in the epithelioid-like cells found in the alveoli and the alveolar walls. The bacilli are extremely fine, long and often slightly bent or tortuous. They are isolated or in groups within the cytoplasm of the cells. In the mediastinal lymph nodes all epithelioid cell foci and giant cells are crowded with bacilli.

No. 213. — Reddish-yellow calf with white patches. Received with No. 212 on August 13. Weight, 103 lbs.

September 13. 9 p.m., receives .4 cubic centimeter human tuberculin V.

September 14. 6 a.m., 102.8; 8 a.m., 102.8; 10 a.m., 102.5; 12 noon, 102 4; 2 p.m., 102.5; 4 p.m., 103.1; 6 p.m., 103.1.

October 16. Weight, 150 lbs. Receives into right jugular vein forty milligrams of a suspension of culture Human XXIV. (For details see No. 212 under same date.)

The temperature began to rise October 27, fluctuating between 105° and 103°.

October 30. Weight, 161 lbs. Slight diarrhea.

November 16.   Respiration, 120.   Occasional cough.   Weight, 136 lbs.

November 20.   Respiration about 50.   Slight sighing noise with each expiration.   Occasional cough.

December 2.   Calf lying down with head turned towards right flank. Respirations rapid.   As the animal was evidently dying it was killed by a blow and severing vessels of neck.   Autopsy performed at once showed irregular areas of red-flesh consolidation in all lobes involving perhaps a fourth or fifth of the entire lung.   In these areas are seen a few yellowish points one to two millimeters in diameter.   The air tubes in the affected areas are partly filled with a thick, muco-purulent secretion.   Pleura not involved.

Mediastinal and bronchial lymph nodes enlarged to twice normal size. On section soft, succulent, of a slightly yellowish cast.   No distinct tubercle formation visible to naked eye.

General resorption of fat in abdominal cavity leaving a gelatinous-looking mass behind.   Spleen normal in size; somewhat congested.   Liver somewhat swollen, rather firm, surface slightly irregular.   On section the parenchyma shows as a collection of light yellowish-brown areas in a dark red network.   Bile thick, viscid.   Kidneys with base of pyramids deeply reddened.

Smears from the plugs of bronchial mucus showed several colonies of acid-fast bacilli, perhaps fifty to one hundred in a colony.   Smears from the lung tissue showed scattering acid-fast bacilli, singly or in twos, rather thin, and irregularly jointed.

Histological examination: Sections from five different regions of the lungs show in all more or less extensive accumulations of polynuclear leucocytes in alveoli and small air tubes mixed with some alveolar epithelium.   The compact eosin-stained fibrin found in No. 212 also present in this case, but only in small amounts.   The greater portion of the exudation is amorphous and coarsely and finely granular.   The more compact exudate is in many instances associated with masses of endothelial-like cells.   The alveolar epithelium is highly dropsical and resembles cylindrical epithelium.   Extensive cell proliferation as found in No. 212 is absent in this case.   Epithelioid cell masses are seen here and there, and among them acid-fast bacilli are regularly demonstrable.

In the thoracic lymph nodes many isolated giant cells and larger areas of epithelioid cells, with occasional necrotic centers.   Many well-stained, rather slender tubercle bacilli in these areas.

Sections of liver show extensive rarefaction of the lobule, extending from central vein to near periphery.   This rarified area is filled with red cells enclosed in the connective tissue stroma.   Within these areas irregular clumps of eosin-stained material indicate the remains of liver cells, whose chromatin has disappeared.   Tubercle bacilli were not detected in these areas.   Small groups of epithelioid cells are found in various abdominal lymph nodes and in lymphoid tissue from walls of duodenum. In these very slender, feebly-stained tubercle bacilli are occasionally demonstrable.

No. 214. — Red and white calf; received Sept. 7, 1909. Weight, 129 lbs.

September 13. 9 p.m., weight, 130 lbs. Receives .4 cubic centimeter human tuberculin V.

September 14. 6 a.m., 102.6; 8 a.m., 102.2; 10 a.m., 101.8; 12 noon, 102; 2 p.m., 101.9; 4 p.m., 102.6; 6 p.m., 103.

October 16. Weight, 159 lbs. Receives into right jugular vein sixty milligrams of culture, Human XXIV. (For details see No. 212 under same date.)

October 22. Temperature begins to rise to-day.

October 27. Weight, 165 lbs. Slight diarrhea. More or less dyspnea; p.m., temperature 105.8° F.

November 16. Respirations about 54, somewhat labored. Movements of flanks accentuated. Temperature has been above normal since October 22.

November 20. Weight, 146 lbs. Respirations about 50. Occasional cough.

December 12. Weight, 157 lbs. Both eyes partly closed; sensitive to light. A clear, watery discharge from both trickles down cheeks. Cornea clouded. Respiration normal.

December 19. Weight, 173 lbs. Eats well and weight increasing. Both eyes as before. Probably very little vision.

December 20. Temperature low to-day for the first time since October 22.

January 28. Weight, 186 lbs. Watery discharge from eyes has continued. Condition as before. Injected at 9 p.m .3 cubic centimeter mixed tuberculin IX.

January 29. 7 a.m., 103.8; 9 a.m., 103.2; 11 a.m., 102.7; 1 p.m., 101.5; 3 p m., 101.1; 5 p.m., 102.

February 2. Calf chloroformed and bled to death. Eyes removed for more thorough examination. Corneas clouded. No marked conjunctival inflammation. Through the cornea several yellowish spots are visible on iris. On section, that portion corresponding to the ciliary body is occupied by a soft, pultaceous, grayish-red mass crowding upon and distorting the lens. Giant cells and some acid-fast bacilli detected in smears from this mass. Both eyes affected alike. Slight congestion of pia-arachnoid over convex surface of brain. More marked congestion, clouding and edema of membranes at base. Groups of minute grayish bodies along course of vessels at base and on convex surface.

Lungs normal. Thoracic lymph nodes slightly larger than normal but without visible lesions.

In the medulla of kidneys some grayish foci about one millimeter in diameter. Each pyramid shows several on the cut surface.

Histological examination: Extensive cell infiltration into membranes at base of brain. Cells chiefly of the plasma-cell type. Among them epithelioid cell tubercles with giant cells. Some with necrotic center.

Kidney contains small sub-cortical tubercles and larger ones in medulla. They consist chiefly of lymphoid cells. Epithelioid and giant cells practically absent.

Sections of lungs show small, scattering tubercles within parenchyma, occasionally with giant cells. Liver section shows small collections of cells chiefly at central vein. Increase of cells along course of vessels.

Large areas of tubercular tissue in the ciliary body of the eye.

In sections of lungs and eye stained with carbol fuchsin, tubercle bacilli could not be found in the frankly tuberculous tissue. In the meningeal tubercles, however, a considerable number of well-stained, rather long and irregular tubercle bacilli were found.

From one kidney of this case particles of tissue containing tubercles were transferred to tubes of egg media and incubated. The cultures obtained in this way possessed all the characters of the injected strain. Subcultures were used on Calf No. 223 (see page 17). Culture No. XXIV. is shown by the foregoing tests on Calves 212–214 to have been more than normally virulent. The doses were not high, that for No. 212 being the average dose for vaccination. No. 214 received three times this amount, but it survived and when killed was free from pneumonic lesions. The difference in individual resistance to inoculation with tubercle bacilli is again shown in these three calves.

The temperature reaction of No. 214 was highest; next 213; No. 212 did not have a very marked febrile reaction and it was the one to die first. During the highest fever, the respirations were very rapid and shallow. The autopsy showed in two very extensive pneumonia. This pneumonia was probably not an intercurrent affection but due to the vaccination. The injury to the eyes in No. 214, which led to the thickening and opacity of the cornea and destruction of the lens, was due to extensive tubercular new growth in the ciliary body from where it encroached upon the lens. The lesion found in the membranes of the brain of the same calf is also due directly to the injected bacilli. It is frequently seen when bovine bacilli are injected into the blood.[3]

Early in 1910 three fresh calves (Nos. 219–221) were vaccinated with the same strain to determine whether the fatal effect of the former vaccination might have been brought about by causes other than the tubercle bacilli injected.   The calves were slightly over two hundred pounds in weight and each received the same dose, twenty milligrams of moist bacilli, but the cultures differed in age, being respectively fifty-three, thirty-five, and seventeen days old.   The result, however, did not agree with expectations, for the one receiving the oldest culture died, and the others survived, as the following protocols show:

No. 219 — Black and white calf, received Jan. 25, 1910.   Weight, 174 lbs. :

January 28.   9 p.m., injected .3 cubic centimeter mixed tuberculin IX.

January 29.   7 a.m., 102.2; 9 a.m., 101.9; 11 a m., 102.7; 1 p.m., 101.6; 3 p.m., 101.7; 5 p.m., 102.5.

February 4.   Weight, 204 lbs.   Receives into right jugular vein four cubic centimeters of a salt solution suspension of tubercle bacilli, Human XXIV., grown on five per cent glycerine bouillon for fifty-three days. The total suspension injected is equivalent to twenty milligrams moist bacilli.   The temperature began to rise February 13, reaching 104 9° F. February 16.   It remained continuously high until March 2, when it began to fall (see Figure, page 15).   The calf died March 5.   During the fever there was some cough.   The breathing was not accelerated, but there was a slight sighing noise with each expiration.   On March 2 the weight was 201 lbs.; on day of death, 187 lbs.   Before death the respiration was rapid and there was some frothing at the nostrils.

Autopsy made three hours after death.   Lungs very large; almost total consolidation.   Subpleural emphysema extending over considerable areas, giving the appearance of air-injected subpleural lymphatics in some places.   The ventral and cephalic lobes are dark reddish with air-containing pale yellowish lobules scattered through the lobes.   In the large caudal lobes more or less interlobular emphysema.   Here the airless tissue is grayish yellow on section.   Pleura not involved.   No recognizable tubercles anywhere in the lung tissue.

Lymph nodes of thorax all enlarged to three or four times normal size. On section of a mottled, grayish appearance; necroses not seen.   Slight enlargement of abdominal lymph nodes.   Spleen small, flabby.   Congestion of liver and kidneys.

Histological examination: Lung tissue from different regions shows more or less the same lesions.   There is marked distension of the capillaries with red cells.   The alveoli contain a granular substance (fibrin ?) or else a bubbly exudation.   Many of the alveoli are, however, filled with cells fused together and resembling epithelioid cells, i.e., with considerable acidophile cytoplasm.   In some alveoli a giant cell-like body takes the place of the irregular cell-mass.   There are also scattering relatively large foci in an early stage of necrosis.   Tubercle bacilli are quite rare.

In the mediastinal lymph nodes there are large areas where there is a diffuse proliferation of epithelioid-like cells. Many isolated giant cells are found throughout the sections. Bacilli few, quite long and well stained.

In the liver there is extensive vacuolation of the parenchyma (fatty degeneration) extending from the center of each lobule. A few sub-miliary tubercles present.

Kidneys markedly congested but without focal lesions.

No. 220. — Black and white calf. Received with No. 219, Jan. 25, 1910. Weight, 201 lbs.

January 28. 9 p m., injected .3 cubic centimeter mixed tuberculin IX.

January 29. 7 a.m., 103° F.; 9 a.m., 103; 11 a.m., 103.1; 1 p.m., 103; 3 p.m., 102.8; 5 p m., 103.2.

February 4. Weight, 213 lbs. Injected into right jugular vein twenty milligrams tubercle bacilli, Human XXIV., in four cubic centimeters salt solution. The bacilli are from a five per cent glycerine bouillon culture thirty-five days old.

The febrile reaction in this calf was relatively slight. Beginning February 17 the temperature rose to 104° F. and 104.5° F., remaining there until February 25, when it declined, reaching normal February 27.

There was more or less coughing during and shortly after this febrile period.

March 2.	Weight, 236 lbs.		April 17.	Weight, 277 lbs.	
" 9.	"	247 "	May 1.	"	293 "
" 18.	"	258 "	" 21.	"	316 "
" 26.	"	268 "	June 17.	"	385 "

July 12. 9 p.m., injected .5 cubic centimeter mixed tuberculin IX.

July 13. 6 a.m., 106; 8 a m., 105.1; 10 a.m., 104.3; 12 noon, 103.8; 2 p m., 104; 4 p.m., 103.6; 6 p.m., 103.4.

July 23. Weight, 420 lbs.

October 20. 5 p.m., temperature 102.5.

October 21. 6.30 a.m., temperature 101.6; 5.10 p.m., temperature 102 4; 9 p.m., .5 cubic centimeter mixed tuberculin IX. injected.

October 22. 6 p.m., 102.4; 8 p.m., 103.4; 10 p.m., 103.2, drank water; 12 p.m., 102.6; 2 a.m., 102.8; 4 a.m., 103.2, drank water; 6 a.m., 101.8.

October 23. 7.15 a.m., 102.3.

December 6. 8.30 a.m., injected into right jugular vein forty milligrams moist bacilli, Human XXIV., suspended in four cubic centimeters normal salt solution. The suspension had been prepared December 5 and kept at 40°–50° F. over night. The culture had been passed through calf No. 223 (which see) and continued on beef serum and lastly on glycerine agar. The suspension was prepared from a glycerine agar culture seventeen days old.

There was a prompt febrile reaction beginning with 104.8° F. on the same evening. It fell next day to 103°, around which it fluctuated until December 16, when it became normal.

er, in very good condition, was placed in infected

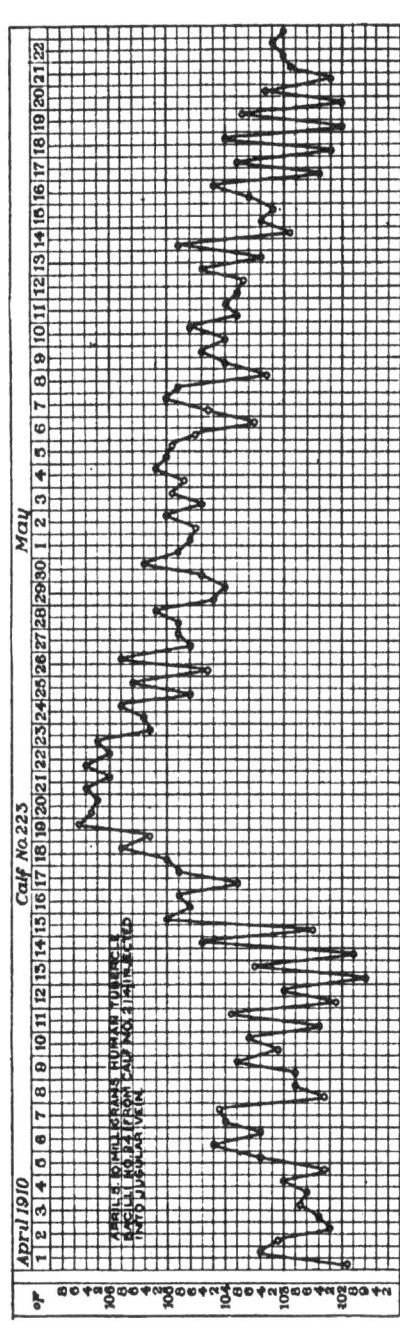

No. 221. — Black and white calf, received Jan. 25, 1910, with Nos. 219, 220. Weight, 230 lbs.:

January 28. 9 p.m., injected .3 cubic centimeter mixed tuberculin IX.

January 29. 7 a.m., 102.4; 9 a.m., 101.6; 11 a.m., 102.1; 1 p.m., 102.2; 3 p.m., 101.7; 5 p.m., 102.5.

February 4. Weight, 229 lbs. Injected into right jugular vein twenty milligrams of moist bacilli from culture, Human XXIV., grown in five per cent glycerine bouillon for seventeen days. The bacilli were suspended in four cubic centimeters salt solution. The febrile reaction was quite severe in this case. Beginning on February 12, it rose above 105.5° in three days and remained high until March 3, fluctuating between 104° and 105°. It touched normal about March 12. During this febrile period there was more or less coughing. The respiration was rapid, up to eighty per minute, during much of this time. From February 5 to March 5 there was a gain of ten pounds, the weight being 239 lbs. March 26, it was 265 lbs. What was probably a slight relapse occurred between April 4 and 15, when there was some coughing, slight decline in weight and some fever. From the latter date the progress was steady.

July 12. Weight, 361 lbs. 9 p.m., injected .5 cubic centimeter mixed tuberculin IX.

July 13. 6 a.m., 104.2; 8 a m., 106.2; 10 a m., 105.6; 12 noon, 105; 2 p.m., 105.2; 4 p.m., 104.8; 6 p.m., 104.9.

October 20. 5 p.m., temperature, 101.7.

October 21. 6.30 a.m., 101.6; 5 p.m., 101.8; 9 p.m., received subcutaneously .5 cubic centimeter mixed tuberculin IX.

October 22. 6 a.m., 102.3; 8 a.m., 103.4; 10 a.m., 103.4; 12 noon, 103.2; 2 p.m., 104.2; 4 p.m , 104.6, watered; 6 p.m., 103.6.

October 23. 7 a.m , 102.7.

Dec. 6, 1910. 8.30 a.m., injected into right jugular vein forty milligrams moist bacilli of same suspension of Human XXIV. used on Calf No. 220 on this date.

The febrile reaction beginning on the evening of the same day at 105° fell during the following day to 103°, touched normal on December 12. This heifer was placed with No. 220 in infected herd B.

Rabbit No. 385. Feb. 12, 1910; weight, 2,880 grams. Injected into ear vein about one-half milligram moist bacilli, Human XXIV., the same suspension used on Calf No. 221.

April 12. Weight, 2,980 grams. May 6. Weight, 3,100 grams.
" 20. " 3,020 " " 11. " 3,120 "
Chloroformed.

In lungs, translucent foci two to four millimeters in diameter. In some a minute opaque center. Small number of minute tubercles in cortex of kidneys. Large fat deposits in abdomen. Other organs normal.

The relatively high virulence of strain No. XXIV. suggested the desirability of passing it through calves to determine whether or not it could be made to approach the bovine type. Although past evidence did not promise any success in this undertaking, yet the great importance of the subject from a general standpoint and the possibility that only certain strains may become transformed from human to bovine type and the reverse made the present attempt desirable. The following calf was therefore inoculated with this strain, after isolation from Calf No. 214:

No. 223. — Red calf, rather graceful build, weighing 245 lbs. when received March 29, 1910:

April 1.   9 p.m., receives .5 cubic centimeter mixed tuberculin IX.

April 2.   6 a.m., 103.1; 8 a.m., 102.7; 10 a.m., 102.6; 12 noon, 102.7; 2 p.m., 102.7; 4 p.m., 102.2.

April 5.   The culture used upon this calf has the following history:

Oct. 16, 1910.   Culture, Human XXIV injected into Calf 214.
|
Feb. 2, 1911.   Calf 214, kidney on
egg media
|
February   28.   Beef serum
|
March   16.   Beef serum
|
April   5.   Injected into Calf 223

A suspension in salt solution was prepared from the beef serum culture so that one cubic centimeter contained two milligrams moist bacilli. Of this suspension five cubic centimeters or one centigram was injected into the right jugular vein of No. 223. The temperature began to rise about eight days after injection, going as high as 106° F. with accelerated respiration and coughing. It touched normal about May 19, forty-four days after injection, but rose again after several days, fluctuating between 103° and 105° F. (see Figure, page 15). Eyes became affected and sight was partially lost. Corneas opaque and slight watery discharge.

June 4. Calf lying down with head and neck retracted. Breathing normal. Chloroformed and bled to death.

The weight of this animal reached a maximum of 277 lbs. on April 15. It slowly fell to 254 lbs. on May 20, and then rose again to 260 May 27.

Autopsy immediately after death.

Thorax: The free half of right cephalic, the whole of right and left ventral lobes hepatized. Scattering consolidated areas in azygos and in large caudal lobes, reddish in color. The consolidated lobes not specially enlarged, delicately mottled with faint gray-yellow points. The dependent portions of both ventral lobes show larger grayish-yellow, ill-defined areas. On section a creamy, semi-fluid mass exudes from the cut air tubes, consisting wholly of cells (polynuclear and alveolar).

Countless minute, glassy tubercles on pleura of those lobes not hepatized. Similar but less marked appearances on costal pleura. All thoracic lymph nodes enlarged to twice normal size, on surface and section mottled with petechiæ. Other changes not evident.

Beneath endocardium of both ventricles of heart many gray, slightly elongated foci, about one-half millimeter in diameter.

Abdomen: Mesenteric lymph nodes twice normal size and permeated with gray tubercles one-half to one millimeter in diameter.

Adrenals similarly affected.

Kidneys, both cortex and medulla, densely permeated with minute grayish tubercles .5 to one millimeter in diameter. Slight injection of pia-arachnoid over convex surface and base of brain. Minute grayish tubercles in membrane between pons and chiasma.

Histological examination: In the dependent lobes of lungs the alveoli and air tubes are largely filled up with polynuclear leucocytes, among which occasional mononuclear (alveolar?) cells occur. In these lobes there are also cell aggregations about the size of an alveolus consisting of mononuclear elements, among which typical epithelioid elements and more rarely giant cells are found. In these portions of the lungs there is also marked proliferation of the submucous and lymphoid tissue around small air tubes, causing considerable deformation and constriction. In the other affected portions of the lungs these tubercles are more abundant and there is besides a diffuse hyperplasia of the alveolar walls or lung parenchyma, the new cells being largely mononuclear (endothelial) in type.

Submiliary areas of epithelioid cell groups occur in the walls of the heart, liver and Peyer's patches. In the kidney subcortical and medullary cell foci occur, made up in part of epithelioid cell elements. The same is true for the meninges over base of brain. Here the tubercular tissue is firmly attached to the brain tissue and the foci larger and more characteristic than elsewhere. The ciliary body in both eyes is replaced by tubercular tissue. Sections stained with carbol fuchsin for tubercle bacilli showed only rare bacilli, well stained. A rabbit was inoculated with the same suspension kept in a refrigerator from April 5 to April 16.

April 16. Rabbit No. 391, weight, 3,050 grams, receives into a marginal ear vein .5 cubic centimeter of a suspension of tubercle bacilli used on Calf No. 223. The suspension is equivalent to one milligram moist bacilli.

July 13. Weight, 3,245 grams. Has been well since inoculation. On iris of left eye a one-half millimeter gray tubercle. Chloroformed. In lungs many one-half to one millimeter not well-defined gray foci. Organ normal in texture and elasticity. Many grayish, partly softened foci in cortex of both kidneys, some projecting above cortex, others linear and extending towards pelvis.

The relatively high degree of virulence of this stock led the writer to use a much smaller dose on the following animals, Nos. 227 and 229, in order to produce a more prolonged disease and a better opportunity for the bacillus to become modified if this were possible. The two animals reacted with only moderately high temperatures and recovered completely. They were subsequently inoculated with a much larger dose of the same culture and resisted this successfully.

No. 227. — Black calf. Received July 25, 1910:

August 12. 9 p.m., weight, 350 lbs. .4 cubic centimeter mixed tuberculin IX. injected. Temperature 103.

August 13. 6 a.m., 103.6; 8 a.m., 103.4; 10 a.m., 102.8; 12 noon, 103; 2 p.m., 103.2; 4 p.m., 103.4; 6 a.m., 104.3.

August 31. Weight, 370 lbs. 5 p.m., receives into right jugular vein six milligrams tubercle bacilli, Human XXIV., isolated on egg medium from the lungs of Calf No. 223. Culture on beef serum thirty-four days old and very rich when used. The temperature reaction following this injection was somewhat irregular, fluctuating between 103 and 104 and going to 105° several times, probably the effect of insolation. This irregularity lasted until about September 26, when the temperature returned to 102°–103°. The weight had slowly risen and on October 8 was 395 lbs.

October 21. 6 a.m., temperature 103; 6 p.m., temperature 102.

" 21. 9 p.m., .4 cubic centimeter mixed tuberculin IX. injected.

October 22. 6 a.m., 103.8; 8 a.m., 104.7; 10 a.m., 104.6; 12 noon, 103.8;* 2 p.m., 104.6; 4 p.m., 104.8; 6 p.m., 104.2 *

Feb. 14, 1911. 5 p.m., receives into right jugular vein sixty milligrams tubercle bacilli, Human XXIV., isolated from Calf No. 223 (see Calf 226). There was a prompt elevation of temperature lasting in all about four days.

February 15. 6 a.m., 105.6; 8 a.m., 105 6; 10 a.m., 107.2; 12 noon 105.8; 2 p.m., 104.6; 4 p m., 105.4; 6 p.m., 104.2.

February 16. 7 a.m., 104.4; 5 p.m., 104.8.

" 17. 7 " 104.2; 5 " 103.4.

" 18. 7 " 103.2; 5 " 104.

" 19. 7 " 103.2; 5 " 103.

March 10. Calf weighs now between 500 and 550 lbs.

Calf No. 229. — Black. Weight, 252 lbs. Received Aug. 9, 1910:

August 12. 9 p.m., .4 cubic centimeter mixed tuberculin X. injected.

August 13. 6 a.m., 102.4; 8 a.m., 102 5; 10 a.m., 102.2; 12 noon, 102.2; 2 p.m., 102.6; 4 p.m., 103; 6 p.m., 103.8.

---

* Calves drank water just previous to this temperature.

August 31. Weight, 280 lbs. Receives into right jugular vein five milligrams tubercle bacilli, Human XXIV., isolated from one kidney of Calf No. 223 on egg medium and kept on this and on beef serum to date. Culture on beef serum thirty-four days old, rich.

No appreciable reaction followed this dose and it is probable that many bacilli in the culture were dead, *i.e.*, the culture was too old.

October 21. 9 p m., .4 cubic centimeter mixed tuberculin IX. injected.

October 22. 6 a.m., 102.8; 8 a m., 103.6; 10 a.m., 103 3; 12 noon, 102.4;* 2 p m., 103.6; 4 p.m., 104; 6 p.m., 103.4.*

This was evidently a reaction when compared with the temperature record preceding and following the test.

December 6. 8.30 a.m., weight, 320 lbs. Receives into right jugular vein twenty milligrams of culture, Human XXIV., isolated from lung of Calf No. 223 and kept on egg and beef serum until November 19, when a glycerine agar culture was made. The injected dose was prepared from . this culture (see Calves No. 220 and 221). A slight elevation of temperature to a maximum of 103.4 on the same evening lasted several days.

February 14. 5 p.m., weight, 400 lbs. Together with Nos. 226 and 227, this calf receives into right jugular vein sixty milligrams tubercle bacilli of culture, Human XXIV. (see No. 226), isolated from Calf No. 223. There was very little elevation of temperature following this large dose except on the day following, amounting to a tuberculin reaction.

February 15. 6 a.m., 103.4; 8 a m., 104; 10 a.m., 105.7; 12 noon, 105.5; 2 p.m., 103.8; 4 p.m., 104.2; 6 p.m., 102.4.

February 16. 7 a.m., temperature 102.4; 5 p.m., temperature 102.6.

March 10. Weight, 418 lbs.

The much lower virulence of the strain called Human XII. is shown in the following case, No. 226. This calf received fifty milligrams intravenously and responded but slightly. It was, however, influenced by this injection, for a later tuberculin test was positive and a still later intravenous injection of a relatively large dose of Human XXIV. was easily borne.

No. 226. — White calf, received Aug. 9, 1910. Weight, 202 lbs.

August 12. 9 p.m., .4 cubic centimeter of mixed tuberculin IX. injected.

August 13. 6 a.m., 102.8; 8 a.m., 102.8; 10 a.m , 102.8; 12 a.m., 102.4; 2 p.m., 102.2; 4 p.m., 103 2; 6 p m., 104.

Beginning at the close of August 13 a period of high temperature followed, lasting six days. During this time, the temperature twice rose to 106.8. At the decline diarrhea appeared. The cause may have been the change of food and environment.

---

* Calf watered just previously.

August 31. Weight, 212 lbs. 5 p.m., receives into right jugular vein fifty milligrams moist tubercle bacilli from strain No. XII. Eighteen days after this injection there was a slight elevation of temperature lasting about a week. Its relation to the dose of tubercle bacilli is doubtful. In general the reaction to this large dose of bacilli was feeble, confirming what had been learned earlier: that this strain was of relatively low virulence. That the animal was, however, influenced is proven by the following positive tuberculin reaction several months later.

October 21. Weight, 242 lbs. 9 p.m., .5 cubic centimeter mixed tuberculin IX. injected.

October 22. 6 a.m., 105; 8 a.m., 104.8; 10 a.m., 104.2; 12 noon. 102.2;* 2 p.m , 102.6; 4 p.m., 103.2; 6 p.m., 102.6.*

Feb. 14, 1911. Weight, 329 lbs. Receives into right jugular vein sixty milligrams of tubercle bacilli from human strain No. XXIV. isolated directly from the lungs of Calf No. 223, June, 1910, on egg media and continued on egg and beef serum to date. The suspension injected into this calf was made from a mixture of an egg and a serum tube twenty-one days old. The injection was made at 5 p.m. On the following morning, at 6 a.m., the temperature was 105.2. It was taken at two-hour intervals:

February 15. 6 a.m., 105.2; 8 a.m., 105.4; 10 a.m., 105.4; 12 noon, 104.4; 2 p.m., 103; 4 p m., 102.3; 6 p.m., 102 2.

February 16. 6 a.m., 104.2; 5 p.m , 105.6.

"   17. 6 a.m., 103.4; 5 p.m., 103.

"   18. 6 a.m., 102.8; 5 p.m., 103.3.

March 10. Weight, 359 lbs. Thereafter it fell to normal and remained there.

The suspension used for inoculating Calves No. 226, 227, and 229 on February 14 was also used upon a rabbit to test its virulence.

Rabbit No. 418.—February 14. Weight, 1,665 grams. One milligram moist bacilli, suspended in one cubic centimeter of salt solution, injected into an ear vein.

May 12. Weight, 1,750 grams. Chloroformed. Partial grayish-red consolidation of left and right ventral lobe of lungs. No distinct tubercles seen in these areas or elsewhere in the lungs. In both kidneys small firm yellowish nodules one to two millimeters in diameter projecting beyond cortex, about one centimeter apart. No extension of these foci farther into cortex.

This test thus reveals a very low degree of virulence towards rabbits.

---

* The fall in temperature probably due to drinking water.

After a continuous cultivation of the strain, Human XXIV., isolated directly from the lungs of Çalf No. 223, since June, 1910, on egg and serum media, a fresh calf, No. 501, was inoculated on March 21, 1911, with fatal result. From this calf cultures were recovered directly from the lungs and further passages through calves will be made.

Calf No. 501. This calf had been used early in February for the preparation of vaccine lymph. About six weeks later it was tested with tuberculin.

March 12. Temperature, 7 a.m., 101.8; 6 p.m., 102.2.

March 13. 7 a.m., 102; 6 p.m., 103; 9 p.m., receives .4 cubic centimeter mixed tuberculin IX. Weight, 269 lbs.

March 14. 6 a.m., 102.6; 8 a.m., 102.4; 10 a.m., 102; 12 noon, 103; 2 p.m., 103; 4 p.m., 103 4; 6 p.m., 103.2.

March 21, 1911. Weight, 264 lbs. Received into right jugular vein twenty milligrams moist bacilli of culture, Human XXIV., isolated from Calf No. 223. The bacilli were scraped from the surfaces of an egg and a beef serum culture, mixed and suspended in salt solution, so that one cubic centimeter contained five milligrams. The cultures were twenty-one days old.

The calf began to cough and breathe rapidly on April 5. The respirations were shallow, about one hundred per minute. The temperature fluctuated between 103° F. and 104° F. and rose above 104° F. only twice. The continuation of the rapid respiration and the evident rapid failure of the calf made it desirable to kill it. This was done April 14.

The only macroscopic lesions were in the thorax. There was a pale red consolidation of the tips of both cephalic, the whole of both ventral and the adjacent fourth of both large caudal lobes. There was much interlobular and subpleural emphysema. Tubercles not seen with the unaided eye. A few tubercle bacilli detected in smears from the lung tissue.

All thoracic lymph nodes are two to three times normal size. On section no focal lesions detected with the naked eye. Pure cultures of the injected tubercle bacilli were obtained from lung tissue on egg media.

No signs of any preëxisting tuberculosis.

The pneumonic changes are shown in sections of fixed and hardened tissue to be due to marked proliferation of the cells of the inter-alveolar septa and their approximation so that the alveolar lumina are very small and scarcely definable. Focal tubercles absent and only a few scattering giant cells detected. Bacilli very scarce. The enlargement of the lymph nodes of the thorax is due chiefly to a general hyperplasia, for the only characteristic lesions are scattering, very large giant cells. In the liver there are submiliary tubercles.

The strain of tubercle bacilli, denominated Human XII., was used in a herd not far from Boston upon young stock in December, 1909, April and November, 1910. No deaths, joint or eye lesions resulted. In all thirty-three were vaccinated twice, twenty-one with an interval of four months and twenty-one days between the vaccinations and twelve with an interval of six and one-half months. Thirteen had been vaccinated but once up to November, 1910. The vaccine was prepared from glycerine bouillon cultures. The ages of the cultures used in the three vaccines were respectively twenty-seven, thirty-three, and thirty-four days. The doses injected were as follows:

For primary vaccinations in December, 1909, and May, 1910:

    1 centigram   of moist bacilli for 100 lbs.
    2 centigrams "   "   "   " 150 "
    3   "   "   "   "   " 200 "   .
    4   "   "   "   "   " 250 "
    5   "   "   "   " 300 "

In the injections in November, 1910, the following doses were used:

    2   centigrams for 100 lbs.
    2.5   "   " 150 "
    3   "   " 200 "
    3.5   "   " 250 "
    4   "   " 300 "

In the second vaccination six centigrams were injected.

In some of these calves a rise of temperature occurred within a day of the first vaccination, which lasted one or two days. This prompt rise is due to a hypersensitive state of the animal which can only be accounted for on the assumption that these calves were already infected with tuberculosis when vaccinated.

In order to obtain a more accurate record of the febrile reaction following the injection of the living bacilli, the temperature hitherto taken only during four days after the injection was taken over a period of three to four weeks in a lot of twenty-four animals vaccinated in November, 1910. Of these twelve were second and twelve primary vaccinations. In all but two of the twenty-four a rise in temperature occurred on the ninth to the eleventh day, persisting from a few to ten days. The severity of the reaction varied from case to case. Some were reported to have been quite sick during this febrile period but all fully recovered. Any intercurrent digestive disturbances as a possible cause of the fever were ruled out by the attending veterinarian.

The following table contains all the important data bearing on the vaccination of the calves here reported and the result:

TABLE III.

Vaccination of Calves.

No.	Weight in Lbs.	Breed or Color.	Culture Used.	Culture Media.	Age in Days.	Dose in Milligrams Intravenous.	Date of Vaccination.	Result.
				First Vaccination.				
209.	130	Red and white.	Human X.	Glycerine agar.	34 days.	40	May 31, 1909.	In fine condition June, 1910. Placed in infected herd A.
210.	145	Black.	"	"	34 "	60	" " "	In fine condition July, 1910. Killed and examined for surviving tubercle bacilli.
211.	120	Red.	"	"	34 "	20	" " "	In fine condition June, 1910. Placed with 209 in infected herd A.
212.	132	Black.	Human XXIV.	Glycerine agar.	16 days.	20	Oct. 16, 1909.	Dies of acute tuberculosis Nov. 26, 1909.
213.	150	Reddish yellow.	"	"	16 "	40	" " "	Killed in dying condition Dec. 2, 1909.
214.	159	Red and white.	"	"	16 "	60	" " "	Tuberculosis of eyes. Chloroformed Feb. 2, 1910.
219.	204	Black and white.	Human XXIV.	Glycerine bouillon.	53 days.	20	Feb. 4, 1910.	Dies of acute tuberculosis March 5, 1910.
220.	213	" "	"	"	35 "	20	" " "	(See below.)
221.	229	" "	"	"	17 "	20	" " "	" "
223.	270	Red.	Human XXIV.*	Beef serum.	20 days.	10	April 5, 1910.	Chloroformed in dying condition June 4, 1910.
227.	370	Black.	Human XXIV.°	Beef serum.	34 days.	6	Aug. 31, 1910.	(See below.)
229.	280	"	" " °	" "	34 "	5	" " "	" "
226.	212	White.	XII.	Glycerine agar.	39 "	30	" " "	" "
501.	264	Black.	XXIV.°	Egg and beef serum.	21 "	20	March 21, 1911.	Killed in dying condition.

Table III. — *Continued.*

Weight in Lbs.	No.	Breed or Color.	Culture Used.	Culture Media.	Age in Days.	Dose in Milligrams Intravenous.	Date of Vaccination.	Result.
					Second Vaccination.			
	220.		Human XXIV.*	Glycerine agar.	17 days.	40	Dec. 6, 1910.	In fine condition February, 1911. Placed in infected herd B.
	221.		" " " *	"	17 "	40	" " "	In fine condition February, 1911. Placed in infected herd B.
	229.		" " " *	"	17 "	40	" " "	(See below.)
	227.		Human XXIV.*	Egg and beef serum.	21 days.	60	Feb. 14, 1911.	In good condition April, 1911.
	226.		" " *	Egg and beef serum.	21 "	60	" " "	" " " "
					Third Vaccination.			
	229.		Human XXIV.*	Egg and beef serum.	21 days.	60	Feb. 14, 1911.	In good condition April, 1911.

*Culture passed through Calf No. 214.

*Culture passed through Calf No. 223.

DISCUSSION OF DATA AND SUMMARY.

The foregoing experiments show that not every strain of tubercle bacilli clearly belonging to the human type is adapted for the vaccination of calves against tuberculosis under ordinary conditions.   Although strain No. XXIV. was of low virulence towards rabbits and its reaction curve in glycerine bouillon corresponded to that of the human type,[4] yet it was of such virulence towards calves that out of nine first vaccinations four were fatal and one caused blindness and other lesions which would probably have resulted in death sooner or later.

The nature of the lesions produced are of interest.   In the cases here reported, pneumonia, tuberculosis of the eyes and tubercular meningitis developed.   The pneumonia resembled, in its general appearance and primary localization in the small cephalic and ventral lobes, the usual types of pneumonia due to other causes.   A superficial inspection might lead to the inference that the lung disease was intercurrent or secondary.   The histological examination does not, however, warrant this inference.   Besides the lesions directly traceable to the tubercle bacillus, the various kinds of exudates suggest that the disease is due to toxins of the tubercle bacillus set free perhaps by the bacteriolytic forces of the blood.   In the above cases the seasonal influences are ruled out since one died in November, two in March, and one in May.   Tubercle bacilli were readily detected both in the lungs and mediastinal lymph nodes of the acutely fatal cases.   This type of lung disease does not, it is true, occur in the course of the natural disease, but only under the artificial conditions established by the intravenous injection.   It is probable that the injected bacilli which are largely retained in the lungs multiply unhindered for a time and are then suddenly acted upon by the awakening bactericidal forces of the body.   The toxins set free in large quantity then produce the exudative processes suggesting ordinary pneumonia.   The focal histological changes in the lungs and associated lymph nodes were characterized by the presence of epithelioid and giant cells within which sometimes large numbers of tubercle bacilli were demonstrable.

The lesions within the eyeball and in the meninges of the base of the brain were due to the injected bacilli in all cases examined. These lesions are exceedingly rare in the natural disease. We must assume that these localities affected are very susceptible, but that in the spontaneous infection they are pretty thoroughly protected, at least in the earlier stages, from invasion. The same is true of the udder, which is rarely attacked in the natural disease but frequently involved when virulent bacilli are injected into the blood.[3]

The difference between strains of human bacilli used as vaccines upon calves is well brought out by Weber, Titze and Jörn in their studies upon the immunization of cattle.[5] They experimented with the bovo-vaccine of Behring and the tauruman of Koch-Schütz, both representing the human type of tubercle bacilli and sold as vaccines against bovine tuberculosis. Among the calves vaccinated with tauruman there had been reported fatal pneumonia, as well as lesions of joints and disease of the eyes. From the description of the eye lesions given by these writers it is evident that the disease met by me in Nos. 214 and 223 is identical with theirs. The fatal cases of pneumonia following vaccination (in one herd, six out of twenty-two) are not ascribed to the vaccination as a direct cause by these investigators. An autopsy was made on four cases and the diagnosis of calf-pneumonia made. Tuberculosis was excluded. It is evidently assumed that the vaccination simply influenced unfavorably a prevailing disease.

One case of meningeal tuberculosis is reported in an animal vaccinated nearly three years before slaughter, at that time nine days old. Only a few small calcified foci were found in a retropharyngeal and in all mesenteric lymph nodes. The meningeal tubercles were found both on convexity and base of brain and associated with marked vascularity and thickening of the pia.

Slight tubercular affections of joints, notably the carpal joint ("knee"), were observed both after the use of bovo-vaccine and tauruman in several cases.

The relative virulence of Strain XXIV. must be referred to the increased susceptibility of young as compared with adult animals.   The high susceptibility of the very young of different species to infection easily overcome by adults has been brought forward by various observers.   It has also been shown that the blood of the young is deficient in protective substances when compared with that of adults.   The comparative resistance of young and older animals towards the tubercle bacillus has been tested by Bang.[6]   Bang fed the young of goats, horses, and cattle with cultures of avian tubercle bacilli and found them quite susceptible.   For instance, he found the adult horse insusceptible to large doses of avian tubercle bacilli introduced by way of the mouth, but colts and fillies succumbed to this treatment. Similar results were obtained when young and adult cattle and goats were fed.

The occurrence of strains isolated from the human subject which are virulent for calves and even older animals has been frequently noted by investigators, but in nearly all such cases no definite data were presented to show whether the strain was not actually bovine in type.   In our strain, No. XXIV., there was no doubt that it belonged to the human type.   In the very extensive series of inoculations into calves made under the auspices of the Royal Commission of England the subcutaneous method was used with few exceptions and their results are not therefore comparable to ours.   Oehlecker[7] gives a detailed account of one strain isolated from a case of tuberculosis of the right ankle joint which, though belonging to the human type, was quite virulent for cattle.   A steer weighing about five hundred and fifty pounds succumbed to a dose of five milligrams, injected into the jugular vein, in thirty-five days.   A second animal, weighing about six hundred pounds, inoculated with the same dose on a later date, died in twenty-four days.   On the other hand, a dose of one milligram had only the usual temporary effect of human strains on two animals. This culture of Oehlecker's has much in common with No. XXIV:, but it is evidently more virulent for cattle than No. XXIV.

In both strains the inoculation of rabbits has shown itself superior to the cattle test, for the rabbits showed the degree of resistance to both strains which is usually manifested by them towards the human type.

In addition to the definite injuries following the use of certain human strains on young animals which have been described, it has been stated from time to time that calves had been stunted in their development as a result of bovovaccination. Bearing in mind the wide differences in susceptibility among calves towards the same strain we may well credit the statements that certain calves are strongly influenced in their growth by the vaccine while others may be wholly unaffected. Just what functions are disturbed when the nutrition of the animal is affected remains to be investigated.

The introduction of living human tubercle bacilli into the circulation of calves leads to a condition under which tuberculin gives a febrile reaction. For the time being they are like tuberculous animals. The period of time following vaccination during which a positive tuberculin reaction is obtainable varies with the nature of the culture introduced. According to Weber, Titze and Jörn[5] calves treated with bovo-vaccine may react positively towards tuberculin seven months after inoculation, but not one year after. Calves treated with tauruman gave in some cases a positive reaction up to two years and ten months after treatment. Autopsy in these cases failed to show any tubercular lesions. In the cases described in this paper, the tuberculin test was applied at irregular intervals. The results are tabulated below. It will be seen from the dates there given that for the culture denominated Human X. a positive reaction is obtainable at least eight months after vaccination and, in case of No. 210, was obtainable a year after vaccination. Owing to the deaths following the use of Human XXIV. as a vaccine the records of only two animals are available. The animal upon which the treatment acted most severely maintained a sensitized condition for at least eight and one-half months.

TABLE IV.

No.	Time between Last Vaccination and		Culture Used as Vaccine.
	Last Positive Tuberculin Test.	First Negative Tuberculin Test.	
209	7 months, 29 days.	10 months, 2 days.	Human X.
210	12 " 7 "	13 " 13 "	" "
211	7 " 29 "	10 " 2 "	" "
220	5 " 8 "	8 " 18 "	" XXIV.
221	8 " 18 "	....	" "

After an animal has been sensitized to tuberculin by an injection of living bacilli in the form of a vaccine a second injection of vaccine calls forth an immediate febrile reaction as if tuberculin had been administered. This is shown in the notes on Nos. 227 and 229, where the temperature taken every two hours on the day following the injection is given. This prompt rise of temperature following a first vaccination may be regarded as a sign that the animal is already infected. As a rule a rise of temperature unless due to digestive derangements or pneumonia does not occur until at least ten days after the first vaccination.

The attempt to increase the virulence of Human XXIV. by passages through calves has not progressed far enough to be fully considered here. The history of the passages is as follows:

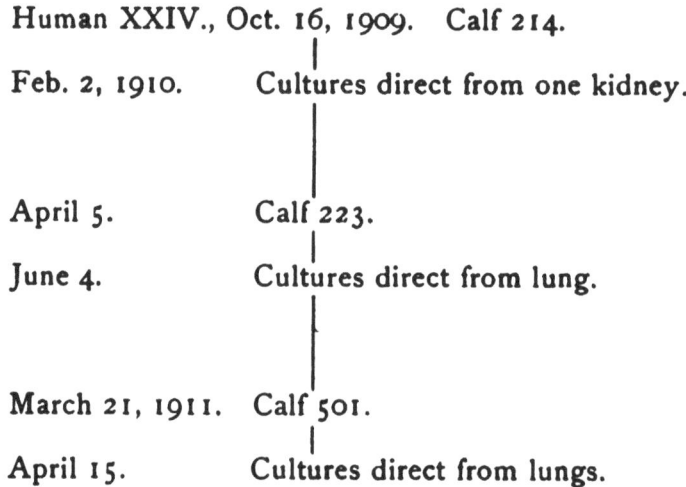

Human XXIV., Oct. 16, 1909.   Calf 214.

Feb. 2, 1910.   Cultures direct from one kidney.

April 5.   Calf 223.

June 4.   Cultures direct from lung.

March 21, 1911.   Calf 501.

April 15.   Cultures direct from lungs.

There was noticeable in the cultures from the first calves a marked increase in the vigor and rapidity of the growth on the usual media.   This is contrary to what might have been expected if the strain was being modified towards the bovine type at this time.   It is to be hoped that there will be opportunity for continuing this series through a number of calves.   The experiments made by others in this direction have been discussed by the writer in a former publication and will not be reviewed here.[8]

CONCLUSIONS.

1.   Calves may succumb to a tubercular pneumonia, not seen in the spontaneous bovine disease, after an intravenous injection of certain cultures of the strictly human type. The initial rise of temperature usually appears within ten to fifteen days and death may ensue after one or two months. Tuberculosis of both eyes associated with complete blindness may be a result.

2.   The culture used in the foregoing experiments, which proved fatal to calves, was rather below than above the average virulence of the human type for rabbits.

3.   The sensitiveness to tuberculin may persist in calves

to from eight to twelve months after an injection of living bacilli of the human type.

4. A second, and a third, larger dose of the same strain is quickly and easily disposed of by the calves which have survived the first dose.

REFERENCES.

1. Journ. Med. Research, 1905, xiii, 253.
2. Journ. Med. Research, 1907, xvi, 435.
3. Journ. Med. Research, 1908, xviii, 451.
4. Journ. Med. Research, xxiii, 185.
5. Tuberkulose-Arbeiten a. d. Kaiserl. Gesundhtsamt., 10. Heft, 1910, 157.
6. Centralbl. f. Bakt. Erste. Abth., 1908, xlvi, 475.
7. Tuberkulose-Arbeiten a. d. Kaiserl. Gesundhtsamte, 6. Heft, 163.
8. Trans. Intern. Congress on Tuberculosis, 1908; Boston Med. and Surg. Journ., 1908, clix, 707.

ORGANIC MATTER IN THE EXPIRED BREATH.[*]

MILTON J. ROSENAU AND HAROLD L. AMOSS.

(From the Department of Preventive Medicine and Hygiene, Harvard.)

The presence of organic matter in the expired breath has long been suspected but never demonstrated. These substances have eluded chemical tests; and animal experimentation has heretofore furnished contradictory and inconclusive evidence. We here present results which we believe demonstrate that the expired breath contains organic matter and that this organic matter is specific in nature.

In the early experiments of Claude Bernard (1857) animals were confined in atmospheric air and in mixtures both richer and poorer in oxygen than atmospheric air. He explained the poisonous effects of carbonic acid when respired to be due to the fact that it deprived the animal of oxygen. Similar results were reported by Valentin and by Paul Bert. Richardson in 1860–61 found that a temperature much higher or lower than 20° C. had the effect of shortening very considerably the lives of animals confined in an unventilated jar. Pettenkoffer shortly thereafter (1860–63) believed that the symptoms observed in crowded ill-ventilated places were not produced by the excess of carbonic acid nor by a decrease in the proportion of oxygen in the air. He further did not believe that the impure air of dwellings was directly capable of originating specific diseases or that it was really a poison in the ordinary sense of the term, but that it diminished the resistance on the part of those continually breathing such air.

The animals exposed by Brown-Séquard to the expired breath of other animals died and he believed they died as a result of poisonous matters in the expired breath. Whether these poisons were of organic nature or not could only be surmised; in fact their very presence was vigorously denied by Billings, Mitchell, Bergey, and others who repeated

---

* Received for publication June 22, 1911.

Brown-Séquard's experiments with contradictory results. The ill effects from breathing air contaminated with the expired breath is now generally assumed to be due to the increased temperature and moisture rather than to the poisons which have so long been suspected and sometimes taken for granted.   Thus Benedict * has kept persons in his calorimeter breathing and re-breathing the same air with a CO_2 content as high as two per cent for twenty-four hours without discomfort, the only precaution being to keep the temperature down and to remove the moisture.   It is to be noted that in these experiments some of the air was passed over lime and sulphuric acid every two hours, and the greater part of the moisture was removed by condensation, which may also remove other substances than the carbonic acid and moisture.

One of the notable achievements of bacteriology was to show that the expired breath during normal quiet respiration is practically sterile.   The moist mucous membranes of the upper respiratory passages act as a bacterial trap.   Before this demonstration it was assumed, and commonly believed, that many of the communicable diseases were transmitted from person to person through specific poisons contained in the expired breath.   The fact that the expired breath usually contains no bacteria robbed it of much of the horror with which it had long been regarded.   When science demonstrated that the expired breath was not particularly dangerous so far as the specific viruses of the communicable diseases are concerned it was entirely acquitted and given a free bill of health in the minds of many sanitarians. Flügge's demonstration of droplet infection and its possibilities partly restored the expired breath to a position of possible danger.

Our changing views towards the dangers in the air are well illustrated by the attitude of the surgeon.   At first Lister and his followers in antiseptic surgery attempted to sterilize the air coming in contact with the wound.   The

* Bulletin 175, Office of Experiment Stations, U.S. Dept. Agric., p. 235.

work of Pasteur, Tyndall and others had fostered the belief that the air was full of dangerous bacteria. The surgeon now largely disregards the air of a well appointed surgical clinic. He, however, ties pieces of sterile gauze about his mouth and nose and over his hair to prevent contamination by these sources.

Since it has been shown that the air under ordinary circumstances is not usually the vehicle by which the communicable diseases and the infections are conveyed, the pendulum has swung to the other extreme, especially as no harmful substances of a chemical nature could be demonstrated in the expired breath. The common experience that a vitiated atmosphere is harmful has recently been explained by its increase in temperature and increase in humidity. Some sanitarians have gone so far as to state that if the temperature and moisture can be kept down and the air kept in motion, say by an electric fan, it may be re-breathed. To this view we cannot subscribe, for we have always felt that a vitiated air must contain substances which are harmful even though not demonstrable to science. This question of ventilation is somewhat foreign though closely related to our present work, but will be discussed in another paper.

An excellent review of the literature up to 1895 is found in the monograph of Billings, Mitchell and Bergey on " The Composition of Expired Air and its Effects upon Animal Life " (published by the Smithsonian Institution). We therefore give only a brief summary of the more important references :

In 1863 Hammond demonstrated the presence of organic matter in vitiated air by experiments upon mice and also by passing the air vitiated by respiration through potassium permanganate. He confined a mouse under a large jar in which the carbon dioxide was taken up by baryta water as fast as it was formed and the moisture absorbed with calcium chloride. Nevertheless the mouse died in forty minutes. The observation was repeated a number of times and death

ensued invariably in less than one hour. Brown-Séquard
and D'Arsonval condensed the moisture in the exhaled
breath, and the liquid thus collected was injected into the
veins of rabbits. Death usually took place in a few days;
sometimes in a few weeks. They believed from this that
they had discovered a volatile organic poison of the nature
of an alkaloid similar to Brieger's ptomaines. These experiments were repeated with variable results, but in 1889 they
reported ingenious experiments in which they obtained
additional evidence in support of their former statements.
Rabbits were confined in a series of jars connected with
rubber tubing permitting a constant current of air to be
passed. The animal in the last jar received the air from the
lungs of the animals in the other jars. This animal died
after an interval of some hours and the animal in the next
died next. The first and second animals usually remained
alive. When they placed absorption tubes containing concentrated sulphuric acid between the last two jars the animal
in the last jar remained alive while the one in the jar just
before was the first to die. This confirmed their belief in
the existence of a volatile poison absorbed by the sulphuric
acid. Haldane and Smith repeated the experiments of
Brown-Séquard and D'Arsonval, using five bottled mice.
They continued the exposure for fifty-three hours without ill
effects to the mice. Beu in 1893 also repeated these experiments and came to the conclusion that acute poisoning
through the organic matters contained in the expired air was
not possible and that the death of the animals was due to
changes of temperature and accumulation of moisture in the
jars. Rauer in 1893, also Lübberd and Peters, concluded
from similar experiments that there are no organic poisons
in the expired air.

Lehmann and Jessen in 1890 collected from fifteen to
twenty cubic centimeters of condensed fluid per hour from
the breath of a person exhaling through a glass spiral laid in
ice. This fluid was always clear, odorless, neutral in reaction
and contained slight traces of ammonia from persons with
good teeth; more from those with poor teeth. Inoculation

of this condensed fluid into animals gave negative results. In 1894 Brown-Séquard and Davis reported further experiments in which they inoculated over one hundred animals with the condensed fluid of respiration and not only confirmed their former statements but were unable to understand the failure of other experimenters, and emphatically reaffirmed that the breath contains a volatile poison and that the death of animals under experimental conditions is not due to an excess of carbon dioxide nor a deficiency of oxygen. This question was studied by Billings, Mitchell and Bergey in 1895, who came to the conclusion that the ill effects of a vitiated atmosphere depend almost entirely upon increased temperature and moisture and not to an excess of carbon dioxide or bacteria or dust of any kind.

It will be seen from this brief review that this question of the presence of organic matter in the expired breath is in confusion.

The present experiments were taken up in order to demonstrate, if possible, the presence of organic matter in the expired breath by means of the reaction of anaphylaxis. It is now generally recognized that the reaction of anaphylaxis is exceedingly delicate, for by its means we are enabled to distinguish so small a quantity as $1/1,000,000$ cubic centimeter of blood serum or $1/20,000,000$ of a gram of purified egg-white — amounts far too small to detect by any chemical method. The reaction of anaphylaxis has the further advantage of being specific, so that if organic matters are found through this reaction their nature may be predicated.

The method adopted consisted, briefly, in condensing the vapors from the expired breath of man, injecting the liquid so obtained into guinea-pigs and, after an interval of several weeks or more, testing the guinea-pigs to determine whether they have become "hypersusceptible" to normal human blood.

The fact that the guinea-pigs showed a definite response clearly indicates that they were sensitized with a protein substance of human origin. This statement is based upon the fact now well established that guinea-pigs may be sensitized only with albuminous matter higher in structure than peptones, and the specific nature of the phenomenon makes it reasonably certain that the protein matter in this case must have been of human origin. At first sight it seems almost incredible to believe that such a complex molecule as protein may be volatile. Nevertheless, such appears to be the natural conclusion to draw from the results of our experiments. After all, the question of volatility may have much similarity to the question of solubility. Theoretically, all substances are soluble, although some in minute amounts; in the same sense all substances may be volatile. Volatility does not mean necessarily a change to the gaseous state in the sense that simple substances are volatile. Thus we may assume that solid and liquid substances may pass into the air in a state of "colloidal suspension." The simplest conception would be to regard the protein as passing off in solution in the watery vapor. Whatever the physico-chemical conception the inference is forced upon us that protein may pass into the air in the expired breath and be again collected in sufficient amounts to produce a definite biological reaction in susceptible animals.

The fact that organic substances, specific in nature, have been definitely demonstrated in the expired breath does not prove that these substances are poisonous. The physiological effect of such organic matter must now be studied.

The following tables show the work in some of its details. The negative as well as the positive results are reported. Only those series are omitted in which the experiments were spoiled by some accident.

SERIES II.

(Some of the series have been omitted on account of the death of the animals or on account of some accident in the experiments. All the negative results are included.)

Oct. 29, 1909.—Liquid condensed from the expired air of Amoss. Collected in accordance with method A. Time of collection, 11 hours. Amount of liquid obtained, 15 cc. Cultures of the liquid developed a spore-bearing organism of the subtilic group. The condensed liquid was warmed to 37° C. immediately after it was obtained and injected at once into the following guinea-pigs:

No.	Date of Second Injection.	First Injection.	Site.	Interval.	Second Injection.	Site.	Result.	Remarks.
1...	12/4/09	5 cc. liquid condensed from expired air of Amoss.	Subcutaneous.	36 days.	.2 cc. normal human serum (Amoss).	Brain (through optic foramen).	Cough once in 5 minutes. Irregular respirations. Convulsions and general paralysis. (Chloroformed 3 hours after injection.)	Reaction ... by pressure ... Autopsy showed basal hemorrhage.
2...		5 cc. same.			.5 cc. same.	Heart.	Agitation in 1 minute. Cough, respirations irregular, jerky, wheezing, with expiration. Fur ruffled, shivering, wobbly, and down. (Chloroformed.)	Autopsy showed no ... one of symptoms.
3...		3 "			.5 "		Coughing and scratching in less than 1 minute. Respiration labored, irregular, and jerky. Down.	Recovered.

Summary of Series II.: Numbers 2 and 3 of this series showed a definite reaction.

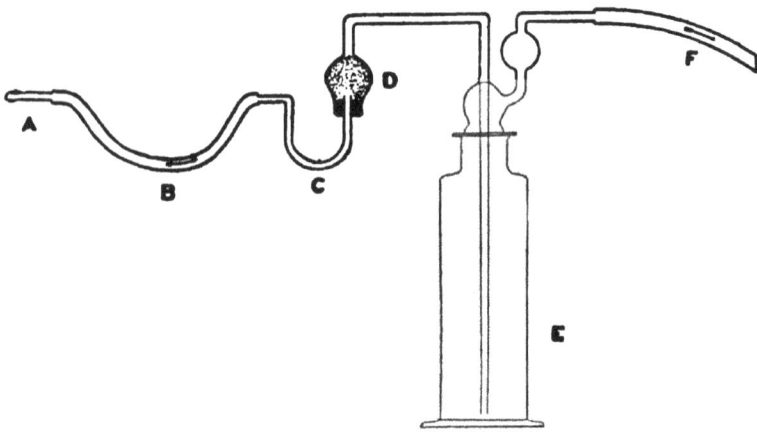

FIG. 1. — METHOD A.

A. Mouth-piece.
B. Rubber connection.
C. Trap.
D. Filter of glass-wool.
E. Drexel bottle which stands in an ice and salt freezing mixture.
F. Rubber connection to vacuum pump.

SERIES III.

Nov. 1, 1909.— Liquid condensed from the expired air of Amoss, 6 hours, then Kenney, 3 hours, into same Drexel bottle. Collected in accordance with method A. Time of collection, 9 hours. Amount of liquid obtained, 17 cc. Cultures from the liquid remained sterile. Remarks: Amoss had rhinitis with coryza. The liquid thus obtained was warmed to 31° C. for 10 minutes and injected at once subcutaneously into the following guinea-pigs:

No.	Date of Second Injection.	First Injection.	Site.	Interval.	Second Injection.	Site.	Result.	Remarks.
1...	12/4/09	8 cc. liquid condensed from expired breath of Amoss and Kenney.	Subcutaneous.	33 days.	.5 cc. normal human serum (Amoss).	Heart.	No symptoms.	
2...	"	3 cc. same.	"	"	.5 cc. same.	"	Ruffling and scratching.	Symptoms suggestive but not positive.
3...	"	5 " "	"	"	.5 " "	"	No symptoms.	

Local reactions: Series III., No. 1, Fluid slowly absorbed, slight edema and inflammation at site of injection. No. 2, No edema; slight induration at site of injection. No. 3, No edema; slight induration at sight of injection.

Summary of Series III.: Only one of the three guinea-pigs in this series gave a suggestive reaction, but not regarded as positive in our conclusions.

SERIES IV.

Nov. 4, 1 99.— Liquid ... air of Amss, 8 urs, thn Kenney, 2 ... ... from the ...d de the chance of ... ... ans, into the same ...e. ...d in ... ...ance with ...thd A. ...d so as to ... ...al ont ... ...in, viz., ...l l ayrs of ... ...ge being wrapped around the neck of the ...tte. Time of ...tn, 10 hours. ...d, 24 ...c. Agar slants of one drop of this liquid showed growth in ...ne ...be; a ...ther ...ilarly. ...d ...d sterile. The ...rd ...id was ...zen as ...n as it was ...id ...d ...ept frozen ...dring the entire ...ic of ...e ...din, then the contents of the Drexel ...te were ...l ...d to ...lt at ...om t ...pe, ...d then warmed for 6 min ...es in a water bath at 30° C. The ...id thus ...d was ...id into four parts, viz.:

(a) Without further treatment i ...jd at once into guinea-pigs. ...d to ...d over ...ght at 5° C., then injected into guinea-pigs.

(b) ...d at once into guinea-pigs. (b) ...r night at 5° C., ...d ...thn injected into ... ...igs. (d) A ...ll

(c) A ...ll ...unt of alkali ...d ...d ...thn ...ll ...d to stand ...r night at 5° C., and ...thn injected into ...

amount of ...id ...d and ...thn ...d to stand ...r night at 5° C., and ...thn injected into ...gpigs.

SERIES IV. A.

The following five guinea-pigs were sensitized with the liquid obtained at once after thawing and without further treatment:

No.	Date of Second Injection.	First Injection.	Site.	Interval.	Second Injection.	Site.	Result.	Remarks.
1....	12/7/09	5 cc. liquid condensed from expired breath of Amoss and Kenney.	Subcutaneous.	33 days.	.5 cc. normal human serum (Amoss).	Heart.	Expiratory difficulty, scratching, coughed in 4 minutes, drowsy, fur ruffled, shivering. Symptoms subsided in 30 minutes. (Chloroformed.)	Autopsy: no mechanical cause of symptoms found.
2...		4 cc. same.			.4 cc. same.		Shivering, breathing expulsive, coughing motions (no sound). Quiet in ½ hour. (Chloroformed.)	Autopsy: no mechanical cause of symptoms.
3...		3 "			.1 "		Agitation and shivering, scratching. Symptoms subsided. (Chloroformed.)	Autopsy: clot in pericardium.
4....	12/7/10	2 "			.1 "		Typical reaction: scratching, coughed in 6 minutes, down, fur ruffled, etc. (Chloroformed.)	Autopsy: small clot around base of heart.
5...		1 "			.2 "		No symptoms.	Autopsy revealed very small clot in pericardium.

SERIES IV. B.

2.6 cc. of the liquid was made up to 3 cc. with sterile water and allowed to stand for 14¾ hours at 5° C. when it was warmed to 20° C. for 5 minutes and then injected into the following guinea-pigs:

(The liquid was diluted in this experiment to serve as a control for Series IV. c. and IV. D.)

No.	Date of Second Injection.	First Injection.	Site.	Interval.	Second Injection.	Site.	Result.	Remarks.
6...	12/7/09	3 cc. liquid condensed from expired breath of Amoss and Kenney and allowed to stand at 5° C. for 14¾ hours.	Subcutaneous.	32 days.	.1 cc. normal human serum (Amoss).	Heart.	Coughing, scratching, expiratory difficulty, agitation, typical convulsions, hiccough. Typical reaction. Recovered. (Chloroformed.)	Autopsy: no mechanical cause of symptoms.

SERIES IV. C.

2.6 cc. of the liquid was placed in a sterile test-tube, with .3 cc. of normal sterile sodium bicarbonate solution, and .1 cc. of sterile water (this makes a solution of 3 cc. of N/10 sodium bicarbonate).

The test-tube was allowed to stand over night (14¾ hours) at 5° C, then warmed at 20° C. for 5 minutes and injected subcutaneously into guinea-pig No. 7.

No.	Date of Second Injection.	First Injection.	Site.	Interval.	Second Injection.	Site.	Result.	Remarks.
7...	12/7/09	3 cc. liquid allowed to stand in alkaline solution.	Subcutaneous.	32 days.	.1 cc. normal human serum (Amoss).	Heart.		Died of pericardial hemorrhage.

SERIES IV. D.

2.6 cc. of the liquid was placed in a sterile test-tube with .3 cc. of N/100 sterile hydrochloric acid and .1 cc. sterile water added (this makes a solution of 3 cc. N/1,000 HCl).

The tube was allowed to stand for 14¾ hours at 5° C. when it was warmed at 20° C. for 5 minutes and then injected subcutaneously into guinea-pig No. 8.

No.	Date of Second Injection	First Injection.	Site.	Interval.	Second Injection.	Site.	Result.	Remarks.
8...	12/7/09	3 cc. liquid allowed to stand in acid solution.	Subcutaneous.	32 days.	.1 cc. normal human serum (Amoss).	Heart.	Coughing, convulsive expiration, agitation, scratching. Typical reaction. Recovered. (Chloroformed.)	1 Apsy: no mechanical cause of symptoms.

Summary of Series IV.: The experiments in Series IV. were evidently designed to test the effect of time, and the action of acids and alkalies on the organic matter in the condensed liquid from the expired breath.

The results show that guinea-pigs may be sensitized even though the liquid is kept for 14¾ hours, cold, and in the dark; and further, that a minute amount of acid had no effect so far as may be judged from the limited data.

The effect of the presence of the alkali was not determined on account of the accident to the animal.

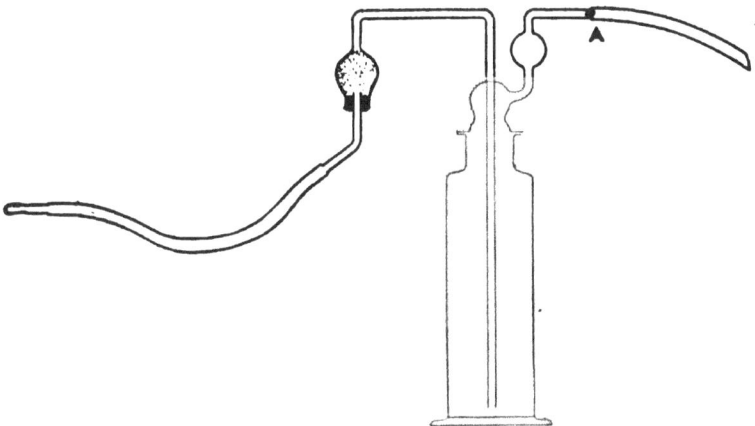

FIG. 2. — METHOD B.

Similar in all respects to method A, excepting the form of the trap, and an additional cotton filter at A, to prevent contamination from the outlet of the Drexel bottle. As a further guard against contamination, the neck of the Drexel bottle in this and all subsequent methods was swathed with several layers of sterile gauze.

SERIES V.

Dec. 17, 1909. — Liquid condensed from the expired air of Amoss. The material was collected in accordance with method B. Time of collection, 7½ hours. Amount of liquid obtained, 50 cc. The condensed liquid was kept frozen in this experiment except for a short interval when it thawed, but was again frozen. The liquid was kept in the Drexel bottle for 15 hours at 5° C., then warmed and injected into guinea-pigs as indicated in the following table. Cultures showed the liquid to be sterile. Biuret and Millon's tests were negative. The glass wool in this experiment was not discolored, as it was in many of the experiments. All tubing and connections were of larger bore than those used in previous experiments, thereby allowing more expired air to pass through in a given time.

No.	Date of Second Injection.	First Injection.	Site.	Interval.	Second Injection.	Site.	Result.	Remarks.
1..	4/1/10	5 cc. liquid condensed from expired air of Amoss.	Subcutaneous.	103 days.	.1 cc. normal human serum (Amoss).	Brain (through optic foramen).	Negative.	
2..	"	4 cc. same.	"	"	.1 cc. same.	Brain (through optic foramen).	"	
3..	"	3 " "	"	"	.1 " "	Brain (through optic foramen).	Convulsive movements. Down, sneezed, coughed.	
4..	"	2 " "	"	"	.1 " "	Brain (through optic foramen).	Negative.	
5..	"	1 " "	"	"	.1 " "	Brain (through optic foramen).	Symptoms masked (?).	
7..	"	4 " "	"	"	.2 " "	Brain (through optic foramen).	"	
8..	"	3 " "	"	"	.2 " "	Brain (through optic foramen).	"	
9..	"	2 " "	"	"	.2 " "	Brain (through optic foramen).	"	

SERIES VI.

Dec. 17, 1909.— Liquid condensed from the expired air of Kenney. Collected in accordance with method B. Time of collection, 6¼ hours with 1½ hour interval after the first 4 hours. Amount of liquid obtained, 18 cc. The warmed glass wool used to filter the expired air was not discolored. Cultures of the liquid were sterile. The liquid was allowed to remain in the Drexel bottle at 5° C. for 15 hours, then warmed to about 30° C. and injected into guinea-pigs subcutaneously as indicated in the following table:

No.	Date of Second Injection.	First Injection.	Site.	Interval.	Second Injection.	Site.	Result.	Remarks.
1...	5/14/10	5 cc. liquid condensed from expired breath of Kenney.	Subcutaneous.	147 days.	.1 cc. normal human serum (Amoss).	Brain (through optic foramen).	Symptoms masked.	Reaction masked.
2...	"	4 cc. same.	"	"	.2 cc. same.	Brain (through optic foramen).	Restless in 15 minutes.	
4...	"	3 " "	"	"	.5 " "	Brain (through optic foramen).	Symptoms of pressure.	
5...	"	2 " "	"	"	.5 " "	Brain (through optic foramen).	Symptoms masked.	
6...	"	1 " "	"	"	.3 " "	Brain (through optic foramen).	" "	

Summary of Series VI.: The results of this series were regarded as negative.

SERIES VII.

Dec. 17, 1 99.— Liquid ... ed from the expired air of Miss J. Collected in ar ie with tbd B at the Deel ttle was smaller (16 es tud of 32). Time of collecti n, 6¾ 1 urs. at of liquid t ad, 40 c. Be liquid was frozen &r the br of in. Culture mia inoculated with the liquid remained sterile. Be gus wol ed as a filter &r the expired air was nt ail ed. Be fan ad liquid was ill ed to remain at a temperature of 5° C. for 18 hours, thn warmed to abt 30° C. in a water-bath and injected subcutaneously into gia-pigs as follows:

No.	Date of Second Injection.	First Injection.	Site.	Interval.	Second Injection.	Site.	Result.	Remarks.
2...	5/14/10	4 c. l iqd con-ad form ex-pired bath of Miss Jones.	Subcutaneous.	147 days.	.2 cc. normal human serum (Amoss).	Brain (through right optic foramen).	Symptoms masked.	
9...		2 cc. same.	"	"	.2 cc. same.	Brain (through right optic foramen).		

Summary of Series VII.: The results of this series were regarded as negative.

Control. — At this point in the work a control was made of the instruments, apparatus, and methods used. This was done to eliminate any faults in the technic by which traces of human protein might have been overlooked. It is well known that it requires exceedingly small amounts of protein to sensitize a guinea-pig.

Dec. 18, 1909.—Using same kind of pipettes, syringes, apparatus, etc., with conditions as nearly alike as possible to those of Series V., VI., and VII., salt solution was injected instead of liquid condensed from the expired air as follows:

No.	Date of Second Injection.	First Injection.	Site.	Interval.	Second Injection.	Site.	Result.	Remarks.
1...	5/14/10	5 cc. salt solution.	Subcutaneous.	72 days.	.2 cc. normal human serum (Amoss).	Brain (through right optic foramen).	Brain injury. Killed by chloroform.	
2...	"	5 cc. same.	"	"	.2 cc. same.	Brain (through right optic foramen).	Negative.	

SERIES VIII.

May 26, 1910.— Liquid condensed from expired air of Amoss. Collected in accordance with method B. Liquid did not freeze. Time of collection, 6 hours. Total amount collected, 26 cc. The liquid was allowed to stand for 25 minutes at room temperature and while still cold was injected into guinea-pigs as follows:

No.	Date of Second Injection.	First Injection.	Site.	Interval.	Second Injection.	Site.	Result.	Remarks.
1...	6/2/10	10 cc. liquid condensed from expired breath of Amoss.	Subcutaneous.	26 days.	.2 cc. normal human serum (Amoss).	Brain (through optic foramen).	No reaction.	
2...	"	6 cc. same.	"	"	.2 cc. same.	Brain (through optic foramen).	"	
3...	"	5 " "	"	"	.2 " "	Brain (through optic foramen).	"	

SERIES IX.

May 26, 1910. — Liquid condensed from the expired air of Miss J. Collected in accordance with method B. Time of collection, 5 hours. Amount of condensed liquid, 40 cc. Liquid did not freeze. Considerable foam in the condensed liquid during collection; this foaming started at about the beginning of the second hour when the condensed liquid reached the height of the inlet tube. The liquid was allowed to stand at room temperature for 25 minutes and injected into guinea-pigs as follows:

No.	Date of Second Injection.	First Injection.	Site.	Interval.	Second Injection.	Site.	Result.	Remarks.
1...	6/21/10	10 cc. liquid condensed from expired breath of Miss Jones.	Subcutaneous.	26 days.	.2 cc. normal human serum (Amoss).	Brain (through optic foramen).	Scratched ' vigorously; coughed, agitation, dyspnea, down, cyanotic.	Typical reaction.
2...		10 cc. same.	"	"	.2 cc. same.	Brain (through optic foramen).	Explosive coughing, scratching, down in 8 minutes.	Positive.
3...		10 " "	"	"	.2 " "	Brain (through optic foramen).	Agitation and down.	

Guinea-pig No. 1 in this series had a small slough at the site of the first injection, and guinea-pigs Nos. 2 and 3 each had a large slough.
Summary of Series IX.: All three guinea-pigs in this series reacted.

SERIES X.

May 27, 1910. — Liquid condensed from the expired air of Amoss. Collected in accordance with method B. Time of collection, 5 hours. Amount of liquid condensed, 50 cc. Air began to bubble through the liquid, which condensed by the beginning of the third hour. Injected at once after collection.

No.	Date of Second Injection.	First Injection.	Site.	Interval.	Second Injection.	Site.	Result.	Remarks.
1....	6/21/10	10 cc. liquid condensed from expired breath of Amoss.	Subcutaneous.	25 days.	.2 cc. normal human serum (Amoss).	Brain (through optic foramen).	Wheezing, expulsive and jerky respiration, down.	Suggestive symptoms masked by central irritation.
2....	"	10 cc. same.	"	"	.2 cc. same.	Brain (through optic foramen).	Agitation.	?
3....	"	10 " "	"	"	.2 " "	Brain (through optic foramen).	"	?
4....	"	10 " "	"	"	.2 " "	Brain (through optic foramen).	No reaction.	
5....	"	7 " "	"	"	.2 " "	Brain (through optic foramen).	Down in 4 minutes.	

Guinea-pigs Nos. 1 and 2 had no slough at site of first injection. Guinea-pigs Nos. 3, 4 and 5 had a slough at site of first injection.

Summary of Series X.: Whether the reactions in this series were those of anaphylaxis or not may be considered doubtful.

SERIES XI.

May 27, 1910.— Liquid condensed from the expired air of Miss J. Collected in accordance with method B, using a 16-ounce Drexel bottle. Time of collection, 5 hours. Amount of liquid obtained, 33 cc. Injected at once after the liquid was obtained.

No.	Date of Second Injection.	First Injection.	Site.	Interval.	Second Injection.	Site.	Result.	Remarks.
1...	6/21/10	10 cc. liquid condensed from expired breath of Miss Jones.	Subcutaneous.	25 days.	.2 cc. normal human serum (Amoss).	Brain (through optic foramen).	Great agitation, down in 14 minutes, convulsions.	Positive.
2...	"	10 cc. same.	"	"	.2 cc. same.	Brain (through optic foramen).	Cough, agitation, down in 13 minutes.	"
3...	"	10 " "	"	"	.2 " "	Brain (through optic foramen).	Cough, down in 10 minutes.	"
4...	"	3 " "	"	"	.2 " "	Brain (through optic foramen).	Agitation, cough, scratching, down in 9 minutes.	"

Summary of Series X.: All four of the animals in this series developed slough at the site of the first injection. All four gave a definite reaction of anaphylaxis.

SERIES XII.

May 27, 1910.— Liquid condensed from the expired air of Kenney. Collected in accordance with method B. Time of collection, 5½ hours. Amount of liquid obtained, 16 cc. The liquid thus condensed was injected at once into guinea-pigs as follows:

No.	Date of Second Injection.	First Injection.	Site.	Interval.	Second Injection.	Site.	Result.	Remarks.
1...	6/21/10	10 cc. liquid f r o m expired breath of Kenney.	Subcutaneous.	25 days.	.2 cc. normal human serum (Amoss).	Brain (through pic foramen).	Negative.	Chloroformed.
2...		6 cc. same.			.2 cc. same.	Brain (tágh optic foramen).		

SERIES XIII.

May 28, 1910.— Liquid condensed from the expired air of Amoss. Collected in accordance with method B. Time of collection, 3 hours. Amount of liquid collected, 30 cc. Injected immediately (after having been warmed) into guinea-pigs as follows:

No.	Date of Second Injection.	First Injection.	Site.	Interval.	Second Injection.	Site.	Result.	Remarks.
1...	6/21/10	10 c. li qd 6m pied breath of Ams.	Subcutaneous.	25 days.	.2 cc. normal human serum (Morse).	Brain (through optic foramen).	Agitation, down and up, jerky respiration.	Positive.
2...		10 cc. same.			.2 cc. same.	Brain (through optic foramen).	No reaction.	Negative.
3...		10 " "			.2 " "	Brain (through optic foramen).	Down, jerky respiration, agitation, cough.	Positive.

Summary of Series VIII.: Guinea-pigs No. 1 had no slough; Nos. 2 and 3 developed slough at site of first injection. Two of the three animals of this series gave a definite reaction of anaphylaxis.

SERIES XIV.

May 28, 1910.— Liquid condensed from the expired air of Miss J. Collected in accordance with method B, using 16-ounce Drexel bottle. Time of collection, 3½ hours. Amount of collection, 26 cc. Warmed and injected at once into guinea-pigs as follows:

No.	Date of Second Injection.	First Injection.	Site.	Interval.	Second Injection.	Site.	Result.	Remarks.
1....	6/21/10	10 cc. liquid condensed from expired breath of Miss Jones.	Subcutaneous.	24 days.	.2 cc. normal human serum (Morse).	Brain (through optic foramen).	?	Positive.
2....		10 cc. same.			.2 cc. same.	Brain (through optic foramen).	Respiratory difficulty, down, convulsions, paralysis. Died in 4 hours.	
3....		6 "			.2 "	Brain (through optic foramen).	Cough, respiratory difficulty, convulsions, paralysis. Died in 3 hours.	

Summary of Series XIV.: Guinea-pigs Nos. 2 and 3 each developed a slough at site of first injection. Guinea-pig No. 1 did not develop a slough at site of first injection. Two of these three animals developed definite symptoms of anaphylaxis.

SERIES XV.

June 1, 1910.—Liquid condensed from the expired air of Kenney. Collected in accordance with method B. Time of collection, 5¾ hours with a 1-hour interval after the 3d hour. Amount of liquid obtained, 48½ cc. Liquid thus collected was brought to room temperature and injected immediately into the following guinea-pigs:

No.	Date of Second Injection.	First Injection.	Site.	Interval.	Second Injection.	Site.	Result.	Remarks.
1...	6/21/10	10 cc. liquid condensed from expired breath of Kenney.	Subcutaneous.	20 days.	.2 cc. normal human serum (Morse).	Brain (through optic foramen).	Jerky respirations and apparently ill.	Recovered.
2...	"	.9 cc. same.	"	"	.2 cc. same.	Brain (through optic foramen).	Injury at base of brain.	Chloroformed.
3...	"	.9 " "	"	"	.2 " "	Brain (through optic foramen).	Jerky respirations and apparently ill.	Recovered.
4...	"	.9 " "	"	"	.2 " "	Brain (through optic foramen).	?	
5...	"	.9 " "	Subcutaneous. (Probably some intraperitoneally.)	"	.2 " "	Brain (through optic foramen).	Agitation, cough, scratching, down.	
6...	"	2.5 "	Subcutaneous.	"	.2 "	Brain (through optic foramen).	Agitation, scratching, convulsions running movements, down.	Recovered.

Summary of Series XV.: No. 4 in this series was the only guinea-pig that developed a slough at site of the first injection. Four of the six guinea-pigs in this series showed definite indications of anaphylaxis.

SERIES XVI.

July 12, 1910.—Liquid condensed from the expired air of Amoss. Collected in accordance with method C. Time of collection, 6 hours. Total amount collected, 46 cc. The liquid was warmed to about 20° C. and injected immediately into guinea-pigs as follows:

No.	Date of Second Injection.	First Injection.	Site.	Interval.	Second Injection.	Site.	Result.	Remarks.
1....	8/13/10	10 cc. liquid condensed from expired breath of Amoss.	Subcutaneous.	32 days	.2 cc. normal human serum (Amoss).	Brain (through optic foramen).	Restless, washed face, fur ruffled, scratching.	
2...	"	10 cc. same.	"	"	.2 cc. same.	Brain (through optic foramen).	Appeared uncomfortable, fur ruffled, shivering.	
3...	"	10 " "	"	"	.2 " "	Brain (through optic foramen).	No symptoms.	
4...	"	10 " "	"	"	.2 " "	Brain (through optic foramen).	"	
5...	"	6 " "	"	"	.2 " "	Brain (through optic foramen).	"	

Summary of Series XVI.: Guinea-pig No. 3 developed small slough at site of first injection, the others none at all. Of this series only the first two animals developed mild symptoms that may have been indications of anaphylaxis.

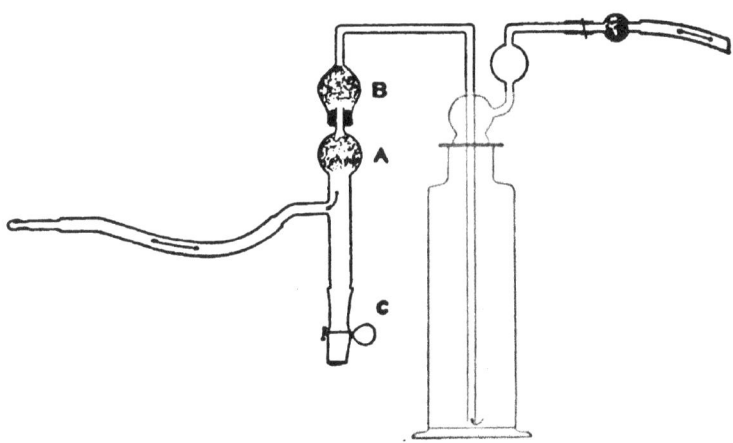

FIG. 3. — METHOD C.

The apparatus in this method contained two glass-wool filters, A and B, and an improved trap controlled with a Mohr's pinchcock. The bulbs A and B were surrounded with a box warmed with a 16-candle power electric globe and the temperature controlled by a thermometer between 37° and 40° C.

SERIES XVII.

July 13, 1910. — Liquid condensed from the expired air of Amoss. Collected in accordance with method C. Time of coll___ ___in, 7½ hours; 1-hour interval after the fourth hour. Amount collected, 51 cc. The liquid was warmed and injected immediately into ___ ___ as follows:

No.	Date of Second Injection.	First Injection.	Site.	Interval.	Second Injection.	Site.	Result.	ths.
2...	8/13/10	10 cc. liquid condensed from expired breath of Amoss.	Subcutaneous.	31 days.	.2 cc. normal human serum (Amoss).	Brain (through pic ___).	Negative.	
3...	"	10 cc. same.	"	"	.2 cc. same.	Brain pic (___ ___).	"	
4...	"	10 " "	"	"	.2 " "	Brain (through pic foramen).	"	
5...	"	9 " "	"	"	.2 " "	Brain pic (___ ___).	"	
6...	"	2 " "	"	"	.2 " "	Brain pic (___ ___).	"	Slight slough at site of first injection.

SERIES XVIII.

July 14, 1910.— Liquid condensed from the expired air of Amoss. Collected in accordance with method C. Time of collection, 6 hours. Amount of liquid obtained, 34 cc. The liquid was allowed to stand at room temperature for 2 hours and injected into guinea-pigs as follows:

No.	Date of Second Injection.	First Injection.	Site.	Interval.	Second Injection.	Site.	Result.	Remarks.
1....	8/13/10	10 cc. liquid condensed from expired breath of Amoss.	Subcutaneous.	30 days.	.2 cc. normal human serum (Amoss).	Brain (through optic foramen).	No symptoms.	Negative.
2....	"	10 cc. same.	"	"	.2 cc. same.	Brain (through optic foramen).	"	"
3....	"	10 " "	"	"	.2 " "	Brain (through optic foramen).	"	"
4....	"	4 " "	"	"	.2 " "	Brain (through optic foramen).	"	"

Number 2 of this series had a considerable slough at site of first injection.

SERIES XXI.

July 20, 1910.— Liquid condensed from the expired air of Amoss and Kenney into same apparatus. Collected in accordance with method C. Time of collection, 4 hours Amoss, and 1 hour Kenney. Amount of liquid obtained, 40 cc. Liquid was warmed to about 30° C. and at once injected into guinea-pigs as follows:

No.	Date of Second Injection.	First Injection.	Site.	Interval.	Second Injection.	Site.	Result.	Remarks.
1....	8/13/10	10 cc. liquid condensed from expired breath of Amoss.	Subcutaneous.	24 days.	.2 cc. normal human serum (Amoss).	Brain (through pic fmen).	No symptoms.	
2...	"	10 cc. same.	"	"	.2 cc. same.	Brain (tl gh pic foramen).	"	
3...	"	10 " "	"	"	.2 " "	Brain (t r ogh optic foramen).	"	

No sloughs following first injection in this series.

SERIES XXII.

July 22, 1910.—Liquid condensed from the expired air of Amoss. Collected in accordance with method C. Time of collection, 5 hours. Amount of liquid obtained, 40 cc. The liquid was allowed to come to a temperature of 30° C. and at once injected into guinea-pigs as follows:

No.	Date of Second Injection.	First Injection.	Site.	Interval.	Second Injection.	Site.	Result.	Remarks.
1...	8/13/10	10 cc. liquid condensed from expired breath of Amoss.	Subcutaneous.	22 days.	.2 cc. normal human serum (Amoss).	Brain (through optic foramen).	No symptoms.	No slough.
2...	"	10 cc. same.	"	"	.2 cc. same.	Brain (through pic foramen).	" "	Some slough.
3...	"	10 " "	"	"	.2 " "	Bain (through pic men).	" "	No slough.
4...	"	10 " "	"	"	.2 " "	Brain (through optic men).	Convulsive movements; jerky respirations.	" "

Series XXXIV.

May 19, 1911. — I inj... in ... with ... thd C. ... 6m expired ... th of K., Miss P., S., A., Ms D., nd R., ... ne using ... t of ... iid ... t... 8, 9 ... ; Miss P., 30 cc.; S., 23 c.; A., 11 cc.; ... ; A., 1 ... ; Ms P., 4 ... ; S., 2 ... ; R., 2 ... Ms D., 3 ... ; Ms D., 3 ... was ... m; the ... wool ... nt discol ... ; Ms D., 34 c.; R., ... c. K. ... a 32-oz. Drexel bottle; liq ... yellow tinge; the ... wool ... nt discol ... ; ... clear; no ... Ms P. ... a 32-oz. Drexel ... d ... ever frozen; had ... ; ... was ... t ... the ... nsg. S. ... d a 64-oz. Drexel ... liquid ... ; ... t ... d ... ; no ... ; d; no ... nt of saliva in the sp. A. ... d a 32-oz. Drexel ... ; liquid ... ; ... ; no ... ; ... el ... ... nt ... dil ed nd ... Ms D. ... d a 64-oz. ... jid was never ... ; ... d; no ... ; ... el ... was nt ... ed. R. ... d a (6-oz. ... ; clear; there ... ... br ... d the ... ... was ... ghly ... d (t ... bao ?). One-half cubic ... r of the ... sh was ... to pl ... in both; the ... was no ... gth ... t in the ... se of S. (staining ... d the ... to be of ... ts of liquid ... ly ... d to 35° C. ... d at ... ne ... no the ... ing ... ; ... (liary origin). The

No.	Date of Second Injection.	First Injection.	Site.	Interval.	Second Injection.	Site.	Result.	Remarks.
1....	6/15/11	9 c. liquid n- ... 6m ... bath of K.	5 cc. into peritoneum. 4 cc. subcutaneously.	27 days.	.2 cc. normal human serum (Amoss).	Heart.	Fur ruffled.	Negative.
3...	"	10 c. li ... 6m ex- ... bath of P.	Subcutaneous.	"	.2 cc. same.	"	Shivering.	"
4...	"	10 cc. same.	Peritoneum.	"	.2 " "	"	Shivering; sudden expiratory movements.	"
5...	6/16/11	10 cc. liquid condensed from expired air of S.	"	28 days.	.5 " "	"	Washed face; agitative; down; looks sick. Chloroformed.	Autopsy; clot in pericardium. Spurted when opened.

No.	Date	Dose	Site		Amount	Symptoms	Result
6...	"	10 cc. same.	Subcutaneous.		.5 "	Negative.	Negative.
7...	"	3 " "	Peritoneum.	27 days.	.5 "	"	
8...	6/15/11	11 cc. ljd on- nd dn cr- rpd nth of A.	Subcutaneous.	"	.35 "	Shivering.	Negative.
9...	"	10 c. ljd n- nd dn cr- rpd nth of D.	"		.4 "	Negative.	
10...		10 c. me.	Peritoneum.		.4 "	Sh in 1 mute; agion; jerky respirations; cyanotic. Down in 4 minutes and up. Limp. Recovered. Chloroformed.	Autopsy: no mechanical cause of symptoms. Lungs O.K. Pericardium O.K.
11...		10 "	Subcutaneous.		.4 "	Sh in 1 minute; respiratory difficulty; down. Convulsions. Died in 4 minutes.	tapsy: no mechanical ose of symptoms. No lesions in pericardium. Mt continued to beat after cessation of respirations. Lungs full when and do not collapse.

SERIES XXXIV.— *Continued.*

No.	Date of Second Injection.	First Injection.	Site.	Interval.	Second Injection.	Site.	Result.	Remarks.
12..	6/15/11	4 cc. same.	Subcutaneous.	27 days.	4 cc. same.	Heart.	...gh in ½ minute; ...; respiratory difficulty. Convulsions in 2 ... ms. ...ed in 4 ... mutes. ..., ...id to beat after respira... ...ed.	Autopsy: slight amount of blood in pericardium. Heart beat at least 10 minutes after being exposed.
13..	6/16/11	10 cc. liquid condensed from expired breath of R.	Peritoneum.	28 days.	.5 "		Scratching; agitation; respiratory difficulty; ...gh in 2 minutes. Down in 5 minutes. ...ks sick; irregular 1 ...as. ...al ...s. Chloroformed.	Autopsy: pericardium O.K. No mechanical cause of symptoms.
14..		10 cc. same.	Subcutaneous.		.5 "		Negative.	

Summary of Series XXXIV.: Of the animals in this series 4 responded typically; 2 died.

SERIES XXXV.

May 27, 1911.—Liquid condensed from the expired breath of A. Collected in accordance with method C. Time of collection, 5 hours. Amount of collection, 35 cc. Liquid frozen during entire experiment. No foaming or bubbling. The glass wool was not discolored. One-half cubic centimeter of the liquid inoculated into broth showed no growth. The liquid was warmed 35° C. and injected immediately into a guinea-pig.

No.	Date of Second Injection.	First Injection.	Site.	Interval.	Second Injection.	Site.	Result.	Remarks.
1...	6/16/11	35 cc. liquid condensed from expired breath of A.	Subcutaneous.	20 days.	.5 cc. normal human serum (A.).	Heart.	Negative.	

Saliva ⁞⁞⁞. — One possible source of ⁞⁞⁞ ⁞⁞⁞ ⁞⁞ kept in mind ⁞⁞⁞ut the ⁞ ⁞⁞⁞⁞, that is, that ⁞⁞⁞ of saliva ⁞⁞ght pass the barrier of the ⁞⁞⁞ ⁞⁞⁞l and ⁞⁞⁞ in the ⁞ ⁞⁞⁞ ⁞⁞⁞. If guinea-pigs can be sensi⁞⁞ed with saliva the least fault in technic in this regard would ⁞⁞⁞ the ⁞⁞⁞s misleading. Every ⁞⁞⁞ ⁞⁞⁞ ⁞⁞en to prevent this accident. The bulbs ⁞⁞ tightly stuffed with g⁞⁞s ⁞⁞⁞l and the ⁞⁞ that bacteria did ⁞⁞ get ⁞⁞gh is a fair indication that ⁞⁞⁞ particul⁞ ⁞e ⁞⁞⁞ did not pass.

It ⁞⁞⁞ ⁞⁞⁞nt to ⁞ ⁞⁞ow ⁞ ⁞⁞r gui ⁞⁞-pigs ⁞⁞sitized with saliva will respond after a proper interval to a second injection of human ⁞⁞⁞ ⁞⁞.

In the following ⁞ ⁞⁞⁞s, Series XXIX., XXX., and XXXI., the guinea-pigs were given a first i ⁞⁞⁞n of an ⁞⁞⁞nt of saliva far in ⁞⁞s of any ⁞⁞⁞ntity ⁞⁞t could possibly ⁞⁞e ⁞⁞en contained in the liquid ⁞ ⁞⁞ ⁞⁞m the expired breath ⁞ ⁞⁞r the con ⁞⁞ns of our ⁞ ⁞⁞⁞s. At the ⁞⁞⁞ ⁞⁞⁞ they ⁞⁞ given normal ⁞ ⁞⁞an ⁞⁞m in the ⁞⁞⁞ ⁞⁞⁞ts and by precisely the ⁞⁞⁞ methods used to ⁞⁞⁞te the presence of ⁞⁞c ⁞⁞r in the ⁞⁞⁞ breath.

The guinea-pigs injected with human ⁞⁞⁞ g⁞⁞ practically no response to a ⁞⁞d injection of ⁞ ⁞⁞an ⁞⁞d ⁞⁞.

SERIES XXIX. — *Saliva.*

Nov. 30, 1910. — The mouth was thoroughly cleansed with boric acid solution, then with sterile salt solution and a few cubic centimeters of saliva were collected into a sterile test glass. The saliva was diluted with sterile salt solution in the ratio of 1 to 10 when it was divided into three portions: (1) without further treatment, (2) passed through a Berkefeld filter, and (3) chloroformed. The first portion (without further treatment) was injected immediately into guinea-pigs as follows:

No.	Date of Second Injection.	First Injection.	Site.	Second Injection.	Site.	Result.	Remarks.
1...	12/29/10	1 cc. (representing .1 cc. saliva).	Subcutaneous.	.1 cc. normal human serum (Amoss).	Brain (through optic foramen).	No symptoms.	
3...	"	1 cc. same.	"	.1 cc. same.	Brain (through optic foramen).	"	"
	"	"	"			"	"

SERIES XXX. — *Filtered Saliva.*

Nov. 30, 1910. — The second portion of the saliva described in Series XXIX. was passed through a Berkefeld filter and injected into guinea-pigs as follows:

No.	Date of Second Injection.	First Injection.	Site.	Interval.	Second Injection.	Site.	Result.	Remarks.
1...	1/12/11	5 cc. filtered saliva (representing .5 cc. saliva).	Subcutaneous.	43 days.	.1 cc. normal human serum (Amoss).	Brain (through optic foramen).	Wheeze, scratched head.	Negative.
2...	"	5 cc. same.	"	"	.1 cc. same.	Brain (through optic foramen).	Down and up immediately.	"
3...	"	5 " "	"	"	.1 " "	Brain (through optic foramen).	Restless, agitation.	"
4...	"	5 " "	"	"	.1 " "	Brain (through optic foramen).	No symptoms.	Negative.
5...	"	5 " "	"	"	.1 " "	Brain (through optic foramen).	Respirations rapid.	"
6...	"	5 " "	"	"	.1 " "	Brain (through optic foramen).	Back up, fur ruffled.	"
7...	"	1 cc. filtered saliva (representing .1 cc. saliva).	"	"	.1 " "	Brain (through optic foramen).	Jerky movements.	"

SERIES XXXI. — *Saliva Chloroformed.*

Nov. 30, 1910. — The third portion of the saliva was allowed to stand in contact with chloroform, 1 cc. of chloroform to 50 cc. of diluted saliva, for 20 hours, at which time the chloroform was evaporated under reduced pressure and the saliva thus treated injected into guinea-pigs as follows:

No.	Date of Second Injection.	First Injection.	Site.	Interval.	Second Injection.	Site.	Result.	Remarks.
1....	1/12/11	5 cc. saliva treated with chloroform (representing .5 cc. saliva).	Subcutaneous.	43 days.	.1 cc. normal human serum (Amoss).	Brain (through optic foramen).	No symptoms.	Negative.
2....		5 cc. same.	"	"	.1 cc. same.	Brain (through optic foramen).	" "	–
4....		5 " "	"	"	.1 " "	Brain (through optic foramen).	Injury to brain.	Chloroformed.
5...		5 " "	"	"	.1 " "	Brain (through optic foramen).	" " "	"
6...	"	5 " "	"	"	.1 " "	Brain (through optic foramen).	No symptoms.	Negative.
7...	"	5 " "	"	"	.1 " "	Brain (through optic foramen).	Jerky respirations.	"
8...	"	5 " "	"	"	.1 " "	Brain (through optic foramen).	No symptoms.	"

SERIES XXIX., XXX., AND XXXI.

The following experi⸺ ⸺ts were ⸺ i ⸺d to ⸺ ⸺e thr albumi⸺ as subst⸺ as ⸺h as ⸻gs, milk, or bl⸺od ⸺ ⸺dd be "volatil⸺" ⸺", ⸺lly to ⸺e protein in the ⸺d fluid. Dry, fil⸺ted air ⸺s ⸺d t⸺h the milk ⸺d over the eggs ⸺d ⸺dd, ⸺n ⸺d in a Drexel b⸺ttle ⸺t in an ice ⸺d ⸺lt mixture. The li⸺d ths ⸺d ⸺s injected into gui⸺ ⸺pigs, ⸺d ⸺r a sui⸺ble interval the ani⸺ls ⸺e tested to ⸺mine their ⸺ ⸺bility.

R⸺s XXIII. — Eggs.

July 23, 1910. — Li⸺d ⸺d from the air ⸺h ⸺d ⸺d over / ⸺le eggs. A⸺ ⸺s al ⸺ed to ⸺s t ⸺gh ⸺ated sulp⸺ ⸺ic acid, th⸺n through a ⸺g ⸺r ⸺ ⸺g line, the to a ⸺tal ⸺e the ⸺ ⸺bs in ⸺r ⸺d the / ⸺t long ⸺lf filled with ⸺le ⸺d eggs where it ⸺s allowed to pl⸺y ⸺r the surface, ⸺d ⸺s th⸺n ⸺n into the ⸺s ⸺y ⸺y ⸺ld as being ⸺d for collecting the liquid f⸺m the expired air. Glass-wool ⸺s ⸺t in the large ⸺be ⸺h the eggs in ⸺r to ⸺t in the c⸺n, ⸺t found to be worthless for this purpose. The tube ⸺s l ⸺t warm to 35° C. by oils of ⸺ ⸺br ⸺g, t⸺h ⸺th ⸺r at this ⸺ ⸺e ⸺s ⸺n at 7.30 A.M. and ⸺n until 6 P.M. 1 ⸺t 36 c. of liquid ⸺s ⸺d in the Drexel ⸺le. I⸺s was ⸺l⸺d to ⸺e to ⸺n ⸺le ⸺d injected at 7.30 P.M. into the following guinea-pigs:

No.	Date of Second Injection.	First Injection.	Site.	Interval.	Second Injection.	Site.	Result.	Remarks.
2...	8/18/10	10 cc. liquid condensed from eggs.	Subcutaneous.	26 days.	9 c. ⸺ted solu-tin white of egg.	Intraperitoneum.	Negative.	
3...	"	10 cc. same.	"	"	9 c. saturated solu-tion white of egg.	"	"	
4...	"	10 " "	"	"	9 c. ⸺ted ⸺lu-tion white of egg.	"	Jerky respirations. Looks sick; sprawl-ing.	

The negative results with eggs may in part be accounted for by the fact that the egg mass in the horizontal tube C (see fig.) soon became dry and hard upon the surface, thereby preventing further evaporation. It is not practical to bubble the air through the egg mixture as in the case of milk on account of foaming.

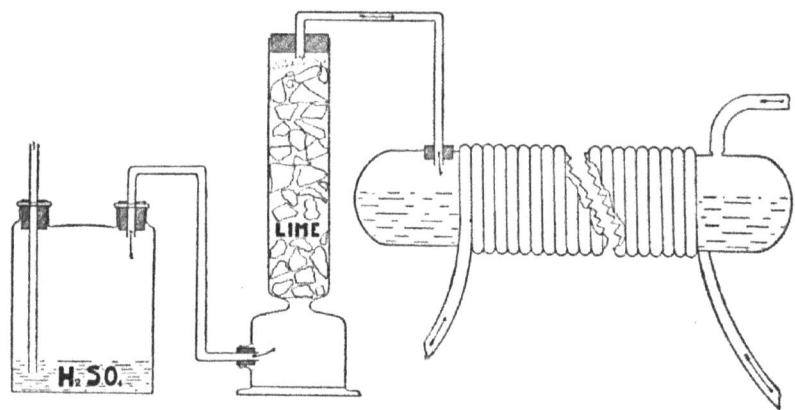

FIG. 5. — METHOD E.

Method used in the Experiments with Egg and Blood.

The air was first dried, being passed through concentrated sulphuric acid in a Woulf's bottle and then through a tower containing lime. The dried air was conducted over the surface of the eggs or the blood in a horizontal glass tube 3″ in diameter and 3′ long, thence to a Drexel bottle (not shown in the diagram) standing in a freezing mixture, where the moisture was condensed. The horizontal tube was surrounded with a coil of rubber tubing through which was circulating warm water at 37° C.

SERIES XXVI. — *Egg white.*

July 28, 1910. — Liquid condensed from the air which had been allowed to pass over the surface of the white of eggs. Collected in accordance with method E. The whites of eighteen fresh eggs were put into the horizontal tube. The air was allowed to pass over the surface of the whites of eggs from 7.30 A.M. until 3 P.M. Amount of liquid obtained was about 36 cc. This was allowed to come to a temperature of about 30° C. and injected immediately into guinea-pigs as follows:

No.	Date of Second Injection.	First Injection.	Site.	Interval.	Second Injection.	Site.	Result.	Remarks.
1...	8/18/10	10 cc. condensations from air passed over whites of eggs.	Subcutaneous.	22 days.	9 cc. saturated solution white of egg.	Intraperitoneum.	No symptoms.	
2...	-	10 cc. same.			9 cc. same.		"	
5...	-	3 " "			9 " "		"	

SERIES XXIV. — *Milk.*

July 26, 1910. — Liquid condensed from the air which was passed through whole milk. Collected in accordance with method D. 200 cc. of fresh whole milk from the "Warelands" was used in the experiment which began at 10 A.M. July 26, and continued until 10 P.M., when the liquid condensed in the second Drexel bottle was warmed to 30° C. and injected into guinea-pigs as follows:

No.	Date of Second Injection.	First Injection.	Site.	Interval.	Second Injection.	Site.	Result.	Remarks.
1...	7/26/10	10 cc. condensations from air bubbled through fresh milk.	Subcutaneous.	23 days.	6 cc. fresh milk.	Intraperitoneum.	No symptoms.	
3...	"	10 cc. same.	"	"	6 " " "	"	"	
4...	"	3 " "	"	"	6 " " "	"	"	

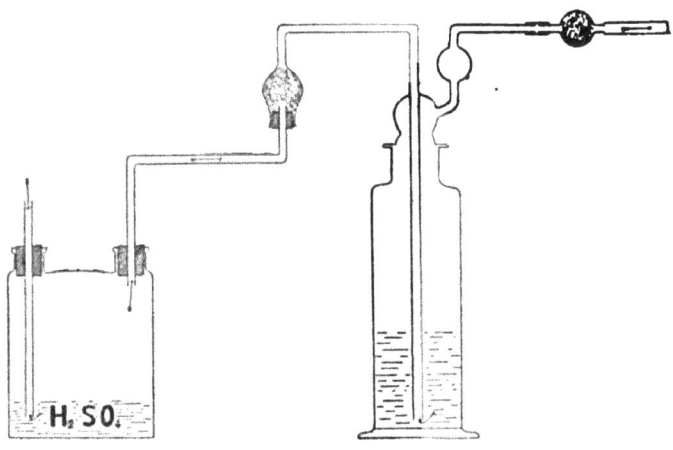

FIG. 4. — METHOD D.

Apparatus used for the Experiments with Milk.

The air was first passed through a Woulf's bottle containing concentrated sulphuric acid, then filtered through glass-wool and bubbled through the milk in the first Drexel bottle and the vapors thus collected were condensed in a second Drexel bottle not shown in the diagram.

SERIES XX. — *Milk.*

July 18, 1910. — Liquid condensed from the air which had passed through milk preserved with boric acid. At the same time that Series XIX. was being run another similar series was carried out under conditions as nearly the same as possible except that the 200 cc. of milk was preserved with 4 grams of boric acid. Amount of liquid collected, 30 cc. Time of collection, 7 hours. It was injected into 3 guinea-pigs as noted in the following table. The second injection consisted of 6 cc. of fresh milk as in Series XIX.:

No.	Date of Second Injection.	First Injection.	Site.	Interval.	Second Injection.	Site.	Result.	Remarks.
1...	8/13/10	10 cc. condensation from milk preserved with H_3BO_4.	Subcutaneous.	25 days.	6 cc. fresh milk.	Intraperitoneum.	No s pigs.	
2...	"	10 cc. same.	"	"	6 " "	"	⬛d; looks sick; down.	
3...	"	10 " "	"	"	6 " "	"	Fur ruffled; looks sick.	

The results are definite enough to suggest positive reactions.

SERIES XIX. — *Milk.*

July 18, 1910. — Whole fresh milk — 200 cc. (milked into sterile bottles at the "Warelands" 6 A.M. July 18) was placed in sterile Drexel bottle at 7.40 A.M. The bottle was placed into a water-bath maintained at 35° C. Condensed liquid collected in accordance with method D (which see). Duration of experiment, 7½ hours. Amount of liquid obtained was 70 cc. The liquid was warmed at room temperature and injected into guinea-pigs as follows (the second injection consisted of 6 cc. of fresh milk from the same dairy taken under the same conditions as the 200 cc. used in the first part of the experiment):

No.	Date of Second Injection.	First Injection.	Site.	Interval.	Second Injection.	Site.	First Injection.	Remarks.
1...	8/13/10	10 cc. condensation from milk.	Subcutaneous.	25 days.	6 cc. fresh milk.	Intraperitoneum.	No symptoms.	
2...	"	10 cc. same.	"	"	6 " "	"	"	
3...	"	10 " "	"	"	6 " "	"	Fur ruffled; looks sick.	
4...	"	10 " "	"	"	6 " "	"	Slight scratching; looks uncomfortable.	
5...	"	10 " "	"	"	6 " "	"	Fur ruffled; looks sick.	

While some of the animals in this series showed very slight symptoms, the reaction was not definite enough to be considered as positive. It will be remembered that milk is not as toxic at the second injection for guinea-pigs as blood serum or egg white.

SERIES XXV. — *Blood.*

July 27, 1910. — Liquid condensed from the air which had been allowed to pass over the surface of fresh defibrinated blood of a calf. Collected in accordance with method E (fig.) The defibrinated calf blood was put into the horizontal tube and the dry air was allowed to pass over its surface from 3 P.M., July 26, to 8 A.M., July 27. Amount of liquid obtained was about 40 cc. This was brought to a temperature of about 30° C. and injected into guinea-pigs as follows:

No.	Date of Second Injection.	First Injection.	Site.	Interval.	Second Injection.	Site.	Result.	Remarks.
1....	8/18/10	1 cc. condensation from blood.	Subcutaneous.	22 days.	1 cc. calf serum.	Intraperitoneum.	No symptoms.	Negative.
2...	"	10 cc. same.	"	"	1 " same.	"	"	"
4...	"	10 " "	"	"	1 " "	"	"	"

SUMMARY AND DISCUSSION.

Through the reaction of anaphylaxis the presence of organic matter has been demonstrated in the expired breath.

In brief the method consisted in condensing the moisture from the expired breath of man and injecting the liquid so obtained into guinea-pigs. After a suitable interval (at least two weeks) the guinea-pigs were injected with normal human blood serum. Ninety-nine guinea-pigs were thus tested in twenty-five different experiments. Of the ninety-nine guinea-pigs twenty-six gave definite symptoms of anaphylaxis; this does not include the animals showing suggestive or mild symptoms. In four of the animals the reaction was so severe that death ensued from anaphylactic shock.

The expired breath of eight persons was tested; of these five gave positive reactions; upon each of the remaining three only one experiment was made.

This organic matter must, according to the interpretations of our knowledge of anaphylaxis in the guinea-pig, be protein in nature. Guinea-pigs may be sensitized only with protein substances higher in the scale (that is, more complex) than the peptones.

Further, this protein substance is specific and for the present it is assumed to come from the blood.

The indications are that this organic matter is probably present in variable amounts, although the reaction is a qualitative and not a quantitative test.

The fact that a number of our experiments resulted negatively may mean either that the organic matter is present in the expired air in exceedingly small amounts or that the guinea-pigs with which we worked did not come from a very sensitive race. There are indications in our work which suggest that the expired breath from certain persons contains more organic matter than from other persons; also that the amount varies with conditions. We obtained a greater percentage of reactions in the guinea-pigs injected with the liquid condensed from the expired breath of females than those injected with the liquid condensed from the expired breath of males. Whether this is a mere coincidence or not may be determined only by collecting more extensive data.

We record a few experiments to determine the effect of time, temperature, acids and alkalies upon the organic matter in the expired breath, but the data are too limited to draw conclusions. It is of practical importance to collect further information along these lines. It would also be interesting to study the organic matter in the expired breath in health and disease; in different ages, etc.

Every possible precaution was taken to prevent contamination with human protein. All glassware was cleaned with chromic and sulphuric acids, and baked at a high temperature for several hours. In most instances the syringe barrels and needles were new. Rubber gloves were worn at the first injection and other precautions taken. Special measures such as traps, filters, etc., were provided to eliminate saliva and particulate matters. That these devices were successful is proven by the fact that the condensed liquid usually showed no growth when inoculated into culture media. Control experiments noted in the text were used to check the technic.

A comparatively large number of the guinea-pigs inoculated subcutaneously with the condensed liquid from the expired breath developed sloughs at the site of the injection. It is not certain whether this was due to the pressure of the relatively large amounts of liquid injected or to some irritating principle contained in the liquid. Occasionally the local effects may have been due to the fact that the liquid was cold when injected. The injection of the condensed liquid caused no other untoward symptoms upon the animals, which is quite contrary to the observation upon rabbits of Brown-Séquard and others.

The best results were obtained when the second injection was given directly into the heart. With a little practice this operation upon the guinea-pig is easily performed and unattended with any ill effects. As human blood serum was used it was not convenient to obtain enough of it to test the hyper-susceptibility of the guinea-pigs by subcutaneous or perito-neal injections. When the second injection was placed under the dura through the optic foramen the results were some-times clouded by the appearance of symptoms which were interpreted to be the result of central irritation.

The logical conclusion from our results is that protein sub-stances under certain circumstances may be volatile. It seems unlikely that such a complex molecule should possess the power of passing into the air in a gaseous form. The volatility, however, now in question may resemble that solu-bility which deals with particles in suspension in a physico-chemical state (colloidal suspension). The protein may simply be carried over in "solution" in the watery vapor.

Our experiments are too few to state that albuminous sub-stances such as egg-white, milk, or blood serum in vitro is "volatile." However, they are sufficiently suggestive to stimulate further work along this line.

The fact that organic matter is present in the expired breath does not mean that these substances are poisonous. The physiological effects should now be studied. It is evi-dent, however, that the air contains many substances which we cannot at present discern, some of which may have an important bearing upon health. Thus it is well known that most of the cases of sudden death following the first injection of horse serum (diphtheria antitoxin) occur in adults. Chil-dren rarely give a severe immediate reaction at the first injection. How the adults become sensitized has always been a mystery. It may now be assumed that some suscep-tible persons may absorb, through the lungs, enough horse protein, from close association with horses, to become sensi-tized. This hypothesis will be tested by exposing guinea-pigs to the expired breath of horses and then testing their power of reaction to horse serum. The expired breath of other animals will be similarly tested and the condensed breath of animals other than man will also be studied.

The fact that the expired breath contains definite amounts of specific organic substances will also have an immediate bearing upon the problems of ventilation and the effects of vitiated air. There has recently been a growing tendency to regard the ill effects of vitiated air as due to the increased temperature and moisture, but it is now apparent that there are other factors which must be taken into account.

A STUDY OF PRIMARY INTIMAL ARTERITIS OF SYPHILITIC ORIGIN.[*]

FRASER B. GURD, M.D., AND H. W. WADE.

(*From the Laboratories of Pathology and Bacteriology, Tulane University of Louisiana.*)

In general, the lesions of the vascular system resulting from the presence of the Treponema pallidum in the body, in addition to those which are frankly gummatous in character, fall into two groups. In the large vessels, especially, there is an involvement of the vasæ vasorum resulting in degenerative changes in the media. Such lesions are those characterizing the so-called mesaortitis syphilitica. The other type is known as endarteritis obliterans. Andrews'[1] describes the essential vascular lesion of syphilis as a perivascular infiltration with round cells, especially plasma cells, together with a hyperplasia of the intima. He gives several excellent plates, but in none is it clear that the occlusion of the lumen is due to endothelial proliferation. Although this condition is usually spoken of as characteristic of the disease, typical lesions are very infrequently found, excepting in the smallest arterioles and venules, so much so that many authors have doubted the correctness of the term, considering that the thickening of the intima is due in part to a compensatory hyperplasia following degenerative changes in the media, but that in advanced cases it results from the organization and canalization of thrombi formed upon the inner aspect of the vessel.

Klotz[2] gives as his opinion, that rarely, if ever, does obliteration of a vessel take place as the result of an intimal proliferation simply. He along with others considers the complete occlusion to be due to thrombus formation.

We have been fortunate in having been able to study the tissue from a case of syphilis in which the size of the vessels involved in the lesion is sufficiently large to make

---

* Read before the American Association of Pathologists and Bacteriologists, Chicago, 1911. Received for publication May 10, 1911.

easy the study of the changes in the three coats. The distinctness with which the changes are demonstrated, in view of the difficulty usually experienced in procuring tissue presenting early and simple syphilitic vascular lesions, warrants, we believe, the publication of the results of our study. As will be seen, the primary endothelial nature of the intimal change is proved, also the possibility of complete vascular occlusion as the result of simple intimal proliferation. In addition, a certain amount of evidence is brought to bear upon the blood supply of the smaller vessels.

In the lesions commonly found, in which there is a narrowing of the lumen due to the thickening of the intima, degenerative areas are usually present in the media; whereas the thickening of the intima is due, apparently, not so much to a proliferation of the endothelium but to an increase in fibrous, hyaline or calcareous material between the internal elastic lamina and the endothelium. So constantly are the degenerative changes in the inner portion of the media, with frequent thinning of the whole wall, found, that usually it is difficult to state with any degree of certainty whether the intimal change is primary or of the nature of a compensatory lesion purposed to give support to the weakened media.

Mallory[3] describes the reaction in the vessels, as the result of the action of the treponema, as a proliferation of fibroblasts which leads to narrowing and more or less complete occlusion, especially in the intima.

In other instances of so-called endarteritis obliterans, notably those occurring in middle-aged individuals in the vessels of the lower extremity, such as have been extensively studied by Buerger,[4,5] the lesion presented has led him as well as others to consider the obliteration of the lumen to be due entirely to the organization and canalization of thrombi formed in situ within the lumen of the vessel. Inasmuch as the evidence as brought forward by Buerger is against the syphilitic origin of these lesions, there is perhaps no reason why the changes discussed in the present paper should be compared with those found in the anterior tibial and other branches in this condition; since, however, the picture presented by many of the vessels in our sections, showing the more advanced lesions, corresponds closely to many noted in the vessels studied by Buerger, it may be justifiable to reopen the question as to the thrombotic or proliferative nature of these lesions. Although Buerger explains the general thickening of the vessels by a canalization of organized thrombi, the cause of the primary thrombosis is by no means clear. There does not appear to be sufficient reason, especially in view of the appearance presented by the vessels in our sections, for excluding a true endarteritis obliterans as the primary cause of the thrombosis which, as Buerger rightly explains, is the cause of the obliteration of large portions of the vessels.

That it is possible to induce the production of experimental exudative and proliferative lesions of such a grade as to completely occlude the vessel has been amply proved by the work of Klotz, Duval, and others. Experimentally, however, only the smallest vessels have been successfully obliterated by an intimal proliferation.

The patient from whom the tissue studied in the preparation of this paper was procured is a male of thirty years with a history of syphilis twelve years ago. Upon Aug. 12, 1910, he was operated upon, at which time a superficial ulcerating area over the sacrum was excised. This lesion was not considered luetic; no specific treatment was therefore used. Three months later he returned showing enlarged inguinal glands upon the right side and hard nodular areas in the erector spinæ on the same side. Sarcoma with metastasis was diagnosed and the new tissue removed.

Upon removal the material consisted of several small and one larger piece of tissue, the latter being composed of a firm mass of grayish glistening material with a reddened periphery, made up of two distinct oval-shaped portions, each measuring two to 2.5 cubic centimeters in diameter, the whole being surrounded by about .5 cubic centimeter of fat. The smaller pieces consist of firm deep red irregular masses, measuring from .5 to 1.5 cubic centimeters in diameter, showing numerous small grayish glistening pin-head sized areas. From the appearance of the gross and microscopic preparations a diagnosis of syphilis was made.

One month later treponemata were demonstrated in the scrapings from a sub-mucous gummatous ulcer in the hard palate.

The tissues were fixed in Zenker's solution and imbedded in paraffin.

Sections were stained with eosin methylene blue, phosphotungstic acid hematoxylin, Mallory's connective tissue stain and Weigert's elastic tissue stain. Formalin-fixed specimens were also prepared after the method of Levaditi and cut after paraffin imbedding.

No treponemata were identified by the last mentioned method, although several collections of granular masses which may have represented degenerated organisms were found. That the treponemata were not found may have been due to the fact that the tissue was not fixed until after a period of three hours following excision had elapsed, this length of time being usually sufficient, in warm weather, to lead to the destruction of the organisms.

Description of sections. — The examination of one section composed of a portion of a lymphatic node surrounded in part by fibrous and fatty tissue presents the following appearance: It is composed of two parts, the major consisting of necrotic tissue, along the border of which lies a narrower zone of tissue which stains well. The tissues everywhere are infiltrated by lymphoid and plasma cells with a moderate number of eosinophiles; in places, too, are collections of apparently proliferated fibroblasts.

The necrotic area is composed in part of a mass of granular débris, but is chiefly made up of fatty and fibrous tissue in which are seen the remains of inflammatory cells and in which are several moderate sized blood vessels and many smaller ones. The larger blood vessels in the area contain thrombi.

The most striking lesion is that found involving the blood vessels in the area which is apparently alive. Several vessels are seen, two of which show lesions characteristic of all. The smaller measuring about four hundred and fifty microns from the outer border of the media demonstrates an early lesion. In this vessel the adventitia is thickened and infiltrated with lymphoid and plasma cells and large mononucleated cells of the lymphoblast type with vesicular nuclei and non-granulated protoplasm. There is a slight increase in collagenous fibrils and an increase in elastic fibers in this area.

The outer layers of musculature are also the site of a few lymphoid cells, the greater portion, however, of the media appears normal, there being no evidence of any degenerative change in the muscle cells nor increase nor decrease in collagenous or elastic tissue fibrils. The internal elastic lamina is normal as is also the sub-endothelial fibrous tissue layer.

The endothelium has proliferated, encroaching upon the lumen of the vessel upon all sides. From four to six cells are heaped upon one another, the later lying next to the vessel wall being loosened and attached only by fine projections of the endothelium. The majority of the cells are club-shaped, the larger portion of each cell being toward the lumen. The individual cells aside from their unusual shape are practically normal in appearance; several, however, contain moderate sized vacuoles. Several lymphoid cells are seen lying between the first and second layer of endothelial cells. No mitotic figures can be found.

A second vessel lying close to the one just described, being about twice the diameter, presents a more advanced lesion. The adventitia presents a similar appearance to that described for the smaller vessel; there is, however, a greater increase in elastic tissue. All but the inner layer of musculature is broken up, the result of an infiltration by lymphoid and plasma cells and lymphoblasts. The musculature shows no evidence of degeneration nor is there an increase in connective tissue.

The internal elastic lamina of this vessel is normal. The lumen of the vessel is occluded excepting for a fine slit in the center, which is but slightly wider than the diameter of a red blood cell. Immediately surrounding this opening is a layer of endothelial cells which are somewhat swollen but otherwise appear normal. Between these cells and the internal elastic lamina is a loose network of young connective tissue cells with a few fine collagenous fibrils scattered about. Between these fibrous tissue cells are large numbers of lymphoid and plasma cells and lymphoblasts.

Other sections showing the same two vessels demonstrate in them an appearance that is very instructive. Both show a somewhat more advanced lesion than those just described.

In the smaller vessel the condition of the wall is similar, the lumen is, however, more nearly occluded. The central opening is somewhat triangular in shape and about twenty-five microns in diameter. Toward the center are endothelial cells three or four cells deep. These cells are swollen, but show no other change except for fine prolongations which protrude into and across the lumen joining those from cells situated opposite. Between the endothelial cells and the lamina are seen a few connective tissue cells and numerous lymphoid and plasma cells. The external elastic lamina is normal in appearance.

The larger vessel is completely occluded, no endothelium being made out with certainty. The mass of obliterating tissue is composed of connective tissue cells and numbers of lymphoid and plasma cells and lymphoblasts. The muscula-ture of this vessel is broken up by numbers of cells of the lymphoid series, but otherwise shows no change. Near the elastic lamina the cells situated in the lumen are separated in places, and minute spaces filled with red blood cells are seen. These spaces are apparently lined by endothelium.

Since the changes present in the two vessels described are characteristic of all, it is unnecessary to mention others in detail. It may be pointed out, however, that several vessels are present in which complete occlusion has taken place and whose walls have almost entirely disappeared as the result of cellular infiltration but not of degenerative changes. In one vessel is a single split or space in the internal elastic lamina at which point a small capillary extends from the media into the occluded lumen.

That the lesions are, in truth, due to the action of the treponemata themselves or their toxins is sufficiently well attested, even though no organisms were demonstrated, by the presence of active syphilis as shown by the detection of the treponema from the lesions in the patient's mouth and by the presence of typical gummata in the lymph nodes and in the muscles surrounding them. As no ulceration was present for several weeks prior to operation it can hardly be conceived that organisms other than the treponema pallidum had been active in the production of the lesion.   Cultures, moreover, made from the excised material remained sterile.

Although the picture presented by these sections shows the lesion usually described as being characteristic of syphilis, namely, perivascular cellular infiltration and intimal proliferation, we consider them to be unique, since very infrequently, at least, are such typical appearances found.   The size of the vessels involved is especially unusual.   We have been unable to find adequate description of similar proliferative changes as the result either of syphilis or any other lesion whatever.

The experimental lesions produced by Duval[6] in rabbits and guinea-pigs using attenuated strains of Bacillus mallei demonstrate that the toxin of the bacillus in the blood stream will so act upon the endothelial cells as to stimulate mitosis on the part of these cells resulting in the complete obliteration of the vessel in the case of very minute vessels.   In these experiments, however, only the smallest capillaries, namely, those devoid of distinct muscular and elastic walls, became occluded, all others were followed by an early and in many instances marked degeneration of the inner layers of the media.

Duval reasons, and apparently justifiably, that this medial degeneration is due to the interference with the nutriment of the inner layer of the media as the result of the thickened intima, explaining this occurrence by the hypothesis that although the outer muscular coat is nourished by the vasæ vasorum and the lymph supply arising in the adventitia, the inner layers are nourished by transudation of fluid or by osmosis from the lumen of the vessels.

Klotz by means of the injection of B. typhosus and attenuated streptococci was able also to produce proliferative changes in the intima.   Such lesions also involved the inner medial layer with proliferation of fibroblasts and fatty degeneration of the subendothelial tissues.   In lesions of the Jores type the internal elastic lamina splits, part entering the proliferative tissue.   As is seen in our sections (Fig. 3) no such rupture of the elastic layer has taken place.

That such a nutrient supply is not essential in man nor in animals is abundantly proved by the rapid involvement of the infected thrombus following a phlebitis, for instance, by polymorphonuclear leucocytes and new blood vessels arising from the vessel wall itself. In the case of the veins also it is difficult to conceive that the vessel wall can be nourished to any considerable extent by the venous and deteriorated blood coursing through these vessels.

The complete obliteration of the vessels described in this paper, unaccompanied by any degenerative change in the muscular or elastic tissues, demonstrates that although perhaps normally such an osmosis or transudation may take place such a method of nutrition is not essential.

In addition to the vascular lesions presented by these sections an excellent opportunity is given to study the two types of necrosis occurring as the result of the syphilitic process. In the striped muscles surrounding the lymph nodes are found numerous minute, typical gummata, composed of small necrotic, homogeneous, more or less granular looking areas, presenting no evidence of being composed of dead musculature nor other fixed tissue, surrounded by moderately large cells with vesicular nuclei of the type known as epithelioid cells, which in turn are surrounded by lymphoid and plasma cells, the latter being in preponderance. Within nodules such as these no blood vessels are found. Such lesions represent new tissue formation with necrosis of those parts farthest removed from the blood supply. Other areas of necrotic tissue are evidently the result of the cutting off of the blood supply by the obliteration of vessels. In these areas, which are of the nature of infarcts, the normal fatty fibrous and lymphatic tissue is readily made out together with the blood vessels passing through them, although all take the acid stain and show every evidence of being necrotic.

With regard to the cells whose necrosis results in the formation of the gumma no discussion will be entered into in this paper. That, however, necrosis may result in syphilitic areas in two different ways is evident from the appearance presented by our sections, such types of lesions are by no means infrequent in lues. We believe, however, that in general an insufficient amount of attention is paid to the manner of formation of so-called gummata since there is no evidence for believing that either one or other of these lesions is usually referred to when the term gumma is used. Recently Mallory[3] has stated that gummata are the result of the necrosis of tissues as the result of vascular changes cutting off nutriment to certain areas. He writes further that these focal lesions in syphilis differ materially from those of tuberculosis in that the latter tend to spread indefinitely. That necrosis in syphilis may occur as the result of the cutting off of the blood supply by vascular occlusion is evident, but that this is the usual manner of formation of gummata we are unwilling to admit nor do we recognize the essential difference between tuberculosis and syphilis referred to by him.

CONCLUSIONS.

Total occlusion of vessels having a complete anatomical structure does result from the simple proliferation of the intima. Syphilis must be recognized as one of the factors potent to induce this lesion.

Although nourishment for the inner layers of the media may be derived from the circulating blood in the lumen of the vessel, our preparations show that such a method of nutrition is by no means necessary.

In occlusion taking place in this manner, endothelial proliferation is primary, this being followed by sub-endothelial fibrous tissue hyperplasia. The vascularization of this proliferated tissue takes place by the growth inward of vessels from the adventitia.

Necrosis in syphilis results both by the formation of tubercle-like nodules by the proliferation of cells with breaking down in the center and also as the result of obliteration of vessels as the result of an intimal proliferation.

BIBLIOGRAPHY.

1. Andrews, M. A. In Power and Murphy's Syst. of Syphilis, i, 111.
2. Klotz, Oskar. British Med Jour., 1906, ii, 1767.
3. Mallory. Johns Hopkins Hosp. Bull., xxii, 69.
4. Buerger, Leo. Am. Jour. Med. Sci., cxxxvi, 567.
5. Buerger, Leo. N.Y. Medical Record, Oct. 15, 1910.
6. Duval, C. W. Jour. Exp. Med., 1907, ix, No. 3.

DESCRIPTION OF PLATE I.

FIG. 1. — (High power.)   Photograph showing early lesion; proliferation of endothelium.

FIG. 2. — (Low power.)   Vessel showing almost complete obliteration. Note absence of degenerative changes in media.

FIG. 3. — (Low power.)   Elastic tissue stain.   Two vessels, one showing early proliferative change, the other obliterated with the exception of a minute slit.   Note absence of splitting of internal elastic lamina.

FIG. 4. — (High power)   Portion of vessel presenting complete obliteration showing the type of cells forming the greater part of the occluding material.

FIG. 5. — (Low power.)   Elastic tissue stain.   Note the entrance of capillaries from the media into the lumen penetrating the internal elastic lamina.

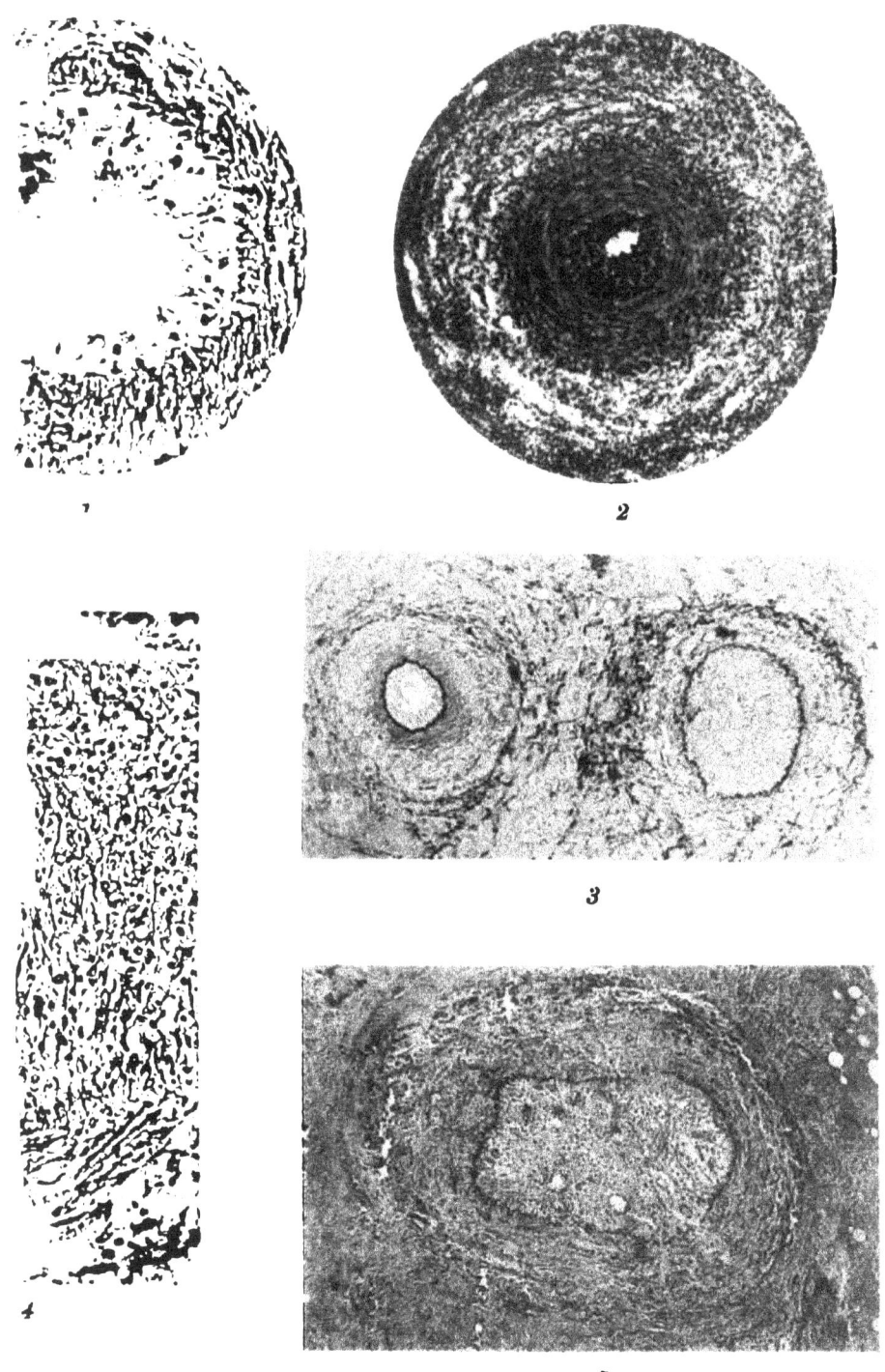

1

2

3

4

5

Syphilitic arteritis

THE RAPID ISOLATION OF TYPHOID, PARATYPHOID, AND DYSENTERY BACILLI.[*]

ARTHUR I. KENDALL AND ALEXANDER A. DAY.

(*From the Department of Preventive Medicine and Hygiene, Harvard Medical School.*)

The marked interest which is being manifested in the methods of spread of typhoid fever and other intestinal diseases of bacterial origin has led logically to a demand for improved and, more specifically, for rapid diagnostic procedures which shall identify these bacteria. The early recognition of the specific organisms concerned in these diseases, furthermore, is a point of special importance, since it has been shown that patients may be sources of infection before the clinical diagnosis can be made, and consequently before appropriate measures are taken to disinfect the excreta. It is well known also that a certain number of apparently healthy individuals may excrete continuously or intermittently considerable numbers of these bacilli, and the ever-growing list of bacilli carriers which have been responsible for definite groups of cases attests to the dangers which these unfortunate carriers present. Without entering into a discussion of the conditions surrounding the early and rapid recognition of typhoid and similar pathogenic organisms of fecal origin, it is apparent that the amount of time which must elapse between the collection of the sample and the formal presentation of the bacterial findings must be a serious handicap in the examination of feces of suspected carriers as well as in the diagnostic work of hospitals and municipal laboratories.

The methods described below, a modification of the well-known Endo medium, together with a procedure for the identification of the organism developing on this medium by agglutinating with a known specific serum of high agglutinating power, can be carried to a definite conclusion in twenty-four hours, or even less.

[*] Received for publication May 29, 1911.

The principles are modifications of well established procedures; consequently no new factors are introduced other
than those leading to the much more prompt growth of the
desired organisms through details to be described.

Briefly, the changes in technic from those usually followed
are:

(1.) The modified Endo medium used by us contains
fifteen grams of agar per liter instead of forty grams. The
colonies developing in fifteen gram agar are much larger in
eighteen hours than those similarly grown in forty gram
agar. The colonies are also more distinct.

(2.) The reaction of the agar is made just alkaline to
litmus, instead of being strongly alkaline to it; consequently
those bacteria forming acid bring about their characteristic
changes very promptly because there is less alkali to
be neutralized before the acid can assert itself.

(3.) After eighteen hours' incubation upon the medium
modified as above, suspicious colonies are removed entire
to small tubes containing one cubic centimeter of sterile
sugar-free broth, which has been kept at 37° C. previous to
use and incubated in them at 37° C. for two hours.

It is obvious that a given initial number of organisms
will be more concentrated, and produce a greater turbidity
in one cubic centimeter of broth than in a much larger
amount, hence the inoculation of the entire colony into this
small volume of broth.

The net gain attained may be summarized as follows:
1, quick growth and differentiation of the organisms sought
for in the modified agar medium; 2, continuity of growth
from agar to broth; 3, rapid growth of bacteria in broth to
such numbers as may be necessary to make a specific microscopic agglutination test with a specific serum. In time,
this gain amounts to nearly three days.

Technic. — Make plain, nutrient sugar-free agar as follows: Tap water (cold), one thousand cubic centimeters;
powdered agar, fifteen grams; peptone (Witte), ten grams;
meat extract (Liebig), three grams. Cook in double boiler

one hour. Make the reaction just alkaline to litmus by the cautious addition of NaOH. Cook fifteen minutes to set the reaction and filter through absorbent cotton.

The tap water should be as cold as possible and the agar should be "dusted" on the surface and allowed to settle into the medium before heat is applied and before the other ingredients are added.

After filtration the medium is stored in flasks containing known amounts, conveniently in one hundred cubic centimeter lots, and sterilized in the autoclave.

To use the medium: (a) Prepare a ten per cent solution of fuchsin in ninety-six per cent alcohol. (b) Prepare a ten per cent solution of sodium sulphite in water.

Add one cubic centimeter of (a) to ten cubic centimeters of (b) and heat in the Arnold sterilizer for twenty minutes = (c).

Add one per cent of lactose (which must be chemically pure) to the agar medium described above, and heat in the Arnold sterilizer until the medium is melted and the lactose thoroughly distributed in it. The decolorized fuchsin solution (c) is then added in the proportion of one cubic centimeter of the mixture to each one hundred cubic centimeters of medium; then thoroughly mixed.

Plates are then poured and allowed to harden (with the covers removed) in the incubator for thirty minutes, after which time they are ready for inoculation.

Preparations of feces for inoculation. — The feces are collected preferably in the small rectal tubes described by one of us.[1] A small portion of feces (about a loopful) is thoroughly emulsified in ten cubic centimeters of sugar-free broth, and preferably incubated one hour at 37° C. prior to the inoculation of the plates. This preliminary incubation does two things: the clumps of bacteria are thrown down, leaving a more uniform suspension of bacteria in the supernatant solution for inoculation, and the bacteria undergo a slight development in a medium particularly suited for their growth.

It should be stated that the transition from feces to

artificial media involves a marked change in the nutritive environment of the bacteria, and experience has shown that cultures grow much more readily if this transition be made from feces to fluid culture media than if the change be made from feces to solid media direct. It is possible, however, to obtain good results even if this preliminary incubation is dispensed with.

Sugar-free broth is essential for this preliminary incubation. If broth containing sugar be used, many of the fecal organisms can form acids, which tend to inhibit the growth of the typhoid, dysentery, or paratyphoid bacilli; the growth of the latter organisms is relatively retarded, and the whole object of the use of broth prior to inoculation of the plates is defeated. The acid formed by this fermentation, furthermore, is frequently sufficient in amount to restore the color to the Endo medium even before the bacteria have grown in it. Hence, sugar-free broth must be used for this purpose.

The thin suspension of the stool having been prepared, and incubated as outlined above, is now rubbed gently but firmly upon the surface of the agar plates by means of sterile glass rods in the usual manner and incubated for eighteen hours at 37° C. At the end of the period of incubation, translucent, colorless " dew drop" colonies, which usually range from one to two millimeters in diameter are sought for and, when found, removed entire to small broth tubes (which contain one cubic centimeter of broth, and which have been kept at body temperature previous to use) and incubated for two hours at 37° C. At the end of this time there will be sufficient growth to make the customary microscopic agglutination tests, using a known serum of high agglutinative power in the usual manner. It is the custom in this laboratory to agglutinate at 1/200 and 1/500, and it has been our experience that organisms agglutinating typically at these dilutions have always been the ones sought for.

Confirmatory cultural characters may be readily obtained by inoculating suitable media from the same tubes as those from which the organisms for agglutination were obtained.

It has been our custom to observe the following order in our work: The samples are brought in about 4 P.M., a thin suspension of each is made as described above and incubated for one hour at 37° C.

While the suspension is being incubated, the plates are prepared and hardened in the incubator.

At the end of an hour, the plates, which are hardened by this time, are inoculated from the suspension, and incubated until nine o'clock the next morning. Suspicious colonies are removed whole to one cubic centimeter tubes of sugar-free broth, and incubated two hours. While the incubation is taking place, the sera are diluted, ready for use.

The cultures are then agglutinated with appropriate sera, and watched for two hours. At the end of that time the results are reported. The whole procedure as outlined requires about twenty hours.

REFERENCE.

1. A. I. K.   Boston Med. and Surg. Jour., clxiv, 1911, 301.

CALIFORNIA

AN INVESTIGATION ON THE PERMEABILITY OF SLOW SAND FILTERS TO BACILLUS TYPHOSUS.[*]

EDWARD B. BEASLEY, M.D., DR. P.H.

(From the Department of Preventive Medicine and Hygiene, Harvard Medical School.)

At the present time a supply of safe drinking water is recognized as an absolute necessity to any community which desires to keep certain well understood types of intestinal diseases at a low level. Indeed the Mills-Reincke phenomenon and Hazen's theorem indicate that the good effects of using a safe drinking water do not cease at the domain of intestinal disorders, but they also extend their influence towards a slight diminution in the mortality from some other diseases.

In this country our main water-borne disease is typhoid fever — a disease which is annually responsible in the United States for a sick list of three hundred and fifty thousand people, thirty-five thousand deaths, a tremendous amount of unrecorded suffering and misery, and a gigantic financial loss. It is no wonder, then, that any system for the purification of water, in order to be considered efficient, must eliminate, as completely as possible, the organisms causing this disease. That good drinking water would surely diminish markedly the many epidemics of typhoid fever is well shown by Schuder's compilation of six hundred and thirty-eight epidemics of typhoid in which he found that seventy-one per cent were attributed to infected drinking water. A very distressing feature of water-borne epidemics of typhoid is the high mortality which usually prevails, while milk epidemics usually form a striking contrast by their characteristic mildness. A possible explanation of this contrast might be the fact that, as multiplication never or seldom takes place in water, an individual would swallow the immediate generation of typhoid organisms from an infected person; whereas in milk multiplication occurs — at times rapidly — and any one drinking such milk would probably receive a somewhat attenuated strain of B. typhosus.

[*] Received for publication June 22, 1911.

It seems justifiable to assume that any scheme for the purification of water which will not permit the presence of typhoid bacilli in the treated supply — when they were known to be in the raw water — is ample protection against a water-borne epidemic of either dysentery or cholera, in case these diseases should ever have the same relative opportunity to infect a water shed. Accordingly, to insure against typhoid infection of a drinking water is to safeguard against other diseases which might be transmitted through the same source.

Ozone, the ultra-violet rays of light (produced by a modified Cooper-Hewitt lamp immersed in the water to be sterilized), "bleach" (Hypo-chlorids), mechanical filtration with or without the use of coagulents, intermittent filtration, slow sand filtration and storage combined with sedimentation have all been used for rendering a polluted water fairly safe for drinking purposes. Each year sees the adoption of one or more of these methods by communities eager to obtain the improved health conditions which usually ensue. Naturally, the relative merits of these processes vary; but, when every factor is taken into consideration, it would be safe now to state that the laurels — for the best and most constant results — belong to the well designed and carefully operated slow sand-filtration plant.

It is, of course, desirable that filters shall have a high bacterial efficiency, but the important point is that their hygienic efficiency (*i.e.*, the ability of a filter to hold back any pathogenic organisms which may be in the raw water) shall be as nearly perfect as possible. Since bacteriological advance, so far, has failed to reveal how many typhoid bacilli are necessary to cause the disease in any individual; and, furthermore, as clinical evidence is lacking to indicate just what conditions must exist in a person in order that the ingested organisms may gain sufficient foothold to produce typhoid fever, it would seem incumbent on all communities to protect their citizens from the most minute "doses" of this organism in the drinking water, and from all other sources when possible.

It was realized about 1885 that bacterial efficiency was an important feature in a sand filter, although prior to this time the filters were used mostly to rid the raw water from as much organic matter as possible. However, an epidemic of typhoid fever in Berlin which was caused by a freeze at the Stralau filters — destroying their hygienic efficiency to a large degree — stimulated Piefke and Fränkel to test the permeability of some experimental filters to typhoid and cholera organisms.

Piefke and Fränkel, in 1890, used two wooden vats for their experimental filters with a height of 2.1 meters, a diameter of .75 meter, and with a sand layer six hundred millimeters deep. The bacteria were grown in nutrient bouillon which was diluted, when applied, in the proportion of one part culture to twenty parts sterile water. The result showed that both typhoid bacilli and cholera vibrios succeeded in passing through the filters. In 1891 Piefke repeated the experiments as some objections had been made to those in the previous year. He found that the results of 1890 were confirmed in all of the important features.

In 1891, 1892, and 1893 similar experiments were carried on at Lawrence, Mass., by means of intermittent and slow sand filters and the results were recorded in "The Massachusetts State Board of Health Reports" for those years. The filters used were tanks lined with galvanized iron and usually having an area equal to one twenty-thousandth of an acre. The experiments were conducted on an extensive scale and with varying thicknesses of the sand-layers, various rates of flow, etc.

It seems well worth while reviewing some results obtained in 1892 with an intermittent filter at Lawrence, although the present paper deals only with slow (continuous) sand filters. The sand layer in the Lawrence filter was five feet three inches deep and the infected water (containing one hundred and forty-five thousand typhoid bacilli per cubic centimeter) was applied continuously for ten hours. The organisms were obtained from bouillon cultures and some sodium chloride was added to the applied water in order to trace the bacteria through the filter. Samples of the effluent were plated each hour and a summing up of the results showed that an average of .05 per cent of the applied typhoid bacilli passed through the filter. Chemical analysis of the effluent was made hourly and at the very hour when the chloride per cent was highest, the greatest number of bacilli appeared in the sample. This coincidence is most interesting as it would seem logical to suppose that after the bacteria succeeded in penetrating the "schmutzdecke" they would naturally have been retarded more readily, in their downward passage, by the mere mechanical action of the sand than substances in solution would have been. However, some work done by Abba, Orlandi and Rondelli in 1899 in the filter galleries at Turin demonstrated the incorrectness of this view. By using water containing a suspension of B. prodigiosus and methyl-eosin and uranin in solution, they showed that, for distances of approximately one-eighth of a mile, the bacteria were able to permeate the soil almost twice as rapidly as the dyes in solution.

Attention is particularly invited to the fact that many of the slow sand filters used in the Lawrence experiments during 1892 and 1893 contained a loam layer which is practically never used in the large filters of to-day, since it keeps the rate down, and in addition requires a more frequent cleaning. In 1892, however, three slow sand filters, Nos. 37, 38, and 39, were used at Lawrence to filter water infected with B. typhosus applied in bouillon cultures by means of small doses every two hours for ten hours with the following results:

No. of Filter.	Depth of Sand Layer.	Effective Size of Sand.	Efficiency.	Rate in Gallons Per Acre Daily.
37..........	5 feet.	.19 mm.	100%	1,540,000
38..........	2 "	.19 "	99.16%	1,500,000
39..........	1 foot.	.19 "	99%	1,500,000

The exact number of typhoid bacilli applied was not stated. The "effective size" is smaller and the rate lower than that used in actual practice in American filters and the results do not apply to our large filtration plants. Although selected from the "Massachusetts State Board of Health Reports," because they conformed to what is now commonly . understood by a slow sand filter, they seem to indicate that the thickness of the sand layer is a very considerable factor in determining the resulting hygienic efficiency. This observation would not hold good unless approximately the same number of typhoid organisms were added to the water on each filter.

In view of the enormous sanitary importance of a filter's hygienic efficiency, it was deemed advisable to undertake again some experiments along this line. At the time of the Berlin and Lawrence work it was an exceedingly difficult problem to differentiate with any degree of certainty B. typhosus from B. coli and several other more closely allied forms. To recognize typhoid colonies in mixed cultures in and on gelatine plates is a rather uncertain task. Even with our present methods for differentiating B. typhosus from some other organisms, obstacles in the road to a correct diagnosis occur. In addition to these cultural drawbacks, it seems likely that a seepage of the infected water might readily have occurred between the sand and wooden walls, and between the sand and galvanized iron walls of the Berlin and Lawrence filters respectively, as the adhesion between the sand and these substances is usually slight. Then, too, the intermittent infecting or "dosing" of the applied water with typhoid organisms, instead of a continuous infection, would probably give the bacilli less chance to permeate the filters. On account of our present crude methods of sewage disposal certainly many of the large cities using surface water supplies must have B. typhosus in their untreated water almost continuously.

It was not until 1896 that the so-called Widal reaction was discovered, and this of course is one of the critical tests for the identification of B. typhosus, although it is by no means infallible. Most of the trouble in working with specimens suspected of containing typhoid bacilli has been in isolating B. typhosus when so many other hardier forms were present. Since 1902 many excellent media — designed for this very purpose — have come into use with varied success. Some of the more important ones are those described by Hiss, Conradi-Drigalski, Endo, and Löffler. Endo's medium was selected from this group for the present work because it had been used with such fine results in Washington for four years during some studies " On the Origin and Prevalence of Typhoid Fever in the District of Columbia," — Hygienic Laboratory Bulletins, P. H. and M. H. Service.

The experimental filters were constructed of reinforced concrete with an aim to make them resemble, as nearly as possible, the ones serving the large cities in the filtration of their supplies. The filters were seven feet high, their walls four inches thick. The filters were built in a basement room and therefore may be considered covered, since a concrete horizontal ceiling was situated about four feet above the filter. The internal dimensions of the filters were exactly two feet one inch square, so that this area was precisely equal to one ten-thousandth of an acre. It was naturally advantageous to have the area of the experimental filters represented by definite fractions of an acre in order to facilitate the calculation and interpretation of results. The bottoms of the filters were sloped gently from the backs forward and from the sides towards their middles, in order that the water would drain readily in the direction of the effluent pipes. A gauge tube was placed above and below the sand in each filter for measuring the " loss of head." All plumbing fixtures, submerged in the water, were of galvanized iron.

Those portions of the walls which were to be in contact with the sand layer were carefully lined with a coarse mixture composed of two parts sand and one part cement. This coarse lining gave an excellent adhesion between the grains of the sand layers and the walls. Grooves (one inch high and extending one-half inch into the walls) were constructed at levels of nine, eighteen, and twenty-seven inches, respectively, below the tops of the sand layers. The object of these grooves was to deflect any possible seepage so that all water which reached the effluent would necessarily have to pass through some sand.

The filters contained a rock and gravel layer one foot in depth which supported a sand layer three feet thick. A rock of igneous origin was used in order to avoid any disintegration which might result from the continuous flow of water through the under-drains. The rocks were placed with considerable care, to obtain a very free drainage. The rocks on the bottom layer were about three inches in diameter and graded upward in such a manner that each succeeding layer was just large enough to keep from falling through its supporting layer. The scheme was similar for the gravel layers — the top one of which was very fine in order to retain the sand layer in place.

Filter A contained sand from the Washington Filtration Plant, while sand from the works at Lawrence, Mass., was used in Filter B. According to the method suggested by Hazen, the sand in Filter A had an " effective size " of .32 millimeter and its " coefficient of uniformity " was 1.78. Filter B's sand had an " effective size " of .31 millimeter and a " coefficient of uniformity " of 3. The sand in both filters was added slowly and water forced up the under-drains through the effluent pipes in order to prevent stratification.

The raw water on the filters stood at a constant depth of two feet nine inches — any excess was taken care of by over-flow pipes. Every day typhoid bacilli (twenty-four-hour cultures) — grown on agar slants — were washed off into a graduated bottle. The bottle was so arranged that its contents were continuously discharged into a thistle tube, which had its distal end situated four inches below the surface of the water. By running the thistle tube diagonally opposite to within about nine inches of the affluent pipe a fairly general distribution of the typhoid organisms in the applied · water was obtained. Plating of samples from the over-flow pipe on several occasions showed that no typhoid bacilli were being carried off the filters.

Agar slants were used instead of bouillon cultures, as it is very desirable to keep the amount of nutriment in the water at the lowest possible level to discourage any excessive growth of water bacteria in the under-drains. These organisms require very little organic matter for their multiplication.

Before infecting the water with typhoid bacilli, samples of both the affluent and effluent were studied in order to see if any organisms were present which in any way resembled B. typhosus. Although the affluent showed a large variety of bacteria when plated, the effluent at times gave almost a pure culture of small, clear colonies. These colonies never attained a large size on Endo's medium and would not grow at all on the ordinary media inoculated with them.

When the infection of the water was commenced, samples were taken three times a day and plated on Endo's medium. The medium was modified by using 1.5 per cent agar instead of four per cent as originally advocated by Endo. Typhoid colonies attain a greater size and grow more rapidly with this modification, but they assume a pinkish tint more quickly owing to the greater ease of diffusion possessed by any liberated fuchsin. The plated samples were placed in the thermostat for forty-eight hours and were then examined. All suspicious looking colonies were "fished off," transferred to bouillon tubes and incubated for twenty-four hours. At the expiration of this time a drop of anti-typhoid serum was added (this gave an approximate dilution of 1 : 300). The tubes which showed either a typical or a suggestive Widal reaction were plated, incubated, and any typhoid colonies were transferred to bouillon. After an incubation of twenty-four hours, inoculations were made from these transfer cultures in gelatine, litmus milk, dextrose, and lactose.

Such a method of procedure was thought necessary in order to be certain that the agglutinated organisms possessed the cultural characteristics of B. typhosus. In addition, somewhat recently, Frost has found a pseudomonas of the proteus group which was present in the raw and filtered water at Washington, D.C.   The interest in regard to this organism (Pseudomonas protea) lies in the fact that the organism grows like typhoid on Endo's medium and is agglutinated by anti-typhoid serum — at times, in rather high dilutions.   Either gelatine or milk would easily differentiate Ps. protea from B. typhosus.   A sharp lookout was kept for it, especially when using Filter A — as the sand in neither filter was sterilized.

The ordinary samples of the effluent were supplemented by a daily one-liter sample which was enriched by adding to it ten cubic centimeters of a concentrated dextrose bouillon. These enriched samples were incubated for twenty-four hours and plated.   Since the effluent commonly contained a small number of organisms, it was hoped that B. typhosus might be discovered by this enrichment method before it was possible to isolate it from the samples directly plated.   A glance at the tables will show that only negative results were yielded from this source.   Possibly the lactose-bile medium might have given more satisfaction.   B. coli was practically never present in the effluent samples and so the use of the caffein-broth, described by Hoffman and Ficker, was not used. Any successful enrichment-method would have shown that the hygienic efficiency was not one hundred per cent, but it would have been of no service in determining the actual numbers of B. typhosus permeating the filters.

The investigation was carried on during the winter months when B. typhosus causes most trouble in the water supplies. No visible "schmutzdecke" was present, but a mere discoloration of the top one-quarter inch of the sand layer.   Records from the Metropolitan Water Board, for the period of the experiments, indicated that the many forms of microscopic life existed in very small numbers in the rain water used (the average for the entire time was: Diatomaceæ, 841; chlorophyceæ, 6; cyanophyceæ, 8; cenothrix, 36; protozoa, 64; total organisms, 955 and total bacteria, 310. Organisms in standard units per cubic centimeter — bacteria in numbers per cubic centimeter). The low count of protozoa and bacteria would probably aid the survival of typhoid bacilli in such a water. Although Whipple cites an example where the numbers of bacteria in Baisely's pond (Brooklyn) varied inversely with the number of clathrocystis organisms present, the condition was very likely due to an entangling of the bacteria by the algæ in a mere mechanical fashion, as it is very improbable that the algæ or any other similar types of plants could actually devour living bacteria.

Some diagrams of one of the filters and a tabulation of the results follow.

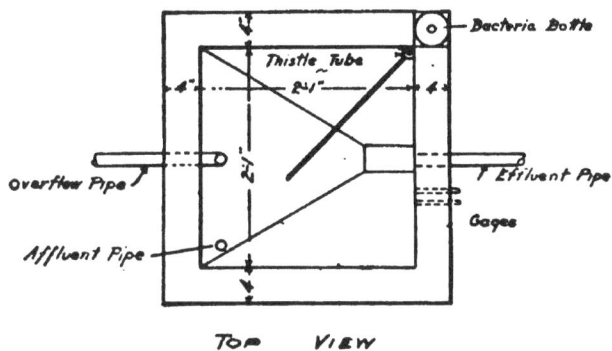

TOP   VIEW

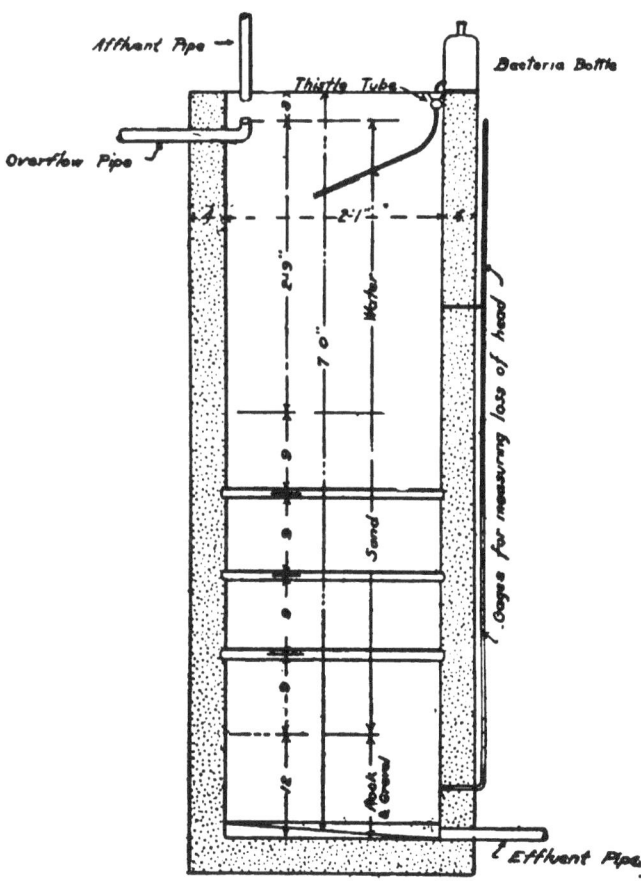

SECTION
Scale ½"=1'.0'

FILTER A.

Day.	Temperature (F.°). Affluent.	Effluent.	Loss of Head (inches).	Rate in Gallons per Acre Daily.	Total Number of Typhoid Bacilli Applied in 24 Hours.	Samples of Effluent Enriched.	Samples of Effluent Plated for B. Typhosus. 1	2	3	Average Number of B. Typhosus per Cubic Centimeter of Applied Water.	Hygienic Efficiency (per cent).
1 .	57.2	62.6	21.5	2,400,000	55,990,000	. . .	O	O	O	60	100
2 .	50.0	60.0	22.2	''	55,300,000	. . .	O	O	O	60	100
3 .	50.2	60.8	22.8	''	60,000,000	. . .	O	O	O	65	100
4 .	58.0	61.0	23.5	''	52,000,000	. . .	O	U	O	55	100
5 .	61.0	63.0	24.3	''	50,000,000	. . .	O	O	O	55	100
6 .	59.5	63.5	24.8	''	47,000,000	. . .	O	O	O	50	100
7 .	59.4	63.0	25.0	''	60,000,000	. . .	O	O	O	65	100
8 .	60.0	63.5	25.5	''	50,000,000	. . .	O	O	O	55	100
9 .	. . .	. . .	. . .	. . .	. . . .	. . .	O	O	O	. . . .	100
10 .	53.0	62.8	26.0	2,400,000	91,000,000	. . .	O	O	O	100	100
11 .	57.2	63.4	26.2	''	114,000,000	. . .	O	O	O	125	100
12	61.4	62.6	26.3	''	126,000,000	. . .	O	O	O	135	100
13 .	58.3	63.4	26.5	''	117,000,000	. . .	O	O	O	125	100
14 .	56.0	62.6	27.0	''	105,000,000	. . .	O	O	O	105	100
15 .	56.2	62.6	27.2	''	126,000,000	O	O	O	O	135	100
16 .	55.4	64.4	27.3	''	41,000,000	O	O	O	O	45	100
17 .	56.0	62.6	26.4	2,700,000	147,500,000	O	O	O	O	140	100
18 .	56.2	62.7	24.0	''	92,000,000	O	U	O	O	90	100
19 .	59.0	62.8	15.0	3,600,000	64,000,000	O	O	O	O	45	100
20 .	56.4	62.7	9.0	3,920,000	60,000,000	O	O	O	O	40	100
21 .	58.4	62.6	4.8	''	41,000.000	O	O	O	O	30	100

FILTER B.

Day.	Temperature (F.°).		Loss of Head (inches).	Rate in Gallons per Acre Daily.	Total Number of Typhoid Bacilli Applied in 24 Hours.	Samples of Effluent Enriched.	Samples of Effluent Plated for B. Typhosus.			Average Number of B. Typhosus per Cubic Centimeter of Applied Water.	Hygienic Efficiency (per cent).
	Affluent.	Effluent.					1	2	3		
1.	55.4	62.6	4.5	2,400,000	90,000,000	O	O	O	O	95	100
2.	59.0	62.6	4.8	2,700,000	104,000,000	O	O	O	O	100	100
3.	59.1	62.6	5.0	2,500,000	110,000,000	O	O	O	O	115	100
4.	57.2	62.6	5.1	2,400,000	175,000,000	O	O	O	O	190	100
5.	68.0	62.8	5.0	"	210,000,000	O	O	O	O	225	100
6.	55.4	66.2	5.2	"	405,000,000	O	O	O	O	440	100
7.	56 0	62.0	5.3	"	544,000,000	O	O	O	O	590	100
8.	56 0	61.2	5.5	"	770,000,000	O	O	O	O	835	100
9.	54.0	61.0	5.6	"	460,000,000	O	O	O	O	500	100
10	60.8	62.6	5.6	"	540,000,000	O	O	O	O	585	100
11.	57.0	61.0	5.5	"	860,000,000	O	O	A	O	935	100
12.	61.2	62.6	5.7	"	880,000,000	O	O	O	O	955	100
13.	60.8	62.0	6.0	"	385,000,000	O	O	O	O	415	100
14.	60.9	62.0	6.2	"	960,000,000	O	O	O	1	1040	99.9
15.	61.8	62.1	6.4	"	825,000,000	O	O	O	O	895	100
16.	61.8	62 7	6.5	"	990,000,000	O	O	O	1	1075	99.9
17.	60.1	62.6	6.9	"	1,100,000,000	O	O	O	O	1195	100
18.	54.2	61.2	7.0	"	1,110,000,000	O	O	1	1	1205	99.9
19.	60.8	60.8	7.2	"	1,300,000,000	O	O	2	O	1410	99.78
20.	53.8	60.8	7.3	"	737,000,000	O	O	O	O	800	100
21.	3.6	60.8	7.4	"	490,000,000	O	O	O	O	530	100
22.	56.0	60.8	7.4	"	638,000,000	O	O	O	O	690	100
23.	57.2	60.8	7.5	"	880,000,000	O	O	O	O	955	100
24.	57.0	60.8	7.8	"	770,000,000	O	O	O	O	835	100

FILTER A (2D TRIAL).

| Day. | Temperature (F.°). | | Loss of Head (inches). | Rate in Gallons per Acre Daily. | Total Number of Typhoid Bacilli Applied in 24 Hours. | Samples of Effluent Enriched. | Samples of Effluent Plated for B. Typhosus. | | | Average Number of B. Typhosus per Cubic Centimeter of Applied Water. | Hygienic Efficiency (per cent). |
	Affluent.	Effluent.					1	2	3		
1.	54.2	57.3	1.7	2,400,000	550,000,000	O	O	O	O	595	100
2.	60.8	59.0	2.0	"	835,000,000	O	O	O	O	595	100
3.	58.0	60.0	2.2	"	1,110,000,000	O	O	O	O	1205	100
4.	54.0	59.0	3.1	"	1,760,000,000	O	O	O	2	1910	99.89
5.	54.3	59.0	4.3	"	3,330,000,000	O	O	2	4	3590	99.88
6.	54.4	59.0	5.2	"	3,278,000,000	O	O	8	6	3565	99.74
7.	55.0	59.0	5.4	"	3,410,000,000	O	O	12	9	3700	99.67

In estimating the total number of B. typhosus applied from the "dosing" bottle, a sample was taken from the bottle each morning after the fresh cultures were added and was set aside to be plated, usually in six hours. By delaying the plating of the sample for several hours, some allowance was made for the decrease in the number of organisms caused by the change from the agar slants at 37° C. to water at a low temperature. The total number of cubic centimeters discharged from the bottle the previous twenty-four hours was read off each morning and later multiplied by the number of colonies per cubic centimeter on the corresponding plate. To obtain the average number of B. typhosus per cubic centimeter of applied water each day, the total number for that day was divided by one ten-thousandth of the rate per acre daily (reduced to cubic centimeters) at which the filter was run for that particular day. The unit figure of the resulting quotient was always changed to either a 5 or a 0 to conform to the methods used in water analysis. The hygienic efficiency was considered one hundred per cent when no sample on that day showed the presence of B. typhosus. When B. typhosus appeared in more than one sample for the same day, the hygienic efficiency was based on the sample showing the greatest number of organisms.

On the ninth day (Filter A) typhoid organisms were applied, but no samples were obtained. The temperatures of the affluent were uniformly lower than those of the effluent, except on the fifth day (Filter B) and on the second day (Filter A " second trial "). The difference in temperature between the affluent and effluent was at times marked and on the seventeenth day (Filter A) an increase in the rate occurred while the gauge tubes showed that the loss of head was receding. Observation at the top of the filter revealed many large bubbles being liberated from the sand which were due to the dissolved oxygen set free as the water in the filter was warmed. As this condition did not subside by the twenty-first day it was decided to abandon the filter temporarily. In spite of the fact that the bubbles made the sand layer much more permeable and the rate greatly increased, no typhoid bacilli were found in the samples for these days.

On the eleventh day (Filter B) a colony looking very much like typhoid was picked off the plate from sample No. 2, and later it gave a suggestive agglutination when the Widal was tried. A trial on the other media soon demonstrated it to be B. alcaligenes.

It may be stated that, so far as these experiments were able to demonstrate, the hygienic efficiency of a well-regulated slow sand filter is one hundred per cent until the number of typhoid bacilli in the raw water reaches the neighborhood of one thousand bacilli per cubic centimeter. Even after this point is passed the typhoid organisms permeate the filters in surprisingly small numbers. It is hardly conceivable that B. typhosus would occur to any such extent in a raw water used for drinking purposes unless the water supply was either of small volume or the source of pollution great, the distance to be traveled by the discharged organisms short and their course comparatively free from a gauntlet of protozoan foes, obstructing algæ, myriads of hardier organisms and a host of other conditions which tend to discourage the existence of typhoid bacilli in large bodies of waters. It seems justifiable to conclude that the installation of slow sand filters, for the purification of a water-supply, is an absolutely safe method if the filtration plant is carefully constructed and supervised.

[DISSERTATION. — Submitted in conformity with the requirements for the degree of Doctor of Public Health at the Harvard Medical School.]

[ACKNOWLEDGMENT. — The writer wishes to express his thanks to Dr. Rosenau for the many suggestions which he made from time to time, and also for the interest he always showed in the work. The writer is also grateful to Mr. E. B. Phelps for suggestions in the construction of the filters ; to Mr. H. W. Clark and to Mr. E. H. Hardy for their kindness in having the sand screened for filters B and A respectively.]

REFERENCES TO THE LITERATURE.

Abba, Orlandi and Rondelli. Zeitschrift für Hygiene und Infection-shrankheiten, 1899.

Conradi-Drigalski. Zeitschrift für Hygiene, etc., 1902.

Endo. Centralblatt für Bakteriologie, 1904.

Frost. Bull. No. 66, Hyg. Lab., U.S. Pub. Health and Mar. Hosp. Serv., Washington, D.C., June, 1910.

Hazen. "The Filtration of Public Water Supplies" (pub. Wiley & Sons).

Hiss. Journal of Medical Research, 1902.

Hoffman and Ficker. Hyg. Rundschau, 1904.

Löffler. Deut. Med. W'ch'nschrift, 1906.

Massachusetts State Board of Health Reports for 1891, 1892, and 1893.

American Public Health Association Reports, 1893. "On the Removal of Pathogenic Organisms by Sand-Filtration," by George W. Fuller. (A review of the work done at Lawrence, Mass., in 1892.)

American Public Health Association Reports, 1894. "Sand-Filtration of Water with Special Reference to Results obtained at Lawrence" (1893), by George W. Fuller.

"On the Mills — Reincke Phenomenon and Hazen's Theorem," by Sedgwick and McNutt. Journal of Infectious Diseases, August, 1910.

Piefke. Journ. für Gas — und Wasserver — sorgung, 1891.

Piefke and Fränkel. Zeitschrift für Hygiene, 1890.

Rosenau, Kastle and Lumsden. "Report on the Origin and Prevalence of Typhoid Fever in the District of Columbia," Bulls. Nos. 35, 44, and 52, Hyg. Lab., U.S. Pub. Health and Mar. Hosp. Serv., Washington, D.C., February, 1907, May, 1908, and October, 1909.

Schuder. Zietschrift für Hygiene, etc., 1901.

Whipple. "The Microscopy of Drinking Water" (pub. Wiley & Sons).

CERTAIN FUNDAMENTAL PRINCIPLES RELATING TO THE ACTIVITY OF BACTERIA IN THE INTESTINAL TRACT. THEIR RELATION TO THERAPEUTICS.[*]

ARTHUR I. KENDALL, S.B., Ph.D., Dr. P.H.

(*From the Department of Preventive Medicine and Hygiene, Harvard Medical School, April 24, 1911.*)

" In the study of the microscopic forms known as bacteria we have what might be fitly called the focal point of the various branches of biological science.   Though their investigation may require careful morphological researches, yet the unmistakable monotony of form, combined with a considerable variation of physiological activity, has compelled the bacteriologist to pay much attention to means by which such physiological variations may be more or less accurately registered in order that they may serve as a supplementary basis for classification.   Again, with unicellular organisms the manifestations of cell activity become the most important phenomena for study.   These manifestations bring together the fields of physiology and chemistry and make bacteriology in one sense a branch of physiological chemistry." — *Theobald Smith.*[1]

" Then again, there was no ulterior interest in the study of bacteria as such, which is a strong impulse in many other departments of biologic science.   It is what bacteria do rather than what they are that commanded attention, since our interest centered in the host rather than the parasite." — *Theobald Smith.*[2]

INTRODUCTION.

The studies recorded here, the logical outcome of work which the writer has engaged in for more than three years,

[*] Received for publication June 22, 1911.

[1] The Fermentation Tube.   Wilder Quarter-Century Book, 1893.

[2] Some Problems in the Life History of Pathogenic Microörganisms.   An address read at the Section of Bacteriology, International Congress of Arts and Sciences, St. Louis, Mo., Sept. 24, 1904.

are illustrative of a new and hitherto practically neglected field of investigation in bacteriology as it is related to curative and preventive medicine.

The present unsatisfactory state of our knowledge concerning the ways and means by which bacteria enter the host, gain and maintain a foothold there, and bring about conditions more or less unfavorable to the patient's well-being was the incentive to the making of these studies.

Notwithstanding that bacteriology as a science is more than a quarter of a century old, and that certain of the best known microörganisms pathogenic for man have been the subject of innumerable investigations, the fact remains that medicine is still uninformed concerning many of even the more general principles which underlie the modes of attack and action of these microbes. The most potent factor which underlies the incompleteness of our knowledge is not difficult to determine: bacteriology, "the handmaiden of medicine," as it has been drolly expressed, besides contributing many of the most brilliant chapters of medicine, enters into so many fields of human activity and interest that it has been neglected as a pure science. Each art or science in which bacteria either directly or indirectly play a part, appropriates to itself only those phases of bacterial activity which appear to be important, leaving the more fundamental problems relating to these organisms to some other branch of learning for elucidation.

The inevitable result has followed this lack of diligence, and bacteriology as it relates to medicine is no exception to the rule. The contributions which bacteriology have made to medicine are largely diagnostic; indeed, with the exception of a very few antitoxins, and a moderate number of sera and vaccines of greater or lesser potency, few, if any, serious attempts have been made to combat pathogenic bacteria through measures directed against the organisms themselves. On the contrary, nature, careful nursing, and supportive treatment are the factors relied upon to bring bacterial infections to a successful termination. It is no argument in defence of this lack of initiative to claim that the time is not yet ripe for fundamental studies of the conditions which enable microörganisms to enter and maintain themselves in the human body because our knowledge of the underlying chemical phenomena is insufficient.   The ultimate cause of this lack of knowledge is the attitude of the medical profession and the medical schools themselves.   Theobald Smith, in discussing Medical Research,[1] says about pathology: " Much that is taught to-day in pathology belongs to clinical medicine and surgery, for it is largely ' special and diagnostic ' in character.   The pathologist is now the servant of the physician and the surgeon in completing and rectifying their diagnoses.   The pathologist of the future will deal with more general phenomena derived from experimental and comparative data   .   .   ."

What is true of pathology is even more true of bacteriology.   In the majority of medical schools in this country, bacteriology is a branch of pathology, and the professor of pathology is also the professor of bacteriology.   Bacteriology practically fills in time, while the greater amount of attention and research is set apart for pathology.   It appears to be a well substantiated fact that a man cannot serve two masters, and it has inevitably followed that bacteriology has been passed by with a minimal amount of attention because the professors have been largely interested in pathology.

From the economic point of view it would appear that a disease such as typhoid fever, with more than thirty-five thousand deaths, and three hundred and fifty thousand cases per annum in the United States alone would be well worth studying, but in spite of the vast loss in time, money, and lives, which typhoid causes yearly, we are still quite in the dark concerning the manner in which the typhoid bacillus succeeds in gaining and maintaining a foothold in the human body in the presence of the intestinal bacteria which under ordinary conditions appear to be able to overgrow this comparatively unresistant organism.   Instead of complacently accepting this state of affairs, and explaining the invasion of the body by typhoid bacilli as " a condition of lowered resistance " (which means ignorance, in the last analysis), it would seem to be more logical to make conditions for the study of this and similar problems attractive to trained observers and to encourage research along these lines.

It must be freely admitted that certain diseases, as, for example, tuberculosis, have been subjected to the most careful scrutiny from the experimental point of view, and the results have been of the utmost value in determining the natural history of the disease. Recently, Schaffer and Coleman, together with Somogyi, Reinoso and Edna Cutler,[2] have applied modern methods to the study of typhoid fever, and have arrived at some very valuable conclusions. This work is mentioned here to call attention to the wealth of material which is untouched, as well as to indicate that even such a well recognized disease as typhoid fever may furnish problems in which much good work may be done, and even as much glory attained, as in the study of the less common diseases, which from time to time suddenly become themes for popular study.

The historical development of the study of the intestinal bacteria will illustrate certain other factors which have detracted from the serious study of bacterial activity in the animal organism, and they will be referred to in some detail in subsequent paragraphs.

The purpose of the work presented here is to discuss certain aspects of the biochemistry and physiology of the intestinal bacteria; to contrast the intestinal flora of man in health and disease; to indicate how specific characters of these bacteria may be made use of both in the attempt to establish conditions in the alimentary tract which shall assist the host to successfully resist certain types of intestinal infection of bacterial origin, and to maintain there conditions which shall tend to prevent infection or reinfection. In carrying out this work, a radical departure has been made from the conventional methods of bacteriological research, and to make clear the premises upon which the results depend, a review of the literature bearing upon this phase of the subject is necessarily included. The literature quoted is not in any sense a complete bibliography of intestinal bacteriology, but rather a summary of certain of the more important monographs which bear directly upon the subject-matter under discussion.

Historical. — Probably the first observer to actually see intestinal bacteria was the Hollander, Anton von Leeuwenhoek, who, in 1675, examined various putrid infusions, drops of water, scrapings from the mouth, and his own diarrheal discharges with lenses of his own grinding. In his collected writings, edited by Robert Hooke,[3] he says: "With great astonishment I observed everywhere through the material which I was examining animalcules of the most minute size which moved themselves about very energetically." From his descriptions, and particularly from his sketches, there can be little doubt but that he actually saw, and drew accurately, motile bacteria. It is possible to recognize cocci, bacilli, and spirilla in his reproduced sketches.

It should be mentioned that previous to von Leeuwenhoek's work on the " animalcules," the learned monk Athanasius Kircher (1659) observed, and described in a very romantic manner, " minute living worms " which he found in various putrefying material. His descriptions, however, are very imperfect and it is doubtful if he is entitled to priority as the discoverer of the bacteria. The fact that he left no drawings leaves one in the dark concerning what he actually may have seen.

For the next century and a half, various observations were made upon what appear to have been in reality bacteria, but no material progress was made and the work of the contributors during this epoch are of historical interest merely.

In 1870 Hausmann,[4] and a little later Woodward,[5] studied the dejecta of various individuals in health and disease, and they independently came to the conclusion that the morphology of the microörganisms in question varied but little in health and disease, although in certain cases they both noticed that the numbers of organisms discernible in the

stools were considerably increased in disease. The purely
morphological relations between the bacteria could not lead
to a definite diagnosis. Woodward was especially keen in
his observations, and described the morphology of certain
bacteria very accurately. He concluded further that the
form of the organism did not establish the identity of the
bacterium, and that bacteria of similar shape and size could
bring about the most varied changes.

In 1877 Weigert and Ehrlich introduced the use of anilin
dyes for coloring bacteria, and in 1863 Davaine actually
demonstrated for the first time the causal relation between a
definite microörganism and a disease. He worked out the
etiological relationship of B. anthracis to anthrax in cattle.
He showed furthermore that the infective agent was in the
blood, and that the blood of an animal dead of anthrax
would reproduce the disease when it was injected into a
susceptible animal.

Koch[6] was the first investigator to actually isolate in pure
culture and then prove the etiological relation of an organism to a human disease. He isolated first the tubercle
bacillus, then the cholera vibrio, studied the latter carefully,
and established the methods of spread and, particularly, the
portal of entry of this organism. Finkler and Prior soon
isolated the spirillum that is named for them from cases of
true cholera, and of cholera nostras, and from this time on
the whole subject of bacteriology occupied the front rank
in medical investigation.

Hoppe-Seyler, Nencki, Brieger and Bienstock studied the
physiology of bacteria and tried to discover the chemical
part which they played both in the human body and in
artificial media, thus laying the foundation for the biochemistry of microörganisms. Nothnagel[7] recognized the fact
that bacteria in the intestinal canal might conceivably set up
various types of fermentations, and he tried to reproduce
these microbic activities in artificial media, inoculating them
with microbes of intestinal origin; his cultures were not
pure, however, and his results are subject to criticism on this
account, although there is at the present time a tendency to

return to this type of experimentation to work out certain forms of bacterial symbiosis. In 1884 Bienstock[8] isolated and described satisfactorily several kinds of bacteria from stools, using the recently discovered solid media to accomplish that purpose which Koch (1881) had shortly before introduced. His organisms were four in number, and one of these, B. putrificus, deserves special mention. It was an obligate anaërobe, which attacked both fibrin and egg-albumen energetically. This organism even to-day is the best known example of an obligate proteolyte.

Other studies of a similar nature were made about this time, many of them of considerable interest, but in 1886 there appeared a monograph by the late Theodor Escherich of Würzburg[9] which was far in advance of any of the previous contributions so far as the real study of the bacteria is concerned. This talented, ingenious, and careful observer studied in great detail the bacteria of the dejecta of babies both in health and disease, noticed the predominant forms, and accurately described both morphologically and culturally some of the organisms best known to bacteriology. He was not satisfied merely to isolate a large number of strains of bacteria. He studied his cultures in great detail upon a variety of media, trying to find out just what they did culturally, and then tried to apply his findings to the intestinal tract. For this purpose he grew his bacteria anaërobically and aërobically, studied the gas formation and gas formula, their action upon the various constituents of milk (a most logical procedure, because milk was the prominent diet of his babies), and in general approached the subject in a thorough, rational, and clever manner. His work, which has stood the test of time, has become a classic, and to-day it is authoritative in many respects. It was not for lack of diligence, insight or thoroughness that he failed to isolate two organisms which are characteristic of the stools of bottle and breast-fed infants. This inability to isolate these organisms, which he clearly saw in his stained preparations made from the freshly passed feces, indeed, is a further indication of his wonderful powers of observation.

In spite of the excellence and far-reaching importance of Escherich's work, comparatively little use was made of it, and the reason is not difficult to discover.

In 1883 Koch discovered the Cholera vibrio, and in 1884 Gaffky [10] succeeded in isolating the Typhoid bacillus, which Eberth had previously (1880) seen in the mesenteric lymph nodes and in the spleens of typhoid cadavers. In 1898 Shiga,[11] a Japanese student, isolated B. dysenteriæ from cases of bacillary dysentery in Japan. The effect of these remarkable discoveries was to detract attention immediately from the study of the normal and abnormal relations of bacteria in the intestinal tract, and to stimulate investigators to turn their full attention to attempts to isolate an organism from some hitherto "refractory" disease which should be the etiological agent. The effect has been what one would predict: at the present time we are practically in the dark concerning the normal intestinal bacteria and their relations to each other, and we are equally uninformed of the laws which underlie their presence and significance in the alimentary tract, points of the greatest importance in the study of the ways and means whereby exogenous or endogenous bacteria gain and maintain themselves and cause disturbances in the alimentary canal.

On the other hand, the literature is overstocked with descriptions of incompletely studied organisms which have been isolated from time to time by careless observers from the intestinal tract.

To repeat: "The very importance of these discoveries has been a potent factor in diverting attention from the studies of the normal intestinal flora with its wealth of problems relating to the principles which govern the activity of these bacteria. Even at the present time the sequence of events, which permits the establishment of these exogenous invaders in the alimentary canal and the exact conditions through which they are able not only to extend and maintain themselves but even to replace wholly or in part the normal flora, are unknown. It is possible to trace the influence of these epoch-marking studies in the subsequent history and development of intestinal bacteriology.

" It appears to be a fact that the majority of bacteria of exogenous origin, pathogenic for man (excluding the anaë-robes) are relatively inert from the standpoint of chemical activity. On the other hand, these organisms grow in more or less distinctive ways in artificial media, and, usually, they may be recognized by their cultural aspect, their inability to bring about deep-seated changes in their nutrient environment, through specific serum reactions or by their power to initiate characteristic lesions in susceptible animals. In these respects these exogenous organisms contrast in a noteworthy manner with many prominent types of the normal intestinal bacteria.

" The more prominent of the latter are distinguished by their chemical or physiological activity, and their identification depends far more upon their ability to bring about well-marked chemical changes in their nutritive environment than upon their cultural properties or serum reactions.

" The lack of appreciation of this fundamental difference which exists between the relatively inert pathogenic microorganisms and the chemical activity of the more important types of the normal intestinal flora, together with the notoriety that attaches to the former, explains the unprogressive attitude which has characterized many researches in intestinal bacteriology.

" While it must be admitted that the purely academic methods of research have resulted in scores of more or less complete morphological and cultural descriptions of bacteria of intestinal origin, this knowledge is fragmentary and unclassified. It is devoid of data which would permit one to correlate the presence of these organisms with the diet or condition of the host, or even to form a judgment concerning their numerical relations with other intestinal organisms.

" This ' bacteriocentric ' conception is not illogical when one is dealing with the exogenous pathogenic microörganisms mentioned above, but it is unproductive of definite results when it is applied in its unmodified form to the study of the normal intestinal flora. It is becoming more and more evident that the problem of intestinal bacteriology must be approached from the dynamical rather than the cultural standpoint."[12]

Bacteria in normal stools. 1. Meconium. Breast-milk stools. — The meconium is sterile at birth, as would be expected from the fact that the uterine cavity is normally sterile (Escherich,[13] Senator,[14] Hochsinger,[15] Moro [16]). Shortly after birth, however, bacteria begin to appear in the meconium, and this early infection may be independent of feeding, as was shown by Breslau as early as 1866.[17] The first bacteria may appear in the meconium as early as the fourth hour, or as late as the twentieth hour, post-partum, although as a rule the organisms are discernible from the tenth to the seventeenth hour, post-partum.

Tissier [18] recognizes three phases in the bacterial infection of infants' stools: 1, period of sterility; 2, period of mixed or promiscuous infection, and 3. period of transition, and establishment of the characteristic nurslings, bacterial flora.

The first stage, the aseptic phase, needs no comment. The period of mixed infection is characterized by the appearance of various types of bacteria, the kinds depending upon circumstances. Seasonal variations play a prominent part in determining the kinds and numbers of bacteria which appear in the semi-meconial stools characteristic of this stage. As a rule, the relatively warm summer months are followed by a more varied and numerous flora than the cool winter months. The various organisms referred to are few in numbers at first, but toward the third day, post-partum, there is a decided increase due, apparently, to the presence of food in the alimentary canal. Theoretically there are three portals of entry for the intestinal bacteria: the mouth, anus, and the blood stream. The latter plays a very insignificant part in infection of the alimentary tract, and it may be neglected in this discussion. The mouth and anus, on the contrary, are very common atria for infection. Moro [19] has traced a not uncommon organism found in infants' stools, B. acidophilus, to the nipples of the mother, and he claims to have found it even in the milk itself as it comes from the breast.

Escherich (loc. cit.), Schild,[20] Sittler,[21] and others have studied the bacteria characteristic of the early stages of infection of the intestinal tract, and they have substantiated the main facts presented above relative to the atria of infection. Escherich has noticed a very common and characteristic organism in the early nurslings' stools, which he called the "Kopfchen" bacillus. This organism characteristically has a large terminal spore resembling closely that of the tetanus bacillus. This Kopfchen bacillus has not been satisfactorily studied in artificial media. Some observers are inclined to identify it with B. putrificus of Bienstock, while others, notably Sittler,[22] believes it is in reality the so-called gas bacillus, Bact. Welchii, or B. perfringens, of Veillon and Zuber. It is probable that these latter two organisms are either identical or very closely related varieties of the same organism.

Besides the Kopfchen bacillus, other spore-forming organisms are very frequently found in the intestinal contents during this period, notably bacilli belonging to the Subtilis-mesentericus group. Certain aërogenic bacilli, notably B. coli, B. proteus, Bact. aërogenes (B. lactis aërogenes), together with coccal forms, chiefly Micrococcus ovalis, Escherich are usually present. (The M. ovalis is almost certainly identical with the Enterocoque of the French writers.) Fluorescent bacteria may rather rarely appear during this stage of mixed infection. It is a fact worth remembering that certain of these bacteria, particularly the proteolytic forms, appear to be able to lie dormant in the intestinal tract for some time, and later on appear in considerable numbers as the conditions become suited for their development. It is equally a significant fact that the majority of the bacteria represented in the early infections of the alimentary tract are proteolytes.

The period of mixed infection merges more or less imperceptibly into the third stage, the period of transformation to the more strictly typical infantile type of bacteria, and the beginning dominance of the " obligate " nursling flora. The bacteria in the intestinal tract become more numerous, and the spore-forming varieties disappear for the most part, and rather abruptly. Coccal forms diminish in numbers more slowly, and never quite disappear. Simultaneously, rather long, thin rods (not infrequently slightly curved, with tapered ends) appear, occurring singly, in pairs, or in groups with their long axes parallel. The pairs frequently show a " geniculate " arrangement. These bacteria are typically Gram positive; in many instances they show a central, Gram-positive granule in an otherwise Gram-negative rod, presenting the so-called " punctate " appearance described by Escherich (loc cit.). Again, under unfavorable conditions, the protoplasm of these rods collects in small, round granules, while the rest of the rod stains faintly or not at all, thus resembling at first glance a series of coccal forms. These organisms are obligate anaërobes, typically fermentative in character, forming moderate amounts of lactic acid, but no gas from the more common sugars. Tissier,[22] who first isolated them in pure culture, called them B. bifidus communis, because of their remarkable property of developing well marked bifid ends when grown in artificial media. The correct name for this organism is B. bifidus [24] since a trinomial is not permissible in botanical nomenclature.

Escherich noticed, in his studies upon the bacteria of normal and abnormal stools referred to previously, that while the majority of organisms which he saw in Gram-stained preparations made from freshly passed stools of normal babies were Gram-positive, and rather long, the dominant forms he succeeded in cultivating in his artificial media were Gram-negative, short, oval rods. He was unable to cultivate the long rods (B. bifidus), and he was inclined to believe that B. coli, the organism he cultivated most frequently, was the organism he actually noticed in the fresh feces, but that somehow it changed its morphology and staining reaction when it was grown in artificial media outside the body. This view was strengthened by an observation made by one of his pupils, Alex. Schmidt,[25] who believed he could transform the Gram-negative B. coli into a Gram-positive organism by growing it in media containing neutral fat. This observation was accepted generally for some time as correct, but the subsequent work of Tissier,[26] Cahn,[27] Rodella,[28] and others has not confirmed these results. Indeed, their work shows almost conclusively that Schmidt was in error. Because B. coli appeared to be the dominant organism in artificial media, Escherich drew the very natural conclusion that this organism was the dominant infantile bacillus, but the results of subsequent observations convinced him that in reality another and very different bacillus was the one typically present in nurslings' stools, namely, B. bifidus.

Early in 1900 two very interesting and important articles appeared, describing two distinct types of bacteria, each of which, it was claimed, was the one typically present in the dejecta of breast-fed babies. One organism, B. bifidus, was described by Tissier.[29] B. bifidus is an obligate anaërobe, very common in the feces of normal nurslings. Tissier believed it was the dominant organism of nurslings' feces, basing his claims for its dominance upon the fact that he was able to isolate it in pure culture, and in considerable numbers from each of several different cases that he studied. The other organism, described by Moro[30] as B. acidophilus, was not an anaërobe, although it was grown with some difficulty when freshly isolated in the presence of oxygen. The latter organism received its name from the fact that it grew well in dextrose broth or acid beer wort, in the presence of an amount of acid fatal to almost all known organisms.

Finkelstein,[31] working independently, isolated an organism indistinguishable from B. acidophilus, which he called "Säureliebende" bacillus, from the fact that he grew it first in acid dextrose broth (acetic acid), the so-called Hayem's medium. Both the "Säureliebende" bacillus and B. acidophilus were morphologically very similar to B. bifidus, as seen in the normal stools of breast-fed babies, and for some time there was considerable difficulty in deciding which type of organism was in reality the one most commonly met with in nurslings' dejecta. Finally, the work of Tissier,[32] Cahn,[33] Moro,[34] Rodella,[35] and Weiss[36] showed that in reality B. bifidus was the dominant organism in breast-fed babies, but that B. acidophilus was relatively uncommon in nurslings' feces, although commonly met with in bottle-fed babies. The latter organism appears to be very common throughout the mammalia, and certain observers have even found acidophili in certain cold-blooded animals, thus indicating that they may be almost as widespread as B. coli in nature. The reason they have not been isolated more frequently is undoubtedly due to the fact that they require rather special media for their isolation. (For a discussion of the distribution and characters of B. acidophilus, see article on Acidoduric bacteria.[37])

Mademoiselle Tsiklinsky [38] has studied the fecal flora of normal nurslings in Warsaw, and finds that B. bifidus is not as common in young babies, particularly those over four to six days old, as appears to be the case in France and Germany, where most of the studies recorded above were carried out; she tacitly asks if B. bifidus is in reality the most characteristic organism of nurslings' stools, and in this sense distinctive. The consensus of opinion at the present time appears to be that bifidi are common, and reasonably distinctive in the normal dejecta of nurslings, but that it is by no means the only organism present.   In fact, in a moderate number of infants, otherwise perfectly healthy, it may be present in very small numbers, or even absent.   In addition to B. bifidus and B. acidophilus, other bacteria are commonly met with, although they occur in lesser numbers than those previously mentioned as a rule.   Among these B. coli and Bact. aërogenes (B. lactis aërogenes) are too well known to need further comment, except to state that the latter tends to disappear when the patients' diet becomes more varied, while the former persists throughout life.   MacConkey [39] has studied the distribution of the aërogenic bacilli (the colon group) extensively, and it is to his work that the above facts relating to the distribution of these organisms is attributable.

A well-known organism, a coccus, also occurs in the stools of normal nurslings, and persists in greater or lesser numbers throughout life.   This coccus, variously known as Micrococcus ovalis (Escherich, loc. cit.), Enterocoque (Thiercelin [40]), Streptococcus lacticus (Kruse [41]), and Streptococcus enteritidis (Hirsch [42] and Libman [43]), is suspected to be an important factor in the etiology of certain intestinal disturbances, particularly those of childhood, and it is especially important to recognize its various synonyms, because it is referred to by any or all of the above-mentioned names by various writers.

Nepper [44] has advanced the theory that Mucomembranous enterocolitis may be caused by the activity of Mic. ovalis (or certain other organisms), due to the superficial infection of the mucous membrane of the intestinal tract, and that the virulence of the organisms is exalted by fecal stasis and mucous overgrowth.

It is a very significant fact, one commented upon by Escherich, that putrefactive (proteolytic) bacteria are very uncommonly met with in the dejecta of normal nurslings; in this connection it is important to remember that while lactose is not commonly met with in their feces, yet the reaction is distinctly acid, due to the presence of lactic acid.   In later paragraphs the full significance of this absence of proteolytes will be considered in considerable detail.   The fact is mentioned here for the sake of completeness.

To summarize : The more commonly recognized bacteria which may be designated the normal nursling fecal flora comprise the following : B. bifidus, Mic. ovalis, B. coli, Bact. aërogenes, and B. acidophilus.   'Bact. aërogenes appears in the upper levels of the tract, the duodenum and jejunum: Mic. ovalis in the lower jejunum, the ileum and to the ileo-cecal valve: B. coli and B. acidophilus in the region of the ileo-cecal valve, while B. bifidus appears to dominate the ascending and transverse colon. The remainder of the tract to the anus is rather poorly populated so far as living bacteria are concerned, partly because of the considerable degree of desiccation of the fecal contents of the intestines, partly because of the accumulation of waste products which appear to inhibit the development of bacteria.

While the distribution mentioned above represents in general the optimum numbers of particularly important types of organisms in the intestinal tract, it must be remembered that bacteria may be, and usually are, carried more or less mechanically from the upper levels to the lower, so that one finds representatives of all types of organisms in the feces.

Cows' milk stools (bottle-fed babies). — Escherich (loc. cit.) distinctly called attention to the fact that the stools of bottle-fed babies, bacterially considered, were distinctly less homogeneous, both morphologically and culturally, than those wholly breast-fed. Practically all subsequent observers are in accord with his observations.

Microscopically (Gram stain), the distinctive features of bottle-fed babies are: the relative increase in Gram-negative bacilli of the Colon-aërogenes type, and of coccal forms of the Mic. ovalis type, associated with the diminution of the Bifidus type. Acidophili are relatively more numerous than before, bifidi less numerous. The increase in the coccal forms is a rather striking feature in many instances. Rodella[45] has called attention to the presence of peptonizing bacilli which, it will be remembered, were almost absent from the stools of normal breast-fed babies; Passini[46] has also isolated, in addition to certain peptonizing organisms, three butyric acid-forming organisms from apparently normal bottle-fed babies. These latter organisms were all anaërobes, and one of them, B. perfringens (or, more properly, Bact. Welchii), is of particular significance because of its relation to a not uncommon type of infantile diarrhea, in which it appears to play a prominent part.[47] B. putrificus, the most typical known example of a purely proteolytic organism, was also found in several cases. Sittler [48] has also found Bact. Welchii in moderate numbers in the stools of certain bottle-fed babies, and many other observers have corroborated this fact.

As these artificially fed babies become older, and their diets become more varied, a new distribution of bacteria makes\ its appearance, and this peculiarity may persist even in adults. Fischer [49] was the first apparently to call attention to this phenomenon, and Sittler [50] has made a similar observation. Bacteria in contact with the intestinal epithelium at a given level of the alimentary tract may be of one kind, while the organisms in the intestinal contents not in contact with the intestinal walls may be of an entirely different and unrelated type. Of course this arrangement is subject to many disturbing factors, but the fact remains that this peculiar distribution is maintained to a striking degree in certain instances. Fischer actually hardened, sectioned, and stained sections of the intestinal canal and observed these differences in bacterial distribution. The writer has noticed a somewhat similar condition [51] in pathological cases. Attention was there called to the fact that the mucus shreds frequently contained an entirely different flora from that of the intestinal contents other than mucus. Autopsy studies also have revealed the fact that the same peculiarity, namely, adherent mucus, not infrequently contained a different flora from that of the lumen of the tract at the same level.

In more marked pathological conditions, as in acute bacillary dysentery, it is well known that the mucus shreds are rather more likely to contain dysentery bacilli than the more fecal portions of the stool or, at least, it is relatively more easy to obtain cultures from such pieces of mucus.

In bottle-fed babies, and more particularly in older children, new factors enter to complicate the bacterial picture. The increased variety of food furnishes new combinations of pabulum which are eagerly seized by various kinds of bacteria, and the number of types represented in the intestinal flora may be said to be almost in proportion to

the complexity of the diet. This applies more particularly, as will be explained later, to the addition of protein
and the diminution of carbohydrate to the food of these
children.

We have seen that the flora of the normal nursling is
relatively homogeneous, both microscopically and culturally:
indeed, this very homogeneity is its distinctive feature, as
Escherich and others have pointed out. Conversely, the
significant feature (bacterially) of the intestinal flora of
older individuals is its complexity, morphologically and culturally. In spite of this seeming complexity, however,
certain organisms tend to localize themselves at rather
definite levels of the alimentary tract.

The duodenum is relatively sterile, according to most
observers: Kohlbrugge,[52] Moro,[53] Landsberger.[54] Certain
other investigators, however, have studied the problem more
carefully, and find that while the duodenum, and even the
tract as far as the lower ileum, may be practically free from
bacteria during the intervals between digestion, yet there is a
relatively large population in the small intestine while food
is passing through it. Gessner[55] found large numbers of
streptococci and staphylococci; Tavel and Lanz[56] found
moderate numbers of similar organisms; Klein[57] also found
bacteria during duodenal digestion. Schütz[58] noticed that
"strange" bacteria introduced into the duodenum were
freely demonstrable there for some time, but that they disappeared rapidly in the lower levels of the tract. Mieczkowski[59] made observations upon a case of duodenal fistula,
and found that the secretion was not bactericidal for B.
typhosus, Microspira comma, Streptococci nor B. pyocyaneus. The consensus of opinion, then, appears to be that
the duodenum is relatively free from bacteria, particularly
during the periods between feeding, that the disappearance
of these organisms in the intervals noted above is due probably to mechanical transportation to lower levels. Singularly enough, most investigators have failed to call attention
to the presence of bacilli of the subtilis group, although they
are frequently found in the duodenal region. Undoubtedly,

these bacteria have been regarded as " foreign " organisms, and have thus been overlooked.   Schmidt and Strasburger [40] have, however, called attention to the relative frequency of liquefying bacteria in the upper levels of the small intestine.   In the lower portions of the small intestines the liquefying organisms become relatively fewer in number, while the coccal forms begin to dominate.   Still further down, toward the ileo-cecal valve, Bact. aërogenes may be found, although as a rule it occurs in levels below the liquefying bacteria and above the coccal forms.   It should be remembered that this organism tends to disappear when milk ceases to be a factor in the diet of the host.   In the ileo-cecal valve, and the beginning of the large intestine, B. coli, B. acidophilus, and even B. proteus may be found, although the latter is uncommon except on a regimen potentially protein in character.   The bacterial population of the ileo-cecal region, and including the ascending and transverse colon, is most luxuriant and varied, for reasons discussed above.   Besides the organisms mentioned, both those characteristic of this region and those mechanically transported from higher levels, anaërobes are frequently present, the kinds of anaërobes depending upon the diet, B. bifidus, rather more rarely, Bact. Welchii, and other organisms less well known, and for the most part less important so far as our knowledge at the present time tells us.   The preponderance of B. coli is perhaps the most distinctive and significant feature of this flora.   This organism comprises a very considerable portion of the flora of bottle-fed babies, and it becomes even more prominent in adult life under ordinary conditions.

Upon a soft diet, little meat, and a considerable amount of carbohydrate and milk, the bacterial flora resembles that of the normal bottle-fed baby except that Gram-negative bacilli are relatively more common, and precisely as the bottle-fed fecal flora was more Gram-negative than that of the normal nursling, so the flora of the adolescent under the regimen described above was more Gram-negative than that of the bottle-fed baby.   With the addition of meat, and the gradual restriction of carbohydrate and milk, the flora becomes still more varied. Coccal forms make their appearance in considerable numbers, and liquefying bacteria of the subtilis-mesentericus type appear, in addition to the Gram-negative forms characteristic of the coli-proteus groups. Schmidt [61] and v. Streit [62] estimated that fully ninety per cent of the cultivatable bacteria from such stools are B. coli, but the evidence appears to show that these figures are considerably exaggerated, and that with a judicious selection of media the numbers of B. coli are much less. Matzuschita,[63] using special media, cultivated and studied at least forty-four different "species" of bacteria, and the probability is that there is in reality a very heterogeneous collection of types represented. With the beginning complexity of food, therefore, there is a corresponding complexity of bacterial types represented in the fecal flora, and this complexity has been the stumbling block which has prevented much progress beyond that of an attempt to catalogue the different "species." To attempt to summarize the literature bearing upon the bacterial flora of the intestinal canal in adolescence and adult life would be beyond the scope of this work. The main fact brought out is that there is a multitude of forms represented in the fecal flora, and that in the attempt to make this catalogue complete a hopeless mass of imperfectly prepared and digested descriptions of bacterial "species" has resulted.

Significance and quantitative relations of the intestinal flora. — Woodward (loc. cit.) and Nothnagel [64] noticed that in stained preparations of the feces bacteria were the most prominent objects seen, and Escherich (loc. cit.) made similar observations. Various investigators, impressed with the apparent dominance of bacteria in this intestinal flora, have attempted to actually number these organisms. Eberle,[65] Hellström,[66] Alex Klein,[67] Hehewerth,[68] de Lange,[69] Strasburger,[70] Sucksdorff,[71] Fürbinger,[72] Stern,[73] and many others have actually weighed or counted bacteria in the feces, and while there are wide variations in their figures, the fact remains that the number excreted daily is very great. Not only are there very large numbers of bacteria in feces, but many of them are not viable. Gilbert and Dominici,[74] Salkowski,[75] Kumagawa,[76] Sehrwald,[77] Hammerl,[78] Kuisl,[79] Mieczkowski,[80] and others have shown that while it is a fact that many of the fecal bacteria are undoubtedly dead, yet many different kinds, and even large numbers of them, may be caused to grow, provided the appropriate media are employed.

A conservative estimate of the number of bacteria excreted daily by a normal adult upon a mixed diet is represented by the number thirty-three times ten to the twelfth power (MacNeal, Latzer and Kerr[81]). This very large number of bacteria represents about 5.34 grams of dried bacteria daily, or approximately .585 gram of bacterial nitrogen. This amount of nitrogen comprises about 46.3 per cent of the total fecal nitrogen. Strasburger (loc. cit.) estimates that there are about one hundred and twenty-eight billions of bacteria excreted daily under normal conditions in an adult, representing one-third of the total dry residue of the feces, or .827 grams of nitrogen. Similarly, Klein (loc. cit.) found 8.8 billions of bacteria, comprising .13 per cent of the total dry residue of the feces. Whatever number most closely approximates the truth is of little account in this discussion, since our attention is directed to the significance rather than the frequence of their occurrence.

As MacNeal and his associates have pointed out, it is very improbable that an ordinary man would take into his stomach with his food even a moderately small percentage of this number of organisms, and the supposition is that the bacteria of the feces for the most part are actually those which have developed in the intestinal tract during the twenty-four hours prior to defecation. Additional evidence in favor of this view is furnished by the fact that the fecal bacteria growing in ordinary media are for the most part different from those commonly ingested with the food. Again, in such individuals as those which have been fed upon sterile diets, there is no very great reduction in the numbers of bacteria encountered in their stools. Sucksdorff[82] and Brotzu[83] believed that sterilized food did actually diminish in a noteworthy manner the numbers of bacteria in the intestinal tract, but this assertion has been practically disproved by Adrian,[84] Schmitz,[85] Hammerl,[86] Stern,[87] Peurosch,[88] Albu,[89] and Wang.[90] Levin[91] has observed that the feces of many mammals in the polar regions are practically free from bacteria. One is forced to the conclusion that the intestinal tract is a wonderfully perfect incubator and culture medium combined; in this intestinal incubator there is such a range of reaction and diversity of foodstuffs that the most varied types of bacteria capable of growing at body temperature can conceivably find a place where the environment is particularly adapted for their development. It must be evident that the direction which this flora takes will not be without influence upon the host.[92] Distinctly exaggerated examples of harmful effects produced by, or associated with, a perverted flora, or of exogenous organisms such as the cholera vitrio, need no comment, and the idea is gradually growing that less marked, but nevertheless abnormal developments of the normal organisms of the alimentary canal may also bring about pathological conditions, although the symptoms may be, and usually are, less marked and much less definite than those associated with the presence of frank acute intestinal infections of exogenous origin. It must be self evident that a lack of knowledge of the normal intestinal bacteria and their relations will be a serious handicap in recognizing the abnormal bacteria and their relations, and that this difficulty is increased many fold when one realizes that the response on the part of the host to the bacterial stimulus may appear quite out of proportion to the observed deviation in the normal flora or conversely. Somewhat similar conditions arise in acute infections of the intestinal tract, as, for example, bacillary dysentery, where the severity of the symptoms is of no great value in gauging the numbers of bacteria concerned in the process. It is of course well known that the number of dysentery bacilli which may be recovered from the stools, either ante- or post-mortem, furnishes no index of the clinical course of the disease.

The difficulties encountered in the study of the intestinal bacteria are great and the pitfalls numerous, but the problem is an important one, and will become increasingly so as our knowledge of bacterial activity becomes greater. The solution of the problem must be reached, and it must be evident from what has gone before that the mere isolation and morphological study of microörganisms will add but little to the real question, "what bacteria do, not what bacteria are."

The preceding paragraphs contain a summary of the better known and more important organisms represented in the intestinal tracts of human beings at different ages; while the list is not complete (due partly to our present imperfect methods of study) it is sufficiently so to make the following pages more intelligible. These organisms have been mentioned in order to bring out the relations which they exhibit toward each other, and to their host. These bacteria do not represent an erratic and disorderly development of whatever microörganisms may gain entrance to the alimentary canal, but on the contrary represent a definite microbic response to calculable stimuli, notably the food of the host. That is to say, it is possible to influence in a noteworthy manner the types of bacteria in the alimentary canal of the host through variations in his diet. The nature of this bacterial response to dietary alterations is so marked, and takes such a prominent part in the theory and practice of what is to follow, it will be explained in some detail. The methods employed for this purpose have been fully described in previous publications,[93-99 inc.] and the details need no repetition here. The principles involved, however, are extremely important, and they deserve special mention.

In carrying out these studies, particularly those in which the fundamental principles involved are being investigated, frequently it has been necessary to resort to exaggeration of the factor sought for, because it has been found to be impossible to suppress other factors which also affect the result. It will be argued by some that this indirect method of experimentation is not accurate, but even in such exact sciences as physics, for example, the same procedure is not infrequently employed.

Lord Kelvin and Tate [100] say: "When a particular agent or cause is to be studied, experiments should be arranged in such a way as to lead to results depending upon it alone, if possible; or, if this cannot be done, they should be arranged so as to show differences produced by varying it. . . . When, in an experiment, all known causes are being allowed for, there remain certain unexplained effects, these must be carefully investigated, and every conceivable variation of arrangement of apparatus, etc., tried; until, if possible, we manage so to exaggerate the residual phenomenon as to be able to detect its cause."

Fermentation and putrefaction. — It will be essential to define clearly just what is meant by fermentation and putrefaction. Two most fundamental types of bacterial activity are expressed by these terms, which are in reality essentially distinct, although they have been greatly confused, and used synonymously by the laity and even by professional men. This confusion is attributable partly to the use of terms to designate certain processes which occur in nature before these changes were studied either biologically or chemically.

In reality, fermentation and putrefaction are generic terms, indicating respectively microbic activities upon two entirely distinct types of organic compounds, the carbohydrates (and closely related compounds) and the nitrogenous bodies. It is very probable there are groups of substances intermediate between the carbohydrates and nitrogen-compounds referred to, in which it would be difficult to predict à priori just how definite types of bacteria would react, and, indeed, such is probably the case. This does not, however, militate against the correctness of the general theory that fermentation and putrefaction are entirely distinct processes, but rather emphasizes the fact that bacteria are reagents far more sensitive than those commonly used in chemistry. Turning now to a separate discussion of fermentation and putrefaction, we will first discuss the former: fermentation.

Fermentation is probably the earliest example known to science of the application of microörganisms to accomplish a definite purpose. The fermentation of grape juice to wine was a process known to remote antiquity so far as the results were concerned, and even to-day, although the organisms concerned are fairly well known, the process is not essentially different from that known to the ancients. The essential feature, the action of yeast upon carbohydrate to form alcohol, is a typical example of a fermentation.

It is known, however, that various yeasts possess the power of forming alcohol from various sugars, and, indeed, even bacteria may accomplish the same end, thus indicating that fermentation is a much more general phenomenon than the mere action of yeasts upon sugars.

Coincidently with this production of alcohol from sugars, other products are formed, even by yeasts, although certain of these latter substances are formed in very minute amounts, so that their presence is rather suspected than demonstrated chemically. These secondary products, both those formed in minute quantities, and those more tangibly present, again indicate strongly that fermentation is a more complex process than would appear at first sight, even if it is brought about by but a single "species" of yeast. Without going into details, the effects of various yeasts, and bacteria, upon various carbohydrate media make it apparent that fermentation is a rather general term embracing a variety of phenomena. In other words, fermentation must be used in a generic sense, and it is now possible to distinguish various types of fermentation, brought about by the activities of different kinds of microörganisms : we recognize alcoholic, lactic, acetic, and other fermentations. The use of the term fermentation in the generic sense does not indicate a loss in dignity or comprehensiveness; indeed, it actually gains in importance through the recognition of its more general application.

Alfred Fischer [101] has defined fermentation in the broadest sense, and his definition leaves little to be desired, being straightforward, definite, and comprehensive. "Als Gärung soll hier nach dem Beispiele vieler Autoren, die biochemische zersetzung stickstofffreier organischen Verbindungen, besonders der Kohlenhydrate durch besondere Gärungserreger, Fermentorganismen bezeichnet werden."

It is very probable that Fischer did not separate two very distinct processes which occur in practically every fermentation: the destruction of carbohydrates, containing no nitrogen, and the metabolism of nitrogenous substances which all microörganisms need in their dietary. This distinction between the very noteworthy breakdown of carbohydrates (which is the salient feature of every fermentation) and the limited, and usually not recognized nitrogen metabolism by the fermenting organisms is of great importance, and, as will soon be shown in the discussion of putrefaction, a similar distinction is to be made there. In putrefaction, where the action of the microörganisms concerned is limited sharply to nitrogenous substances, it is very difficult to distinguish between the widespread breakdown of protein, the essential feature of the process, and the very limited utilization of protein for purely dietary purposes.

Turning now to the definition of fermentation, in the light of what has been said above, fermentation may be defined as "the action of microörganisms upon carbohydrates." This is a very general definition, and covers the salient feature of the process, the action of microbes upon carbohydrate. In order to specify the particular type of fermentation under consideration, the terms lactic, acetic, etc., are utilized.

Putrefaction. — The wildest confusion exists concerning the essential features of putrefaction. The term has been used as a synonym for fermentation, and in supposedly authoritative dictionaries and encyclopedias, even to this day, the terms are thus confused. The popular conception of putrefaction associates this condition with the generation of foul odors, and many scientists adhere to this conception as well. To this association of foul odor, many add the

necessity of anaërobic conditions. This definition of putrefaction being associated with the elaboration of foul odors,
under anaërobic conditions is based upon intangible evidence, for it is impossible to discuss odors, foul or otherwise,
in measurable or even definable terms, strictly speaking.
A striking example of the correctness of the contention that
putrefaction is not as simple an entity as the definition would
indicate is shown by the fact that indol and skatol are not
as a rule formed by the action of anaërobic bacteria, at least
in appreciable amounts. Indol and skatol certainly are to
be classed with the "foul odors," and suggest the essence of
putridity, yet it is a singular fact that anaërobic bacteria
(which are supposed to be the *sina qua non* of putrefaction)
form them only in minimal amounts; indeed, many anaërobes,
including B. putrificus, do not form them at all. On the
other hand, bacteria of the facultative type, as, for example,
B. coli and B. proteus, form indol in considerable amounts
under favorable conditions. Yet the majority of authorities
would not classify these organisms as typical putrefactive
bacteria, nor would they admit that they could initiate and
carry out a typical putrefaction. Hydrogen sulphide is
another moderately vigorous odor, foul, and suggestive of
putridity; it may be produced in relatively large amounts by
bacteria not putrefactive in the regularly accepted sense.
Bienstock and his co-worker Wollach were among the first
to study putrefaction from the standpoint of the organisms
concerned, and they were inclined to believe that true putrefaction was brought about by a single organism, an obligate
anaërobe, B. putrificus. Other workers in this field, following their lead, have made similar declarations, until at the
present time the opinion is widespread that this organism,
together with possibly one or two other obligate anaërobes,
are the true agents able to bring about putrefaction.
According to Bienstock,[102, 103, 104] B. putriforms NH_3, H_2S, peptone, amino bases, valerianic and butyric acids, leucin and
paraoxyphenylpropionic acid. It is a significant fact that
not one observer has detected indol or skatol among its
decomposition products. Yet Bienstock and his school

believe these latter substances are very typically found in putrefaction, in the restricted sense in which they define putrefaction.

Nencki,[105] Kerry,[106] and Zoja,[107] have specially studied the chemistry of B. putrificus, and they have also found that it produces no indol or skatol. Seelig[108] and Dauber[109] have found that while B. putrificus forms no indol, organisms commonly associated with it in natural putrefactions, as for example B. coli and B. proteus, form indol in peptone containing media, and they believe that the obligate anaërobe breaks the native protein to the peptone stage, where the facultative anaërobes take up the work and carry the decomposition through the indol stage to lower and simpler compounds.

Indol and skatol represent somewhat advanced decomposition of protein substances, and it appears to be a fact that anaërobes as a class do not bring about enough action upon the protein molecule to reach the indol and skatol stage. In fact, anaërobes, generally speaking, bring about a merely superficial, but widespread destruction of protein, contrasting in this respect with the facultative anaërobes, which take the initial products formed by the anaërobes to their lowest terms. Hueppe[110] recognized this distinction in action between anaërobes and aërobes as early as 1888: "The changes induced in the substratum by anaërobic bacteria differ from the changes taking place in the presence of free oxygen. The maintenance of life without free oxygen depends solely upon the availability of compounds from which oxygen may be split off. The amount of chemical change, therefore, is relatively much less intense than in anaërobic conditions: thus, if one thousand grams of sugar be completely oxidized to CO_2 and H_2O in the presence of free O_2, three thousand nine hundred and thirty-nine calories of heat units are produced. If, however, it is split into butyric acid, H_2 and CO_2, only four hundred and fourteen calories are produced. It follows, therefore, that anaërobic bacteria must superficially disintegrate a far larger quantity of material to obtain this necessary oxygen than aërobic bacteria, a circumstance that has considerable significance in the large production of toxins by organisms growing in the living body."

In nature, at least (and it is here that the conception and term putrefaction arose), the phenomena grouped together as putrefaction represent a series of symbioses in which the initial, superficial break-down is brought about by the anaërobes, while the process is carried to its lowest terms by the facultative anaërobes. Certain Germans have dimly recognized that putrefaction is indeed more than a simple anaërobic decomposition, and without knowing definitely the steps involved, they have made a distinction between Fäulnis, putrefaction in the popular sense, and Verwesung, which apparently is synonymous with Eremecausis, or the terminal stages of nitrification as we know them in the presence of free oxygen, moisture and the activities of many kinds of bacteria.

The principal difficulty met with in defining putrefaction appears to be that the process has been named without knowing much about it. In order to bring out this contention distinctly, a few definitions of putrefaction are appended. They have been gathered from what ought to be authoritative sources.

Flügge.[111] By putrefaction, or putrid fermentation, one understands the rapid and intensive dissociation of nitrogen-containing, chiefly protein, substances by certain bacteria, through which process gaseous, offensive products are produced in considerable amounts.

Migula.[112] Putrefaction: the decomposition of animal or vegetable substances without regard to their chemical properties.

Cornil et Babes.[113] Putrefaction must be considered as the process and result of different fermentations which vegetables and animals undergo after their death. Fermentations of nitrogenous substances are naturally the most important to recognize in the phenomena of putrefaction.

Gunther.[114] Fäulnis, putrefaction: the anaërobic decomposition of protein, with the production of foul smelling products. Verwesung (eremecausis): an oxidization process, with free access of air; a very important factor in the self purification of the soil.

Kolle und Haetsch.[115] Certain anaërobes break down nitrogen-containing organic compounds in a definite manner through a series of reductions resulting in the formation of volatile and non-volatile compounds, many of which are foul smelling.

Encyclopedia Britannica.[116] Putrefaction: the scientific meaning of this term coincides pretty much with its popular conception, except that it must be understood to be exclusive of all forms of oxidization.

Standard Dictionary.[117] Putrefaction: the act or process of putrefying, decomposition of vegetable or animal matter, accompanied by fetid odors; now regarded as a sort of fermentation or breaking up of complex organic compounds into simpler compounds produced by the microörganisms called putrefactive ferments.

With such a variety to choose from it would appear to be time to define putrefaction in a logical manner.

Putrefaction, reduced to its lowest terms, is a generic term quite similar in this respect to fermentation. The phenomena involved are much more complex than those of fermentation because the nitrogenous substances involved are, or may be, far more complex than the carbohydrates. The wide range of intermediary break-down products contrast in a noteworthy manner with the simpler and more direct transformation of carbohydrates, and the very fact that there is such a variety of intermediary compounds presupposes in nature a multiplicity of bacterial " species " to bring about the transformation of the complex to the simple.

At first sight it might seem to be a very difficult if not impossible task to formulate a definition of putrefaction which should be specific enough to sharply define the process to a definite type of microbic activity, and yet general enough to attain the necessary degree of comprehensiveness. It is possible, even in the face of our present comparative ignorance of the chemistry of putrefaction, to formulate such a definition.

Fischer[118] has again grasped the salient feature of putrefaction, and his definition is well worth considering: "Sie (Fäulnis) ist demnach die Zersetzung stickstoffhaltiger Produkte des Tier — und Pflanzenlebens besonders der Eiweisskörper durch Bakterien."

By putrefaction, then, is meant the bacterial decomposition of nitrogenous substances. This definition is a generic one, precisely as the definition for fermentation given above was a generic one, and it may be qualified in precisely the same manner that fermentation may be qualified.   As our knowledge of putrefaction widens, doubtless many types of putrefaction will be recognized.

In the last analysis, both putrefaction and fermentation are enzyme phenomena, differing but slightly from the digestive processes in man; the terms, however, must be specifically limited to the action of microörganisms or enzymes elaborated by microörganisms.

The necessity for distinguishing sharply between fermentation and putrefaction is by no means an academic one; the two phenomena are fundamentally different and diametrically opposed.   Together they represent the most fundamental phenomena of bacterial metabolism.

It will now be possible to enter into a discussion of the significance of putrefaction and fermentation respectively. It was stated in preceding paragraphs that for the most part the studies upon the intestinal bacteria have led to little that is definite concerning their modes of action, their significance and their relations to their host.   These purely academic studies upon bacterial morphology, while necessary, are after all studies in differences (more or less trivial for the most part) between closely related forms; the problem of intestinal bacteriology is much more a study of relationships and biochemistry than of differences of morphology and physiology.   The most satisfactory manner of approaching the problem of bacterial relationships in the alimentary canal is to realize at the start that these organisms depend upon the food of their host for their sustenance and their very existence.   Consequently, it should be a çomparatively simple matter to vary the food of the host in a definite manner, and observe the character of the bacterial response to these dietary changes.   This has been done by the writer,[94, 97, 98, 99] and the results obtained will be referred to briefly, because they illustrate the differences between putrefaction and fermentation in a manner not appreciated when these experiments were published.

If an animal, preferably a monkey (whose physiology of

digestion closely resembles that of man), be fed alternately
upon an essentially protein and a carbohydrate diet, allowing a sufficient interval between dietary changes to permit
the flora to adjust itself in a characteristic manner, the following phenomena will ordinarily be observed.

Protein diet. — The monkey gradually becomes apathetic,
is dull, takes but little interest in its surroundings, responds
slowly to external stimuli, and becomes sluggish in its movements. The urine becomes small in amount, rather dark in
color, of a relatively high specific gravity, and contains considerable amounts of indican and skatoxyl, together with
other products characteristic of protein putrefaction in the
alimentary canal. The feces become dark colored, tarry,
with a decidedly foul odor. Chemically, they contain indol,
skatol, some aromatic oxyacids and other products of bacterial decomposition of protein. Their reaction is alkaline.
Bacterially, the flora represented is distinctly proteolytic in
character. Many organisms which liquefy gelatin are present, agreeing in this respect with Escherich's experiments
upon dogs,[119] as well as considerable numbers of aërogenic
bacteria of the coli-aërogenes-proteus group. Microscopically, with the Gram stain, one sees many rather large,
Gram-positive bacilli, some containing spores, belonging to
the subtilis-mesentericus group, as well as Gram-negative,
oval rods of the colon group. Some coccal forms Gram-
positive, and resembling greatly the Mic. ovalis referred to
previously, are also usually present, and even in considerable
numbers. So far as morphological types of bacteria are
concerned, the flora is very heterogeneous. The bacteria
which may be isolated from such stools reproduce with
considerable readiness, either singly, or in symbioses, the
changes which the mixed fecal flora brings about in artificial
media, and there is little doubt but what the more common
and important bacteria concerned in this protein flora are
easily subcultured and can be made to reproduce the chemical changes characteristic of the mixed flora. The more
striking phenomena seen in artificial media, which the mixed

fecal flora, or selected, dominant organisms from this flora (acting singly or symbiotically) are the following: rapid liquefaction of gelatin, peptonization and gas formation in milk, and the production of large amounts of gas in dextrose, lactose and saccharose broth fermentation tubes.

Carbohydrate diet (milk containing an excess of lactose about seven per cent). — Clinically, the animal becomes more alert, active and bright. It takes a lively interest in its surroundings.

The urine becomes lighter in color, the specific gravity diminishes, and the putrefactive products (particularly the etheral sulphates) diminish considerably in amount, or even disappear.

The feces become lighter colored, more voluminous, less tarry, less odoriferous, and acid in reaction. Indol and skatol diminish, or disappear, as do the aromatic oxyacids.

The fecal flora undergoes a decided change, both with respect to the change in types of organisms represented and in the behavior of those common to the two regimens. Microscopically (Gram stain), the flora is more homogeneous; longer, thinner Gram-positive rods replace the rather stout bacilli seen in the protein feces, while the Gram-negative forms and coccal forms tend to disappear as well. Culturally, the proteolytic bacteria disappear largely, and they are replaced by fermentative bacteria. Those organisms which are able to grow on either protein or carbohydrate become fermentative in their metabolism when the carbohydrate regimen is instituted.

The mixed carbohydrate fecal flora, inoculated into artificial media, act differently from the protein flora. They do not liquefy gelatin rapidly, do not peptonize milk, form little or no gas in it, and but little gas in dextrose, lactose or saccharose broth fermentation tubes. Indeed, the growths are less vigorous in any of these media. These changes are associated with the development of a distinctly fermentative flora: a flora in which acid products predominate when they are grown in suitable artificial media; it is partly (not wholly) due to this formation of acid that the fermentative flora gains the upper hand when the diet of the host contains even a moderate amount of carbohydrate. As will be shown later, the acid formed is the result of the action of the bacteria upon the food in the intestinal tract of the host, and, in this sense, the acid is the cause of the disappearance of the purely proteolytic bacteria, although the carbohydrate is after all the essential factor. This is a distinction of great importance and deserves special mention.

It is a significant fact that the carbohydrate diet results in the establishment of a flora resembling that of a normal nursling. This resemblance is morphological, cultural, and chemical, although it is not absolute since it appears that the composition of the intestinal juices themselves has some effect upon the flora. This influence is not marked, however, but it is a factor to be taken into consideration in discussing the results. It will be remembered that the liquefying forms largely disappeared as the carbohydrate flora became dominant. Escherich [120] and Spiegelberg [121] have commented upon the lack of liquefying organisms in the stools of normal nurslings. This infantile type of flora, then, following logically the infantile type of diet, is evidence of the correctness of the assumption that food largely determines the type of intestinal bacteria. Jacobson [122] has made somewhat similar observations upon dogs; he fed normal breast milk, and found that the normal nursling flora gradually became dominant. What has actually happened as the result of these dietary changes? Attention has already been directed to the more apparent changes which follow alternations in diet: briefly, a proteolytic flora develops when the host is fed protein and a carbohydrate-fermenting diet results when the diet is essentially carbohydrate. This is not the whole story, however, and it is precisely the changes which will be mentioned in the following pages that are most significant in regulating the character of the intestinal flora.

Not only is there a change in the types of bacteria prominently represented in the protein or carbohydrate diets as alternations in regimen are made, but certain organisms accommodate their metabolism now to a protein, now to a carbohydrate diet.

Theoretically, a third method of influencing the intestinal flora might be considered, namely, by intestinal antisepsis. Intestinal antisepsis would be the most satisfactory method of regulating bacterial activity in the alimentary canal, because it is direct and far simpler than other methods.

Wassilieff,[132] working in Hoppe-Seyler's laboratory, believed that calomel did not influence artificial digestion, but that it seemed to inhibit putrefaction. Dogs fed with calomel showed a diminution of indican in the feces after a short period. Baumann [134] also noticed that after the administration of calomel the ethereal sulphates diminished in the urine. Bartoschewitsch [135] noticed a disappearance of ethereal sulphates regularly, and Fr. Müller [136] in a single case corroborated this finding. Fürbringer [137] studied the effect of calomel upon the numbers of intestinal bacteria and found that they were considerably decreased. Schütz,[138] on the contrary, noticed an increase in the numbers of certain organisms which he fed to dogs, following calomel (Spirillum metchnikovi). Strasburger also noticed that the bacteria in the small intestine increased after the administration of calomel. Morax showed that the cause of the diminution of bacteria in the feces was simply the result of the diarrhea which mechanically removed them.[139] Stieff [130] and Biernacki [131] corroborated Morax's observations that the diarrhea was the essential factor. This explanation is not wholly satisfactory, because it is well known that there is, or may be, diarrhea both in typhoid and intestinal tuberculosis, while putrefaction is frequently increased, judging from the ethereal sulphates in the urine.

Rossbach [132] and Albu [133] tried naphthalin, the former with doubtful success, the latter with very moderate success. Sehrwald [134] noticed that the numbers of bacteria discernible in the feces diminished following the administration of naphthalin. Morax (loc. cit.) tried iodoform, Calderone [135] tried both iodoform and calomel, Stieff (loc. cit.) used camphor, Strasburger [136] used salicylic acid, Rovighi [137] tried turpentine, camphor, menthol, and boric acid, and all found a greater or lesser decrease in the numbers of fecal bacteria following their use. Strasburger (loc. cit.) noticed some reduction in fecal bacteria following the use of thymol, but believed it was without significance. Bourchard [138] found some reduction after beta naphthol, Loebisch [139] found some reduction with urotropin (hexymethylamin), while Stern tried calomel, salol, beta naphthol, naphthalin, and camphor.[140]

The general consensus of opinion seems to be that any or all of these substances, used in permissible amounts, are without practical effect other than a more or less mechanical diminution of the numbers of organisms by the diarrhea produced. It seems to be fair to state that the attempt to destroy the intestinal bacteria by chemical means has had but little success up to the present time. The same generalization is true of starvation. Generally speaking, prolonged starvation will not kill off all bacteria, while it is detrimental to the patient.

Realizing, then, that intestinal antisepsis by chemical means is not practical, the importance of modifying the intestinal flora by dieting becomes great, because it appears to be the method available at times to combat the bacteria upon their own ground. The change in the metabolism of bacteria, together with the change in types following definite alternations in diet, becomes more realistic when one considers that the intestinal bacteria are necessary accompaniments of life, at least outside the arctic zones. The question arises — do the intestinal bacteria serve a useful purpose? Are they necessary for life?

Nuttall and Thierfelder [141] reared guinea-pigs delivered by Cæsarian section in a sterile environment upon sterile food for thirteen days and found that they gained in weight. These experiments appeared to confirm the observations of Levin, [142] who examined the feces of arctic animals bacterially and found them sterile, or nearly so. His observations, it is fair to state, were made in the very northern part of Europe, far beyond the lower limits of the Arctic Circle. Somewhat similar observations are recorded by the scientist who accompanied Nansen. The parrot is said to have very few intestinal bacteria.

Schottelius [143] hatched chickens in a sterile environment and reared them in sterile surroundings. He commented on the difficulty of finding sterile eggs, and his work deserves special mention because of the very thorough manner in which it was carried out. His chicks were divided into three lots: (a) kept sterile; (b) kept sterile for several days, then given infected food; (c) controls, fed in the ordinary manner.

Group a did not do very well, particularly after the first ten days. Group b paralleled Group a, but began to gain rapidly after being fed upon ordinary, unsterile food. Group c did well from the start. The chicks were kept as indicated for three weeks, but normal growth was obtained only in those fed upon unsterile food.

Madame Metchnikoff [144] made similar observations upon tadpoles with results practically the same as those of Schottelius, and Moro [146] cor-roborated Schottelius' work using turtles.

These experiments do not prove that the intestinal bacteria are necessary for the well-being of their host, for obvious reasons.   The facts, after all, hinge upon the truism that man has an intestinal flora, and that it is practically impos-sible, if not undesirable, to rear him in a sterile environment. The huge numbers of bacteria excreted daily, as well as their potential capacity for causing harm, make it imperative that something definite be known about their mode of action, the laws underlying their presence and, above all, how to bring about definite changes in an undesirable flora.   From what has been said it is evident that reformation, not anni-hilation, is the end to be accomplished, since so many observers have found it impracticable, or even impossible, to sterilize or even to approach sterilization in the intestinal tract.

With a clear understanding of the necessity for reforma-tion rather than annihilation of the intestinal flora, the signifi-cance of the fermentative flora in normal nurslings will be apparent.   This fermentative flora, as has been shown by many observers, is of vital importance to the young child, and nature appears to have provided specifically for the maintenance of such a condition in the intestinal tract.   Breast milk contains, roughly speaking, from six to seven per cent of sugar, and only one and a half per cent of protein.   The fat is potentially related to the carbohydrate, rather than the protein, from the bacterial viewpoint.   When we remember that bacteria utilize carbohydrate instead of protein, whenever possible, it will be seen that the conditions for maintaining a fermentative flora instead of a putrefactive flora are good. A significant corollary to this generalization is the fact that derangements of digestion, associated with the activity of organisms other than the normal flora, are preceded by decided changes in the normal types.   As the child grows older, the food becomes more complex, the types of bacteria present in the intestinal tract become more varied, and the

young individual enters upon the cow's milk regimen. It is
noteworthy that trouble is more likely to occur during this
period of life, and one of the best remedial measures in many
cases, at least, is to place the patient upon breast milk again,
if this be possible. Again, breast-fed babies under very poor
sanitary conditions do better on the whole than those of the
same age on cows' milk, but under better sanitary conditions.
The reasons for this will be apparent after a discussion of the
differences between the normal breast-fed flora, and the flora
in which there is a decided diminution in carbohydrate.

Carbohydrate prevents the metabolism of body nitrogen
in man to a considerable degree; this protective or sparing
action of carbohydrate for protein also holds for bacteria,
and even to a greater degree.

The above statement expresses what is probably the most
fundamental property of bacteria. So far as the writer is
aware, this generalization has not been made before, although
many facts bearing directly upon it have been described, and
from many points of view. While many observers have
apparently hit upon and explained successfully most of the
factors entering into this generalization, it appears to be true
that the full significance of the phenomenon has escaped
notice. This statement is particularly apt to be true because
it is hardly possible to conceive of any one grasping the principle as a whole and failing to see its vast field of action and
its possibilities both in medicine and the industries.

Evidence supporting this hypothesis. — In the preceding
paragraphs mention was made of the fact that certain bacteria could accommodate their metabolism now to a protein,
now to a carbohydrate regimen. This statement will now be
discussed in its relation to certain types of bacteria where
the facts have been worked out in great detail and in which
the results are particularly striking.

Lyons [146] and Cramer [147] have analyzed bacteria grown upon media with and without carbohydrate and had made the very interesting and important observation that the actual chemical composition, especially the nitrogen content, of these bacteria varies considerably, the nitrogen content being greater in those organisms grown in media containing no carbohydrate, less in media containing carbohydrate. Before discussing their results it should be stated that Dreyfuss,[148] Nashimura,[149] Duclaux,[150] Nägeli,[151] and others have shown qualitatively that bacteria, yeasts, and moulds vary in composition, and elaborate different compounds from different nutrient substrata, the kind of product elaborated depending upon the composition of the medium in which or on which they are grown.

Lyons and Cramer have furnished definite figures, showing in a striking manner just what these differences are. Their figures are fairly in accord, and inasmuch as the observed differences between the selected media are greatly in excess of the probable sources of error, it may be assumed that, at least the generalization involved, the difference in composition of the bacteria is proved.

Lyons used three bacilli for his experiments, and the following table contains his results:

Organism.		Dextrose in Per Cent.		
		1	5	10
Pfeiffer bacillus	Nitrogen-substance	62.75	58.88	45.88
	Ether extract.	1.68	3.50	2.67
	Alcoholic extract	12.17	17.30	29.60
	Ash..	7.16	2.97	3.09
Bacillus No. 28	Nitrogen-substance	71.81	59.12	46.25
	Ether extract.........	3.32	3.84	2.84
	Alcoholic extract	11.39	15.19	22.78
	Ash.................	6.51	3.66	4.18
"Thread" bacillus...	Nitrogen-substance	61.06	44.31	33.25
	Ether extract.........	1.74	2.24	1.87
	Alcoholic extract	18.40	21.80	27.50
	Ash.................	8.09	4.50	3.02

The organisms were grown upon agar (slanted) carefully collected, washed and dried in vacuo. The organisms were tested for purity before they were dried, and found to be typical, culturally.

Cramer used four bacilli, and his experiments are more extensive than those reported above. His organisms included the following: Pfeiffer's bacillus (1), No. 28 (2), " Pneumonia bacillus " (3), and the Rhinoscleroma bacillus (4).

Bacillus.	Nitrogenous Substances.			Extract Ether-alcohol.			Ash.		
	Peptone Per Cent.		Dextrose Per Cent.	Peptone Per Cent.		Dextrose Per Cent.	Peptone Per Cent.		Dextrose Per Cent.
	1%	5%	5%	1%	5%	· 5%	1%	5%	5%
Medium No.	1	2	3	1	2	3	1	2	3
No. 1	66.6	70.0	53.7	17.7	14.63	24.0	12.56	9.10	9.13
No. 2	73.1	79.6	59.0	16.9	17.83	18.4	11.42	7.79	9.20
No. 3	71.7	79.8	63.6	10.3	11.28	22.7	13.94	10.36	7.88
No. 4	68.4	76.2	62.1	11.1	9.06	20.0	13.45	9.33	9.44

The media were made in the following manner: 1.5 per cent agar, common to the three types of media. Medium No. 1 contained in addition one per cent peptone. Medium No. 2, two per cent peptone. Medium No. 3, one per cent peptone plus five per cent dextrose.

Cramer was impressed by the relatively low amount of nitrogen-containing substance in the dextrose media, and he was unable to account for it. He was certain there was no loss of volatile nitrogenous products, and, indeed, he comments on their absence in the dextrose media. The dextrose agar was strongly acid when he removed the bacteria, while the peptone media were alkaline. He concludes that the composition of bacteria is not constant, but varies within somewhat wide limits upon different media.

In the light of the theory proposed at the head of this section, there is little doubt that the composition of the bacteria is directly influenced by the presence or absence of utilizable carbohydrate, and that not only does this carbohydrate spare, to a considerable degree, the protein of the medium in which they are growing, but that it even influences the actual amounts of nitrogenous-substance in their bodies. Reference to the possible significance of this phenomenon will be made in the discussion of typhoid, dysentery and similar diseases of the intestinal tract.

Decomposition products of bacteria grown in media with and without carbohydrate. — Theobald Smith,[151] in his classical studies of the conditions which influence the appearance of toxins in cultures of the diphtheria bacillus, has shown conclusively that the amount of toxin formed in artificial media is dependent, largely, upon the presence or absence of muscle sugar (dextrose). As the result of his work he finds that three distinct types of growth may occur, as follows: in bouillon free from sugar, alkaline reaction, much toxin produced; bouillon containing but little sugar (.1 per cent or less), initial acidity, followed by alkalinity, with less toxin formed; bouillon containing relatively a great amount of sugar (more than the bacilli can ferment, before the reaction becomes too acid) initial acidity, which increases and remains permanent, practically no toxin. This observed difference in toxin production is not due to acid formation as Dr. Smith has shown: "At first I thought that the acid formed combined with the toxin to make it inert. This easy way out of the difficulty was shown to be not probable by a simple experiment. Toxin filtered and containing .2 per cent carbolic acid received as much lactic acid as was necessary to make the acidity 3.6 per cent, at which point toxin is not detected in cultures (containing fermentable sugar). The acid toxin was placed in the incubator to approximate the environment of the cultures, and tested on guinea-pigs from time to time. After several weeks the test showed no decline in toxic power and the experiment was abandoned." These experiments illustrate in a very graphic manner the shielding action of (utilizable) carbohydrate for protein, and inasmuch as certain of the products of protein metabolism by the bacteria bacillus are strongly toxic, it will be apparent that this work is particularly valuable because it is possible to demonstrate even relatively small amounts of protein break-down through the action of the diphtheria toxin on guinea-pigs. The presence of even small amounts of dextrose prevent the diphtheria bacillus from attacking protein. Dr. Smith further states, in discussing the factors which favor the production of toxin — "we have no reason to believe that the action of diphtheria bacilli on mucous membranes differs from that manifested in cultures."[152] Dr. Smith tells me that the tetanus bacillus acts in a similar manner in cultures, and that the same general phenomena obtain in the production of the tetanus toxin.

Brown,[153] working on spore-bearing anaërobes in market milk, has also published a similar observation for the gas bacillus (Bact. Welchii). He says: "When grown in ordinary broth plus tissue, and inoculated into guinea-pigs, this anaërobe produced no lesion beyond a small subcutaneous nodule that was transitory, but if grown in bouillon plus tissue that had been rendered sugar free by fermentation with B. coli, it was pathogenic for guinea-pigs, producing the characteristic lesions of B. aërogenes capsulatus infection. This is a phenomenon discovered by Theobald

Smith and is mentioned here with his permission, though it has not pre-
·viously been described. It applies to some other anaërobes beside the
one under discussion, but not included in this paper." Here again, we
have an example of the protective action of carbohydrate, sparing protein
from bacterial attack. When this sugar is removed by fermentation, the
protein metabolism may take place, in this case with the production of
products inimical to the host (guinea-pig).

Turning now to bacteria which do not form strong extracellular toxins, it will be apparent that there is the same
general protective action of utilizable carbohydrate for
protein.

Both the typhoid and dysentery bacilli form acids when
they are grown in media containing both protein and dextrose, but form alkaline products when they are grown in
media containing protein alone. The conditions which favor
the development of toxins in the dysentery bacillus will be
described in detail in succeeding paragraphs.

B. coli produces considerable amounts of easily recognized
putrefactive products from protein, and this organism well
illustrates the theory.

As early as 1896, Seelig [154] was impressed with the difference in the
nature of the decomposition products produced in media with and without sugar by the growth of B. coli. In media containing no lactose, for
example, he noticed that the organism formed considerable amounts of
indol and phenol, while in media containing lactose, no indol or phenol
were demonstrable. In milk, similarly, he was unable to demonstrate
either indol or phenol; in fact, no putrefaction products were present in
milk inoculated with pure cultures of B. coli.

The writer [155] has made similar observations. Two portions of meat-
juice-peptone broth (freed from sugar) were prepared, having exactly the
same composition. To one portion, one per cent of dextrose was added
(lactose would have been equally good for the purpose), then the two portions were inoculated with the same strain of B. coli and incubated at
37° C. for ten days. At the end of that time, the two portions of broth
were analyzed chemically.

The portion containing no sugar was foul smelling, alkaline in reaction,
and contained ammonia, hydrogen sulphide, indol, some skatol, phenol,
and aromatic oxy-acids. This was a typical protein putrefaction.

The portion containing lactose was not disagreeable to the sense of
smell (on the contrary it had an appetizing odor), and contained carbon
dioxide, hydrogen, lactic acid, smaller amounts of acetic and succinic

acids. This was a typical fermentation. The only difference between the two portions of broth was one per cent of dextrose which was added to one of them. There can be no other explanation than that the organism (B. coli) attacked the dextrose (carbohydrate) in preference to the protein, and in this sense the dextrose spared the protein.

Auerbach [156] investigated the factors which influenced the liquefaction of gelatin, and selected members of the proteus group for his experiments. He noticed that gelatin containing no fermentable sugar was rapidly liquefied, and that the reaction of the liquefied gelatin at the termination of his experiment was strongly alkaline.

Gelatin of the same composition, with the addition of fermentable sugar, was either not liquefied at all or very slowly. The reaction of this gelatin was either strongly acid, in which case liquefaction did not take place, or moderately acid, when the liquefaction commenced after several days. It is again apparent that the bacteria attacked the sugar in preference to the protein.

Heilner [157] studied the physiological effects of injecting sugars of varying strengths into experimental animals subcutaneously. While his results from the purely physiological point of view are of no particular interest in this connection, he made the very interesting observation that his sugar solutions rarely, if ever, became infected after they were introduced under the skin; he did not get abscesses as the results of his injections. In this connection it is well to remember that protein solutions are relatively difficult to inject without obtaining abscesses, and even making allowance for the difference in rapidity of absorption of the sugar and protein solutions respectively by the subcutaneous route, the lack of infections in the sugar experiments is significant. Here, again, the bacteria which infect and cause trouble in protein injections appear to do so through the elaboration of products from the break-down of this protein which are of real benefit to the invading bacteria, aiding them to gain and to maintain a foothold in the animal body. The corresponding products derived from the decomposition of sugar solutions are acid in character, in contrast to the alkaline products of protein break-down, and are of little or no use to the invading organisms, therefore not aiding them to gain and maintain a foothold.

Cohnheim [158] was aware of an antagonistic action between fermentation and putrefaction, although he was unable to explain the cause of this difference. P. Cohnheim has pointed out that protozoa are more readily obtained from stools containing mucus and having an alkaline reaction: [159] "Die Aussichten, beweglische Urthiere nachzuweisen, sind um zu grösser, je flüssiger, schleimhaltiger und alkalischer der Koth ist." This is not to be construed as indicating that an alkaline reaction is more favorable for the development of intestinal protozoa, but the observation is suggestive.

Hirschler [160] made putrefactive mixtures of meat and pancreas. He added dextrose, glycerine and calcium carbonate to some of them, then allowed them to putrefy spontaneously. The lime neutralized the acid as fast as it was formed in the media containing dextrose (and glycerine), but in spite of the neutralization of the acid no indol or phenol were found in the mixtures containing carbohydrate. He concluded that acid production was not of itself sufficient to prevent putrefaction, and he was inclined to believe that the organisms concerned acted upon sugar first, and that possibly the nascent hydrogen was in some manner important in preventing putrefaction.

Winternitz [161] also noticed that milk had " antiputrid " properties. If he inoculated infants' stools into milk it did not putrefy, but became sour instead. He attributed the antiputrid property of milk to lactose, and he invented a series of experiments which indicated strongly that the casein and the fat, alone or together, were unable to inhibit putrefaction. He attributed a prominent part to the lactic acid which was produced, and his explanation of the action of lactose was that the lactic acid, which was formed almost from the start, prevented the proteolytic bacteria from attacking the protein. He made a mixture of meat, milk and calcium carbonate and found that no putrefaction took place, although the meat alone very quickly decomposed. This last experiment shook his faith in the lactic acid theory, but he left the subject at this point.

Gorini [162] has shown that the addition of dextrose to media in which the cholera vibrio is growing inhibits the formation of indol and cholera-red.

Schmitz [163] fed cheese to certain individuals and found that the intestinal putrefaction was increased as the casein (cheese) was increased. Iwanoff [164] and Rubner [165] noticed that yeasts exercised a peculiar, restraining action upon putrefaction when they were present in protein containing media and sugars.

Rovighi and Simnitzki [166, 167] studied the products of fermentation and believed that lactic acid was in reality the agent that prevented putrefaction.

Seelig, [168] Albu, [169] and Blumenthal [170] examined milk that did not putrefy upon standing, and were able to isolate B. coli or Bact. aërogenes from it. Their experiments indicated that these two organisms, either acting singly or together, could prevent the putrefactive decomposition of milk.

Bienstock [171] infected unsterile milk with his obligate anaërobe, B. putrificus, and found that no putrefaction took place, although this organism could bring about energetic decomposition of native albumens in putrefactive mixtures. Sterilized milk, infected with this organism, on the contrary, did putrefy. Sterile milk inoculated with B. putrificus and B. coli did not putrefy, while sterile milk, B. putrificus, and B. prodigiosus, or B. proteus, did putrefy. He concluded that lactose does not prevent putrefaction, but that B. coli and Bact. aërogenes were apparent exceptions to his rule. He did not realize that the salient feature of his experiments hinged on the fact that B. putrificus, B. proteus or B. prodigiosus did not ferment lactose, and that B. coli and Bact. aërogenes did attack lactose. In the presence of lactose-fermenting bacteria no putrefaction occurred, while in the absence of lactose-fermenting bacteria putrefaction went along without any trouble. If Bienstock had selected organisms to prove the theory advanced in preceding paragraphs, he could hardly have made a better choice to bring out the details of the process.

Tissier and Martelli,[172] and Passini [173] studied the action of B. coli and similar organisms, and found that protein-containing solutions in which these bacteria (together with sugar?) did not, as a rule, putrefy. Their conclusions are not clear cut because they apparently did not understand the necessity for fermentable carbohydrate in order to bring about the inhibition of putrefaction.

Maly [174] noticed that protein was attacked after the sugar in the medium was exhausted, but he does not specify the amounts of sugar which can be fermented before the protein-attack is instituted. This is a point of some importance, because the amounts of acid formed by the fermentation of sugars not infrequently renders the medium unsuited for the further development of bacteria within it.

Putrefaction and fermentation in man. — Senator [175] stated that under ordinary conditions toxic substances formed in the intestinal tract are derived from the decomposition of protein within it. Bouchard [176] elaborated a theory of intestinal auto-intoxication, and measures the degree of intestinal putrefaction by the amounts of putrefactive products in the urine. He called his measurements " Urotoxic coefficients."

Baumann,[177] Wassiliew,[178] Ortweiler,[179] Fr. Müller,[180] Nencki,[181] and Hoppe-Seyler [182] found that the aromatic components of the ethereal sulphates of the urine in man are dependent for their formation upon bacterial action in the intestinal tract. In this sense they regarded the ethereal sulphates as rough indices of the degree of intestinal putrefaction. Albu, [183] Jaffe,[184] Salkowski,[185] Brieger,[186] and Sucksdorff [187] studied more closely the factors influencing this intestinal putrefaction. and Müller and Ortweiler [188, 189] discovered that the administration of carbohydrate tends to lessen it. Krause [190] made similar observations on dogs; these observers carried their theories to excess, and other workers, notably Biernacki,[191] Eisenstadt,[192] Bachmann,[193] repeated the work in its essential details, and were much more conservative in their reports. although they, too, were impressed by the anti-putrefactive action of certain carbohydrates.

The work of these observers naturally turned the attention of many chemists to the study of milk. Milk is the earliest food of man, and even during childhood it not infrequently forms a very important portion of the diet. Poehl,[193a] Biernacki,[193a] Winternitz,[194] Rennert,[195] Skorodumow,[196] Albu,[197] Schmitz,[198] Gussarow,[199] Nasarow,[200] Rovighi,[201] Strauss,[202] Matteoda,[203] and Eisenstadt [204] have made careful studies of the effects of milk diets, or diets consisting largely of milk products, upon intestinal putrefaction, and they agree in the essential details of their results, namely, that milk does in fact tend to inhibit the decomposition of protein by bacterial action in the intestinal tract.

Stimulated by these discoveries Solucha,[205] Wereschtschagin,[206] Kopecki,[207] and Strauss[208] tried to discover what constituent or constituents of milk were responsible for the diminution of putrefaction, and their attention was naturally directed toward the lactose, because previous workers had found that this sugar could inhibit putrefaction in regular protein-media. As the result of their work they found that both lactose and dextrose would inhibit intestinal putrefaction, if these sugars were fed by mouth.

Poehl,[209] Brudzinski,[210] Fischer,[211] Rovighi,[212] and Embden[213] then turned their attention to sour milk and found that that, too, would inhibit intestinal putrefaction to a certain extent at least. This work is in reality the foundation upon which Metchnikoff built his sour-milk therapy. His claim to recognition appears to depend largely on the fact that he popularized this form of administration of lactose and lactic acid.

Not only sour milk, whole milk, and carbohydrates (sugars) inhibit putrefaction, but there is evidence that cereals as well have anti-putrefactive properties in virtue of the carbohydrate which they contain. Rothmann,[214] Gottwald,[215] and Krauss[216] have administered cereals in the usual manner, and noticed a diminution in the ethereal sulphates of the urine, which they interpret as an indication of the diminished formation of putrefactive products in the alimentary canal. It is noteworthy that cereals require a longer time before the effects appear than either milk, sour milk or sugars. This is possibly due to the fact that the cereals break up rather gradually in the intestinal tract, so that there is never a great concentration of sugar (maltose or dextrose) at a given period. This relative paucity of sugar naturally is absorbed more rapidly, or disappears more rapidly, than a large amount, so that there is never at any one time enough food to encourage an overgrowth of fermentative bacteria.

On the other hand, there is evidence that the administration of protein, particularly in relatively large amounts, encourages intestinal putrefaction, as the observations of Salkowski,[217] Salkowski and Jaffe,[218] and others have shown.

It is a point of some importance, as will be shown later,

what sugar one employs to bring about fermentation in place of putrefaction in the intestinal tract.   Sugars are absorbed more rapidly, other things being equal, than proteins. Albertoni [219] has shown that lactose is absorbed more slowly than dextrose, consequently it remains longer in the intestinal tract, and theoretically, at least, is useful for that reason, because it is possible to reach the lower levels successfully when the absorption is less rapid.

As a matter of curiosity it may be mentioned that Jacques Bey [220] in 1886 applied sugar syrups in the treatment of burns with good results, judging from his reports.   Fischer [221] preserved pus, transudates, and exudates in strong sugar solutions, and found that he could keep them for quite a long time without obvious change.

In recent studies upon eggs it has been shown that eggs preserved by the addition of cane sugar did not putrefy, even though the numbers of bacteria per gram egg-sugar mixture ran well up into the millions.   Dr. Folin showed chemically that there were nó products indicative of putrefaction in spite of the high bacterial count.

Finally, experiments carried out in Lawrence (Massachusetts), upon sewage filters show that the addition of sugar almost immediately inhibits nitrification.   The bacteria in the effluent increase greatly in numbers, and the efficiency of the filter becomes greatly, even if not entirely, impaired. [222] While these Lawrence experiments are not conclusive the facts are in harmony with the theory, so far as they go.

To recapitulate: it may be stated that a fundamental and general property of bacteria is expressed by the formula: carbohydrate protects protein from bacterial metabolism, to a considerable extent.   It must be remembered that " utilizable " carbohydrate spares protein, or other nitrogenous substances, because carbohydrate that cannot be fermented by bacteria obviously is not to be considered in this connection. Naturally, the kind of carbohydrate, or carbohydrates utilizable, varies greatly with different bacteria.   This point is well illustrated in the experiments which Bienstock carried out in

milk; sterile milk containing B. putrificus undergoes putrefaction.   Sterile milk containing B. putrificus and B. proteus
also putrefies, because none of these bacteria can utilize the
milk sugar, which is lactose.   B. putrificus and B. coli, inoculated into sterile milk, behave very differently; no putrefaction ensues, because B. coli ferments lactose, and the products
of this fermentation, chiefly acids, inhibit the growth of
the alkali-loving B. putrificus.   These experiments illustrate
another important fact.   Ordinarily, a strict proteolyte, such
as B. putrificus, cannot grow if there is an active fermentative
process going on simultaneously.

Returning to the antagonism between bacterial fermentation and putrefaction, the theory advanced above, relating to
these phenomena, may be stated thus: fermentation takes
precedence over putrefaction.   A most striking example of
this antagonism has been presented, the relation of carbohydrate to toxin production in the case of the diphtheria bacillus.   Dr. Smith has shown conclusively that this organism
does not form toxin in media containing muscle sugar.   If
there is an excess of muscle sugar (dextrose) in the medium
in which the bacilli are growing, no free toxin is formed.
The bacilli attack the sugar in preference to protein.   If a
small amount of sugar is present, no toxin is demonstrable
until it is fermented, but the resulting amount of toxin is
rather less than that formed in media quite free from sugar.
This organism furnishes an excellent opportunity to test the
validity of the hypothesis, because even very small amounts
of protein decomposition, resulting in the formation of toxin,
are readily detectable through inoculations into guinea-pigs.
In the presence of utilizable sugar apparently no toxin is
formed.

The results of experiments upon the factors influencing
putrefaction quoted in previous lines at first sight seem
confusing.   Some observers claim that sugars prevent putrefaction because considerable amounts of acid are formed as
the result of the process.   They believe the acid is the
inhibiting factor.   Others answer this contention by adding

calcium carbonate to the media prior to fermentation and yet find that putrefaction does not occur. The liberation of nascent hydrogen is supposed by some to be the salient factor; these workers again explain the facts by a different interpretation of the observed phenomena. In man the same general type of result obtains; milk, milk products, sour milk, and sugar all appear to exercise a distinct inhibitory action upon intestinal putrefaction, as judged by the decrease of ethereal sulphates in the urine. A careful study of the facts presented indicates strongly that two essential features of the problem have been overlooked. The prevention of putrefaction has been regarded wholly from the standpoint of chemistry; the bacteriology of the process, the reagents so to speak, have received no attention. The prevention of putrefaction has resolved itself into a discussion of the end results, where the observed facts are interpreted in the light of the antagonism of one set of chemical products, the acids, to the development of another set of chemical products, the putrefactive products, alkalis. The second, and most essential feature, has been a disregard of the study of the workings of single " species " grown in media favoring fermentation or putrefaction. The solution of the mystery depends upon the antagonism of fermentative processes to putrefactive processes in individual organisms. Bacteria can accommodate their metabolism, now to a protein diet, now to a carbohydrate diet, but in the presence of utilizable carbohydrate they appear to universally act upon the carbohydrate in preference to the protein. The carbohydrate shields the protein from metabolism (except that minimal amounts of protein break down to meet the nitrogen requirements of the organism) or, in other words, fermentation takes precedence over putrefaction.

The disregard of these fundamental principles has led Metchnikoff and Combe to attribute far too much curative value to bacteria in sour milks used for therapeutic purposes. The theory that Metchnikoff has elaborated in his interesting but highly speculative volume, entitled, " The Prolongation of Life," is too well known to need repeating here (a full discussion of the theory, with an exposition of the true action of sour milk, has already been published [223]). What actually happens when sour milk (usually with a dietary reduction of protein, and an increase of carbohydrate) is administered for therapeutic purposes appears to be thus: the diminution of protein and the increase of carbohydrate in the patients' diet changes the type of metabolism of the organisms elaborating indol and similar products in the intestinal tract from the putrefactive to the fermentative state. The bacteria which are responsible for the production of intestinal indol are not, generally speaking, the putrefactive anaërobes as Metchnikoff believed, but organisms of the coli-proteus type which are well known for their ability to form these products. If B. coli and B. proteus act upon carbohydrate it is evident, from what has just been said, they will not at the same time act upon protein to any extent. If they do not act upon protein they do not form protein decomposition products, and if protein decomposition products are not formed in the lumen of the intestinal tract they will not be absorbed and cannot, of course, appear in the urine. The disappearance of these substances, which are present as etheral sulphates in the urine, is the sign relied upon by most diagnosticians who utilize this " sour milk therapy " to indicate improved conditions in the alimentary canal. While the sign is a good one, as far as it goes, it does not indicate what Metchnikoff and others believe it indicates. It merely expresses the fact that certain bacteria have ceased their attack upon protein in the alimentary canal and are fermenting sugars instead.

As an isolated fact the theory outlined above may be interesting from the bacteriological point of view; the writer has endeavored to go further and apply the principle evolved to actual practice. The experiments upon animals, and the few studies of intestinal bacteria in man carried out along these lines, indicate that it is possible to actually influence the nature of the products formed by invading bacteria which have gained a foothold in the intestinal tract.

The underlying principle is to provide utilizable carbohydrate for these invading organisms. If bacteria are fermenting carbohydrate they are not decomposing protein to any considerable degree. The bacteria are forming acid products instead of putrefactive or toxic products.

Before discussing the practical applications of this hypothesis to clinical medicine, it will be well to consider what we actually know about bacterial infections of the intestinal tract.

We are beginning to realize that typhoid bacilli alone are probably unable to cause typhoid fever. Monkeys, whose intestinal conditions are more nearly comparable to man than those of most lower animals, appear to be quite refractory to even large numbers of these organisms. The writer has tried to produce typhoid fever in the ordinary R. macacus by feeding enormous doses of typhoid bacilli, together with a diet calculated to bring about conditions favorable for their development in the alimentary tract, and even to reinforce their action by adding large numbers of B. proteus in the hopes of breaking down the barrier which prevents infection, but without success so far.

Not only are the lower animals apparently refractory to this disease; man himself varies greatly in his susceptibility to it, judging from the distribution of cases in a community where theoretically all are equally exposed to it. The distribution of cases of typhoid fever in an epidemic caused by some universally used commodity as, for example, water, where the whole population may be regarded as being equally exposed to infection, shows conclusively that there is no really well defined reason why some are victims, others spared. The robust, the strong, the delicate, and the weak all fall victims, or escape, without our being able to determine the incidence and the reason for it. Medical men are accustomed to explain the fact that some contract the disease while others escape, by stating that some have "lowered" resistance, others have normal resistance. A closer scrutiny of the term "lowered resistance" indicates that it is merely a name for an undetermined factor (or factors).

It is an interesting thing to realize that a few bacilli, relatively non-resistant, can gain a foothold in the intestinal tract of man, maintain themselves there, and set up the group of disturbances which make up the symptom complex of typhoid fever, and yet be singularly inert in artificial media and unable to cope with the normal intestinal flora. After the typhoid bacilli have gained a foothold, and the invasion of the host commences, the problem becomes simplified.

Do intestinal parasites make the otherwise intact intestinal mucosa vulnerable? Is the lymphoid tissue of Peyer's patches more suited for the development of typhoid bacilli than other tissues? Are putrefactive conditions more likely to favor the growth of these organisms than fermentative conditions? What is the flora in typhoid fever?

The writer has been unable, up to the present time, to study cases of typhoid fever so as to throw light upon certain of these problems, but during the last summer he had unusual opportunities to examine many cases of a somewhat similar disease, Bacillary Dysentery, on the Boston Floating Hospital. These investigations were carried on so as to throw some light upon the associated flora of bacillary dysentery, and to determine the kind of bacterial activity represented in this disease, whether it be putrefactive or fermentative. These investigations have been carried out in a different method from that of previous efforts along this line; the attempt has been made to discover the relationships and symbioses which the dysenteric flora present, rather than to discover differences and isolated facts concerning individual organisms.

Bacillary dysentery, particularly as it occurred in young children during the summer of 1910, is an acute, febrile disease constantly associated with the presence of dysentery bacilli in the intestinal tract. These dysentery bacilli are of at least two distinct types: the Shiga, or alkaline variety, and the Flexner, or acid variety. The distinction relied upon chemically for their separation is mannite, in which the Flexner type produces acid while the Shiga does not. According to Jehle and Charleton,[224] the Shiga variety is more frequently associated with epidemics of bacillary dysentery, while the Flexner variety is more closely related to sporadic cases. Dysentery bacilli differ somewhat from typhoid bacilli in relation to their ability to actually penetrate the body of the host. Dysentery bacilli do not pass beyond the mesenteric lymph nodes, contrasting in this respect with the typhoid bacilli. Consequently, bacillary dysentery may be considered to occupy an intermediate position between typhoid, which is often a true bacteriemia, and cholera, which is practically a disease produced by saprophytes living upon the intestinal contents, and limited to the lumen of the tract.

The harmful effects which these dysentery bacilli bring about are due to the action of toxins elaborated by the organisms during their growth in the intestinal canal. These toxins are intracellular in the case of the Flexner organisms, but apparently more diffusible in the instance of the Shiga variety.

The toxins are said to act specifically upon the intestinal mucosa and the anterior horn ganglion cells. An antitoxin may be prepared for the Shiga toxin, but it has much less affinity for the toxin than most antitoxins have. The toxin has much more affinity for the ganglion cells than for the antitoxin, so that very little curative value can be attributed to it. The tetanus toxin, also, has a greater affinity for nerve cells than for its antitoxin.

The lesions of bacillary dysentery have been admirably analyzed and summarized by Southard, McGaffin and Richards [225, 226] and need not be repeated here.

Conradi [227] obtained a soluble toxin from the Shiga bacillus by aseptic autolysis, and Gay [228] showed that freshly killed cultures increased in toxicity on standing. Lentz [229] found that the filtrates of Shiga cultures in broth were toxic. Todd [230] has done some excellent work upon the dysentery toxin. He finds that a fairly potent poison may be obtained from cultures of the Shiga bacillus if it is grown in broth having a strongly alkaline solution. From what has gone before it must be evident that the maintaining of a strongly alkaline reaction in broth is incompatible with more than minimal traces of muscle sugar (dextrose). He was actually able to immunize horses to this toxin and thus to produce a moderately strong antitoxin. The antitoxin, however, was not satisfactory as a curative agent, for the toxin has a greater affinity for certain nerve cells than it has for the antitoxin, as was stated above. The Flexner bacillus does not produce an extracellular toxin.

The salient feature to remember is that the Shiga toxin is elaborated only in a medium having a strongly alkaline reaction, and that this alkaline reaction can only be maintained if the medium is practically free from dextrose, or other utilizable sugar. Todd did not appreciate the significance of this alkaline reaction, but his results indicate strongly that the production of the dysenteric toxin takes place only in non-fermentable media, thus conforming in every respect to the theory. That is to say, B. dysenteriæ, in common with most other known bacilli, forms putrefactive products (toxic and otherwise) in media containing no fermentable sugar, but that it attacks dextrose (or other fermentable sugar) whenever this is present in the medium in which it is growing, in preference to the protein.

Turning now to the cases of dysentery studied last summer it will be seen that there were thirty-nine in all, and that three organisms were constantly present in each of them: B. dysenteriæ (Flexner or Shiga), B. coli, and the Streptococcus. These three bacteria may be designated the "obligate dysentery flora" so far as these cases were concerned, and they will be so referred to in this connection. Certain other bacteria (almost invariably proteolytes) were found in irregular numbers, and not uniformly distributed in all cases. They will not be considered in detail. The most significant fact derived from the preliminary study of the bacterial flora of bacillary dysentery was that the organisms represented a perfectly typical putrefactive type of activity.

Practically no typical fermentative organisms could be isolated. This state of affairs is so markedly different from that of normal children of similar age (three months to four years) that it calls for explanation.

It will be remembered that normal children and experimental animals fed upon the diet of normal children both showed large numbers of fermentative organisms in their dejecta, while putrefactive organisms and putrefactive activity were at a minimum. The babies suffering from bacillary dysentery, on the other hand, were distinctly abnormal in this respect; their fecal flora was found to be distinctly putrefactive, both in activity and in composition so far as individual organisms were concerned. This putrefactive flora is characterized by the absence of fermentative types (of the more or less obligate kind as, for example, B. bifidus or B. acidophilus), together with the presence of bacteria, such as B. coli, which shift their metabolism to accommodate their mode of action to the diet. Both the Streptococcus and particularly B. dysenteriæ are to be regarded as exogenous invaders.

The routine clinical treatment of bacillary dysentery accounts perfectly well for certain of the observed facts.

The babies are starved, being allowed only sterile water either until the temperature drops or until there is a more or less normal movement. The idea of this starvation appears to be to actually starve out the bacteria which are causing the trouble. This starvation is perfectly logical in many respects; it allows the intestinal tract to rest so far as possible, and it prevents the ingestion of harmful substances which might aggravate the harmful conditions already established. On the other hand, several unnatural and probably distinctly harmful results are effected by this starvation.

The dextrose (and reserve of glycogen) and body fat are used up very quickly, so that the body has to draw upon its nitrogen to maintain the vital functions of respiration, heart beat, and so on. This condition is abnormal, even in health; in disease it must be apparent that the reserve of energy must soon be depleted to an alarming extent for several reasons.   The loss of body nitrogen means in plain language that some part of the body is being piled into the furnace to keep up steam, and the amount of convertible material is limited.   Again, no food is entering the body to make good this waste.   It is, furthermore, a well-known fact that nitrogen metabolism is much more rapid during the acute stages of a febrile disease than later.   This metabolism may actually be several times as great as is normal during starvation in health.   The excess of nitrogen, the "toxic" nitrogen, that is excreted during this active, acute stage of the disease, represents a loss that the body cannot well afford to suffer.

In addition to the unusual loss of body nitrogen, the intestinal bacteria are adding their quota of trouble.   The dysentery bacillus is manufacturing toxins, which are absorbed, and the colon bacillus and streptococcus are forming putrefaction products, which also must be absorbed, in part at least.   These putrefactive products are partly paired in the liver and excreted through the kidneys, thus adding a fresh increment of work to the already overburdened patient. The colon bacillus and streptococcus are in a measure symbiotes of the dysentery bacillus in this sense: they are growing in the lumen of the intestinal canal with it, and their products are helping to overwhelm the host.   Velagussa has carried out experiments which seem to indicate that B. coli is much more pathogenic under putrefactive conditions than otherwise,[21] and Nepper,[22] has concluded that B. coli is likewise greatly increased in virulence whenever there is a marked increase in mucous secretion.   All of these conditions are present in the intestinal tract during bacillary dysentery, and the unusual preponderance of this organism (B. coli) is certainly in accord with these observations.

Bacterially, then, the conditions are these: the obligate dysentery flora is essentially putrefactive in its activities; it is putrefactive because there is no carbohydrate for it to ferment.   The products formed, and in part absorbed by the host as the result of this putrefaction, are not normally present in the intestinal tracts of children of like age.   Certain of these products are toxic (the dysentery toxin) and their action upon the anterior horn ganglion cells may explain certain of the train of nervous symptoms common in bacillary dysentery. The putrefactive products elaborated by the colon bacillus are also absorbed, in part at least, and paired in the liver, to be finally excreted through the kidneys. Here, again, there is an additional burden thrown upon the already weakened organism. It will be seen that this additional " putrefactive burden " is not as a rule present in health, so that the patient is compelled to burn the candle at both ends and in the middle, as well as to take care of all the problems thrust upon him at this time. The process is almost wholly katabolic, with nothing coming in to help balance the account.

Theoretically, the remedy for these conditions is not difficult to figure out: to feed the patient something to restore the normal constituents of the body (dextrose), so that the vital functions shall be carried along in the customary manner, to reduce or absolutely abolish conditions in the intestinal tract so that putrefaction and toxin formation shall be prevented, to provide nitrogen to make good the loss already sustained, and to restore the anabolic processes so far as possible. In other words, to feed the patient, reform his perverted intestinal flora, to prevent further intoxication, and to remove the damage already accomplished.

The question arises — how will it be possible to judge of the individual and collective effect of these remedial measures? It must be evident that the success of the procedures must be measured in three distinct directions: the clinical, bacterial, and chemical. Each of these methods has its advantages, disadvantages, and limitations, and without entering into a discussion of the relative merits of each it may be asserted with considerable assurance that our present state of ignorance of these diseases is partly attributable to the neglect of coördinated efforts to bring the three classes of factors into harmony.

With a clearer realization of the factors entering into the problem of bacillary dysentery, it will now be possible to discuss the measures adopted to correct at least some of the more amenable of these. For very apparent reasons the interest centered upon the bacterial flora and, as has been stated, this was found to be essentially putrefactive in character and composed of three distinctive organisms: the dysentery bacilli, one or both types, B. coli and the streptococcus. Furthermore, it appears that this flora is putrefactive because the conditions in the intestinal tract of the patient are of such a nature as to force this type of metabolism upon the bacteria. What the antecedent causes are which permit the dysentery bacilli to gain a foothold in the alimentary tract we do not know, and this work does not pretend to shed any light upon the matter. All that is attempted here is to realize the fact and to devise means for altering it to a more favorable type of activity.

The method most suited for this purpose is to add carbohydrate to the patients' meager allowance of sterile water. The choice of sugars is a large one, and it is necessary to take into consideration two features of the problem: the effect upon the patient, and the effect upon the bacteria concerned in the morbid process.

The essential features of the theory underlying the use of sugars are the following: [23]

1. Bacteria in the presence of protein, or protein decomposition products, and fermentable carbohydrate act upon the carbohydrate in preference to the protein, because utilizable carbohydrate shields protein from bacterial attack.

2. In the human body, carbohydrate likewise spares nitrogen metabolism to a very considerable extent.

3. The harmful effects of bacteria are associated to a considerable degree with their action upon protein; it can be shown experimentally that the organisms concerned in the production of bacillary dysentery act thus, and it can further be shown that the introduction of utilizable carbohydrate tends to prevent this elaboration of putrefactive or toxic products.

4. The blood of normal human beings contains nearly .2 per cent of dextrose as such; this dextrose (which together

with the reserve supply of glycogen disappears during starvation) represents the immediately utilizable source of
energy for the maintenance of the vital functions of heart
beat and respiration. Only abnormally does the body utilize nitrogen wholly for this purpose.

5. The normal sugar for young children is lactose, a true
animal sugar in contradistinction to other sugars which are
vegetable in origin. With the exception of a very few obligate intestinal bacteria, but few organisms can ferment lactose. Lactose, furthermore, is less rapidly absorbed from the
intestinal tract than are most other sugars. This is a point of
some importance, for this lack of absorption rate, combined
with the diarrhea which is a feature of dysentery cases, practically insures a stream of lactose throughout the range of
the alimentary canal. In the lower levels of the tract the
lactose is undoubtedly broken down into dextrose and galac-
tose, which are utilizable by the dysentery bacilli as well as
B. coli and the streptococci.

Bearing these facts in mind, the proper treatment from the
bacterial standpoint is, theoretically, to alter the intestinal
flora, both with respect to the types of bacteria represented
and in the character of their metabolism. It is desirable to
introduce conditions which shall prevent further toxin formation, elaboration of putrefactive products, and loss of body
nitrogen. Practically, the measures may be carried out in
the following manner:

1. Infuse chemically pure (Kahlbaum's) dextrose, as a
2.5 per cent solution in normal saline subcutaneously. The
amount of dextrose to be infused can be readily calculated
from the following data: amount of blood equals approximately one-thirteenth the body weight; amount of dextrose
in the blood, .2 per cent. The patient benefits in two directions from this initial dextrose infusion: the tissues of the
body are once more bathed in dextrose, the normal constituent of the blood, which is immediately utilizable for the
production of energy, and, theoretically at least, those bacilli
which have invaded the intestinal mucosa are also immersed
in a fluid containing dextrose. While the writer does not

claim that this circulating dextrose will to any extent change
the metabolism of invading bacteria, it is worth trying.   In
any event, the patient receives readily utilizable food without
the expenditure of an atom of chemical energy for its digestion or absorption.   This dextrose is intended primarily to
prevent the normal and "toxic" metabolism of the body
nitrogen as far as possible.

2.   Castor oil, or other purgative, to clean out the intestinal tract thoroughly.

3.   Feed sterile lactose solution, five to seven per cent, by
mouth.   This lactose is intended to provide the patient with
a readily assimilable food, to flood the intestinal mucosa with
sugar to prevent the invading organisms contained within it
from attacking the body protein, thus increasing the toxemia
already present, and to change the metabolism of B. coli, the
streptococcus, and (after its inversion to dextrose and galac-
tose) the dysentery bacillus from the putrefactive (toxic)
type to the acid-forming, fermentative type.   In other words,
prevent further production and absorption of toxins.   In
addition to the reformation of the obligate dysenteric flora it
would appear to be desirable, and even essential, to encourage the development of the vestiges of the fermentative flora
characteristic of normal individuals to a healthy, widespread
growth.   These organisms, forming as they do under natural
conditions considerable amounts of acid, keep down the
relatively poorly acid-resisting types, such as B. dysenteriæ,
and thus automatically tend to drive them out.   This action
is hastened, theoretically at least, by three factors:

1.   B. coli becomes an antagonist to B. dysenteriæ, where
formerly it was a symbiote.

2.   B. coli forms considerable amounts of acid from lactose, or its decomposition products, and these acids are
inhibitory to B. dysenteriæ, for the latter organism cannot
grow in as acid a medium as the former.

3.   B. dysenteriæ itself forms acids instead of toxins,
acting upon the dextrose which results from the break-down
of the lactose.   This increment of acid, itself harmful to the
bacillus, is probably overshadowed by the importance of the

changed metabolism of the dysentery bacillus, because under the new conditions the dysentery bacilli are no longer elaborating toxins which are in a measure to be regarded as their most potent methods of offence and defence. The patient, therefore, gains in several ways. The lactose should be given not later than the eighteenth hour after castor oil, and in small, often repeated doses. If the doses are larger, and repeated infrequently, there are long intervals in which the tract is free from sugar, so that the organisms can revert to the putrefactive type again.

The after feeding is of extreme importance and, unfortunately, is not at all well worked out. It is evident that some nitrogenous food should be administered as soon after the initial treatment with lactose as possible to replace the loss which the patient has already sustained. This nitrogen, furthermore, should be associated with carbohydrate in physiological excess to prevent further putrefaction in the intestinal tract. All that the initial treatment with sugar can hope to accomplish, and all it can be expected to accomplish, is to prevent further toxin formation, to reëstablish a normal flora in the intestinal tract, and to reduce the metabolism of body nitrogen as low as possible.

The cases studied on the Boston Floating Hospital during the summer of 1910 numbered thirty-nine. Of these, nineteen died and twenty survived. Of those that died eight received no sugar; they were among the earliest studied, and it was from a study of the flora of this group that the ideas employed in the attempt to aid the others originated.

Of those receiving sugar, and which died subsequently, two were on a modified milk diet previous to the use of lactose.

One reacted very well both clinically and culturally to the sugar treatment, convalesced, and was doing very well upon a milk modification. Finally she contracted a severe attack of acute bronchitis, and had not strength enough to rally.

Two cases died within three days of admission, too early

to get the results of the lactose treatment, for the lactose was not given as a rule until after two or three days' starvation on sterile water.

One case was placed on lactose for eleven days, did very well, then upon a milk mixture five days, still with improvement, then upon Eiweiss milk. She died very shortly after.

Two cases were pure Shiga bacillus infections, and died rather promptly.

This is the usual rule with pure Shiga infections, where the patients are seen as late in the disease as were many of those at the Hospital last summer.

Of the remainder, one survived nine days on lactose, then died. The patient did not show the expected change in flora, and the question arises — were there vestiges of the fermentative flora to encourage? Obviously, if there were no fermentative organisms to start the new flora it could not develop. One case died after twenty days' illness on the boat: six days of lactose, two of sterile water, then ten of lactose. The first two days were on sterile water.

The last patient was placed upon lactose solution for fourteen days, then upon a milk mixture. She reacted very well, bacterially, and no definite explanation of the cause of death can be advanced (syphilis?).

To summarize: Eleven of the nineteen fatal cases studied were given lactose for two or more days, with the exception of two which died just as the treatment was instituted. Of the nine cases so treated, the treatment was a failure so far as saving the patients' life. The cause of this failure is not apparent, although several of these cases apparently passed through the acute stage, and were progressing moderately well on modified milk before the end came.

Of the twenty cases that recovered:

One received no sugar. Three showed no noteworthy changes in the bacterial flora as the result of the sugar, although it should be said that the numbers of streptococci did tend to disappear, while the other organisms, notably B. coli, were much less in evidence than before the sugar was administered. It is very difficult to formulate an opinion of the disappearance of the dysentery bacilli, because their numerical relations to the disease are so uncertain.

Seven cases showed marked changes in the types of bacteria found in the feces following the administration of lactose; these cases were not, however, any different, so far as clinical evidence goes, than those that responded more slowly, although, generally speaking, the abatement of the acute symptoms appears to have been a little more rapid.

Two patients received whey instead of lactose, the idea being to furnish them with at least a minimal amount of nitrogen from the start. The cases are too few to draw any conclusions, although it may be said that they did not materially differ from those that received lactose alone. It is very possible that whey sterilized in the usual manner is less valuable for dysentery cases than whey that is sterilized by passage through a Berkefeld filter, because the temperature necessary to Pasteurize whey is high enough to coagulate both the lact-albumen and the lact-globulin.

Of the remaining seven cases all showed distinct changes in bacterial flora along the lines indicated, namely, a gradual disappearance of the Gram-negative forms referable to B. dysenteriæ and B. coli, a complete removal of streptococci, and the gradual appearance of B. bifidus and B. acidophilus. Two of these cases reacted very slowly, so that it was quite seven days before the changes were well under way. The remainder progressed without notable deviation from the desired alternation of organisms.

The results presented are too few to warrant any comments other than to say that in no instance did this early administration of lactose cause even the semblance of harm. This, however, is a point gained, for the patients are getting some nutriment from the sugar instead of starving for several days, and it is quite possible that in those cases where " no harm was done " some good may have been accomplished.

This method is in its infancy, the babies cannot speak to let the clinician know how they feel, and the cases were for the most part desperate, because the parents were poor for the most part and only came to the hospital as a last resort. Consequently, it is far better not to attempt to explain either the beneficial or the fatal results, but wait patiently for more cases.

It may be stated that the babies for the most part held their weight fairly well, even when they were on lactose for many days, and this points strongly to the reduction in nitrogen metabolism, which it is so desirable to prevent in these cases.

The results may be stated thus:

1. The use of lactose (and dextrose) is preliminary only; it is in no sense a complete treatment. The object to be attained is to change the character of the intestinal flora in such a manner as to prevent further elaboration of toxin in the intestinal tract, thus allowing the patient to concentrate efforts upon the repair of the injury already inflicted.

2. The after-feeding will in almost every instance decide the fate of the case, and it is precisely the after treatment that at the present time gives the most concern. The proper after treatment, bacterially considered, is one in which the nitrogen, which the patient must absolutely have, is guarded from putrefaction by carbohydrate, preferably by some starchy food, possibly with a small amount of lactose. Other sugars of course may be used but, at present, lactose seems to be better suited because it is less rapidly absorbed.

3. The fate of any particular case is problematical, because we cannot estimate the amount of damage which the toxin may already have accomplished. For this reason, dysentery cases naturally divide themselves into three classes: those that will probably get well in spite of treatment, those that will die in spite of treatment, and, finally, those in which the balance may swing toward death or recovery, depending upon the treatment.

Lactose does no harm in the first group, does not save life in the second (though it certainly does no harm), but it is in the third group that lactose may be called upon to help most conspicuously. In this latter group, lactose may change the flora from an undesirable, even pernicious type, to a harmless, even prophylactic type, and leave the field clear for the clinician to bring the patient safely to health. Lactose in these cases may be looked upon as a remedial measure of as yet undetermined value.   After-feeding will decide the fate of the patient.

Statistics bearing upon the results of the use of lactose are of little value, because there is no method available to differentiate between the action of lactose and of other remedial measures except by observing the changes in the bacterial flora.   When we realize that the changes in the bacterial flora are only one among many factors, it becomes apparent that the whole solution will be obtained when physiological chemistry becomes effective, and can actually measure the influence of these sugars in reducing toxin formation, sparing body nitrogen from metabolism, and in judging of the effects of after treatment.

Before bringing this dissertation to a close, certain extensions which suggest themselves deserve mention for the sake of completeness.   Not every possibility can be discovered, with our present meager knowledge of the facts, nor, indeed, will it be expedient to even mention many instances where the theory elaborated in the preceding pages could conceivably be of assistance.

Typhoid fever deserves special mention because it is a disease of the intestinal tract running a somewhat similar course to dysentery.   Since doing the work recorded here Professor Folin has called the attention of the writer to the work of Shaffer, Coleman and their associates, which has already been referred to.   Their work was along somewhat similar lines, but their interest lay in the purely chemical aspect of the subject.   They have made very careful, painstaking studies of typhoid fever, from the point of view of the nitrogen metabolism.   To prevent the loss of body nitrogen they have fed their patients carbohydrate, and their conclusions are striking.

" As stated earlier in this paper, other experiments have indicated the effect of pyrexia in causing an increased katabolism of protein may be prevented, but so far as we know the experiments here recorded are the first to show satisfactorily in the human subject that the action of toxins, as well as of the pyrexia, may be prevented from causing a loss of body protein by the addition to the food supply of a sufficient quantity of carbohydrate."

" If, as seems probable from our results, the ' toxic' destruction of body protein may be prevented by large carbohydrate intake, the mechanism of this ' toxic' destruction cannot be a direct (poisonous) injury to body cells and proteins."

It appears to the writer that the observed inhibition of "toxic" loss of body nitrogen is explainable on the lines mentioned in preceding paragraphs. Shaffer and Coleman and their associates find that carbohydrate prevents this loss, but have not thought of the possibilities of changing the metabolism of the organisms which probably elaborate this toxic substance. As an explanation of the phenomena which these investigators have shown so clearly, the writer believes that the typhoid bacilli (and probably other putrefactive organisms) act upon carbohydrate in place of the protein, do not decompose protein, and consequently do not form the necessary substance, or substances, to provoke this toxic loss of body nitrogen.

A diet rich in carbohydrate would seem to be desirable in typhoid fever. One point deserves mention; if the carbohydrate diet is instituted early in the disease, and the intestines are perforated later on, the chances are many times greater in favor of recovery after operation if the flora thus suddenly introduced into the peritoneal cavity is strongly fermentative and in which putrefactive organisms are not numerous than would be the case if the flora was essentially putrefactive.

Similarly, it is not at all impossible that intestinal surgery might profit by studies along lines that readily suggest themselves.

The cholera vibrio is singularly sensitive to acids, and similar measures would be well worth trying, both as curative and, more especially, from the point of view of prophylaxis in this disease.

In chronic Bright's disease, where the kidneys are much restricted so far as excretory area is concerned, it may be a great saving in their action to be freed from eliminating products of putrefaction which have been paired in the liver. Undoubtedly, under ordinary conditions, with normal kidneys, considerable amounts of indican and other similar bodies pass through without apparent harm. In those cases, however, where the activities of the kidneys are greatly on the wane, each added burden, however small, counts for something. The trend of modern medicine, indeed, appears to be toward diets containing moderate amounts of carbohydrate. London[294] has studied the rate of absorption of carbohydrate, associated with protein, in experimental animals, and finds that even in the ileo-cecal valve region there is about two per cent of carbohydrate present. These figures, to be sure, are in experimental animals, and can be applied to man only very cautiously, but it appears to be a fact that even with moderate amounts of carbohydrate in the diet, some persists to the ascending colon, where it may serve very effectively to protect the protein from putrefaction, and thus lift the burden from the kidneys. Bearing London's results in mind, together with the results obtained upon the flora by administration of carbohydrate, it appears to be possible to prevent intestinal putrefaction successfully without over administration of carbohydrate.

Finally, the modification of cow's milk for summer feeding in babies and young children deserves mention. Human milk contains about four times as much lactose as protein. Cow's milk contains not twice as much lactose as protein, and the protein of cow's milk appears to be digested with more difficulty than human milk protein. The fecal flora of breast-fed babies is strictly fermentative in type; that developed on cow's milk is much more likely to become putrefactive. As a preventive measure in summer, one which costs but very little to try, it would seem logical to so modify cow's milk that it will approach human milk in composition in order to reduce, if possible, the chances of successful invasion of young children (two to four years) by bacteria which can thrive apparently amid putrefactive surroundings, but which tend to die out in strictly fermentative surroundings. Naturally, care and judgment must be exercised in such a wholesale measure, and it might be tried on a small scale before it is used on a larger one. The idea, however, is wholly in harmony with what we already know about bacterial action in the intestinal tract, and a cautious series of observations would almost certainly add to our knowledge of the causation of summer diarrheas among young children.

[Offered for the degree of Doctor of Public Health.]

BIBLIOGRAPHY.

1. Theobald Smith. Popular Science Monthly, April, 1905.

2. Shaffer, Coleman, Somogyi, Reinoso and Cutler. Arch. of Internal Med., 1909, iv, 538.

3. Robert Hooke. Collected Memoirs of Anton v. Leeuwenhoek. See also Anton v. Leeuwenhoek. Memoirs to Royal Society of London, 1675 and 1683.

4. Hausmann. Inaug. Diss., Berlin, 1870.

5. Woodward. Medical and Surgical History of the Rebellion, i, Part 2, 1879.

6. Koch, Robert. Wien. Med. Bl., 1883, vl, 1245.

7. Nothnagel. Zeit. f. klin. Med., 1882, iv, 422.

8. Bienstock. Fort. d. Med , i, 1883, 609; Zeit. f. klin. Med., 1884, viii, 1.

9. Escherich. Die Darmbakterien des Säuglings. Stuttgart, 1886.

10. Gaffky. Mitt. a. d. kais. Ges. Amt., 1884, ii, 372; Deut. med. Woch , 1884, x, 215.

11. Shiga. Cent. f. Bakt., 1898, xxiii, 599; 1898, xxiv, 817, 870, 913.

12. Kendall. Journ. Biol. Chem., vi, 1909, 499.

13. Escherich. Loc. cit.

14. Senator. Zeit. f. physiol. Ch., 1880, iv, 1.

15. Hochsinger. 8 Verhandl. d. Ges. f. Kinderheilk., 1890.

16. Moro. Jahrb. f. Kinderheilk., 1900, lii, 53.

17. Breslau. Zeit. f. Geburtskunde, 1866, xxviii, 1.

18. Tissier. Ann. Inst. Past., 1905, xix, 109.

19. Moro. Loc. cit.

20. Schild. Zeit. f. Hyg., 1895, xix, 113.

21. Sittler. Die Darmbakterien beim Säugling. Würzburg, 1909.

22. Sittler. Cent. f. Bakt. Ref., 1909–10, xlv, 138.
23. Tissier. Recherches sue la Flore Intestinale des Nourrissons. Paris, 1900.
24. Kendall. Journ. Biol. Chem., 1909, v, 420.
25. Alex. Schmidt. Wien. klin. Woch., 1892, xviii, 643.
26. Tissier. Loc. cit.
27. Cahn. Cent. f. Bakt., 1901, xxx, 721.
28. Rodella. Ibid., 1901, xxix, 717.
29. Tissier. Loc. cit.; also Compt. Rend. Soc. de Biol., 1899, 943.
30. Moro. Jahrb. f. Kinderheilk, 1900, 3 F., ii, 55; Wien. klin. Woch., 1900, xiii, 114.
31. Finkelstein. Deut. med. Woch., 1900, xxii, 263.
32. Tissier. Loc. cit.
33. Cahn. Loc. cit., 724.
34. Moro. Loc. cit.
35. Rodella. Loc. cit.
36. Weiss. Cent. f. Bakt. Orig., 1904, xxxvi, 13.
37. Kendall. Journ. Med. Research, N. S., 1910, xvii, 153.
38. Tsiklinsky. Ann. Inst. Past., 1903, xvii, 217.
39. MacConkey. Journ. of Hygiene, 1905, v, 333.
40. Thiercelin. Compt. Rend. Soc. Biol., 1899, 269, 551.
41. Kruse. Cent. f. Bakt. Orig., 1903, xxxiv, 737.
42. Hirsch. Cent. f. Bakt., 1897, xxii, 369.
43. Libman. Ibid., 1897, xxii, 376.
44. Nepper. New York Med. Journ., 1908, lxxxvii, 980.
45. Rodella. Zeit. f. Hyg., 1902, xli, 471.
46. Passini. Ibid., 1905, xlix, 135.
47. Kendall and Smith. Boston Med. and Sur. Journ., 1910, clxiii, 578.
48. Sittler. Darmflora beim Säugling, 1909.
49. Fischer. Vorlesungen über Bakterien, Zweite Auflage, 178, 283–284.
50. Sittler. Loc. cit., 32.
51. Kendall. Journ. Biol. Chem., 1908, v, 285.
52. Kohlbrugge. Cent. f. Bakt. Orig., 1901, xxix, 571.
53. Moro. Arch. f. Kinderheilk, 1906, xliii, 340.
54. Landsberger. Diss. Königsberg., 1903.
55. Gessner. Arch. f. Hyg., 1889, ix, 128.
56. Tavel and Lanz. Mitteil. a. klin. d. Schweiz, i.
57. Klein. Arch. f. Hyg., 1902, xlv, 117. Cent. f. Bakt., 1899, xxv, 773; 1900, xxvii, 834.
58. Schütz, R. Arch. f. Verdauungskrankh., 1901, vii, 58.
59. Mieczkowski. Mitteil. a. d. Grenzgeb. d. Med. u. Chir., ix.
60. Schmidt u. Strasburger. Faeces d. Menschen, 291.
61. Schmidt. Deut. Arch. f. klin. Med., 1898, lxi, 281.
62. Streit. Inaug. Diss., Bonn, 1897.
63. Matzuschita. Diss., München, 1902.
64. Nothnagel. Zeit. f. klin. Med., 1882, iv, 223.
65. Eberle. Cent. f. Bakt., 1896, xix, 2.

66. Hellstrom. Arch. f. Gynäkologie, lxiii, 643.
67. Klein, Alex. Arch. f. Hyg., 1902, xlv, 117.
68. Hehewerth. Ibid., 1901, xxxix, 321.
69. de Lange. Jahrb. f Kinderheilk, liv, 1901, 729.
70. Strasburger. Zeit. f. klin. Med., 1902, xlvi, 413.
71. Sucksdorff. Arch. f. Hyg., 1886, iv, 355.
72. Fürbringer. Deut. Med. Woch., 1887, xiii, 209, 235.
73. Stern. Zeit. f. Hyg., 1892, xii, 88.
74. Gilbert et Dominici. Compt. Rend. Soc. Biol., 1894, 277.
75. Salkowski. Virchow's Arch., 1889, cxv, 550.
76. Kumagawa. Ibid., 1888, cxiii, 134.
77. Sehrwald. Berl. klin. Woch, 1889, xxvi, 447, 466, 492.
78. Hammerl. 1897, xxxv, 376.
79. Kuisl. Diss. München, 1885.
80. Mieczkowski. Mitt. a. d. Grenzgeb. d. Med. u. Chir., ix.
81. MacNeal, Latzer and Kerr. Journ. Infect. Diseases, 1905, vi, 165.
82. Sucksdorff. Arch. f. Hyg., 1886, iv, 355.
83. Brotzu. Cent. f. Bakt., 1895, xvii, 726.
84. Adrian. Arch. f. Verdauungskrankh, 1895, i, 179.
85. Schmitz. Zeit. f. physiol. Chem , 1894, xix, 401.
86. Hammerl. Zeit. f. Biol., 1897, xxxv, 376.
87. Stern. Zeit. f. Hyg., 1892, xii, 88.
88. Peurosch. Diss. Königsberg, 1877.
89. Albu. Deut. Med. Woch . 1897, xxiii, 509.
90. Wang. Zeit. f. physiol. Chem., 1899, xxvii, 557.
91. Levin. Ann. Inst. Past., 1899, xiii, 558.
92. Kendall. Journ. Biol. Chem., 1909, vi, 499.
93. Herter and Kendall. Ibid , 1908, v, 283.
94. Kendall. Ibid., 1909, vi, 257.
95. Kendall. Ibid., 1909, v, 419.
96. Herter and Kendall. Ibid., 1909, v, 439.
97. Kendall. Ibid., 1909, vi, 499.
98. Kendall. Journ. Med. Research, N. S., 1910, xvii, 153.
99. Herter and Kendall. Journ Biol. Chem., 1910, vii, 203.
100. Kelvin [Lord] and Tait. Elements of Natural Philosophy, 1879, 113.
101. Fischer. Vorlesungen über Bakterien. Zweite Aufl., 206.
102. Bienstock. Zeit. f. klin. Med., 1884, viii, 1.
103. Bienstock. Arch. f. Hyg , 1899, xxxvi, 335.
104. Bienstock. Ibid., 1901, xxxix, 390.
105. Nencki. Gaz. lek. Warzawa, 2 s, ix, 1889, 732-753.
106. Kerry. Wiener Akad., Berlin, 1889, 98.
107. Zoja. Zeit. f. physiol. Chem , 1897, xxiii, 236.
108. Seelig. Virchow's Arch., 1896, cxlvi, 53.
109. Dauber. Arch. f. Verdauungskrankh., 1897, iii, 57, 177.
110. Hueppe. Principles of Bacteriology, 54.
111. Flügge. Die Mikroörganismen, i, 254.
112. Migula. System der Bakterien, l, 302.
113. Cornil et Babes. Les Bactéries, 43.

114. Gunther. Einfuhrung in das Studium der Bakterien. sechste Aufl., 60.

115. Kolle und Haetsch. Die Experimentalle Bakteriologie und die Infektionskrankheiten, 30.

116. Encyclopedia Britannica, ix, ed. 91.

117. Standard Dictionary, Funk and Wagnalls.

118. Fischer. Loc. cit., 172.

119. Escherich. Darmbakterien des Säuglings, 35, 107.

120. Escherich. Loc. cit.

121. Spiegelberg. Jahrb. f. Kinderheilk, 1899, xlix, 207.

122. Jacobson. Compt. Rend. Soc. Biol., 1909, 143.

123 Wassiliew. Zeit. f. physiol. Chem., 1882, vi, 112.

124. Baumann. Ibid., 1886, x, 123.

125. Bartoschewitsch. Ibid., 1893, xvii, 35.

126. Fr. Müller. Berl. klin. Woch., 1887, xxiv, 405

127. Fürbringer. Deut. Med. Woch., 1887, xiii, 209, 235.

128. Schütz. Arch. f. Verdauungskrankh., 1901, vii, 58.

129. Morax. Zeit. f. physiol. Chem., 1886, x, 318.

130. Stieff. Zeit. f. klin. Med., 1889, xvi, 311.

131. Biernacki. Deut. Arch. f. klin. Med., 1892, xlix, 87.

132. Rossbach. Berl. klin. Woch., 1889, xxvi, 520.

133. Albu. Berl. klin. Woch., 1895, xxxii, 958.

134. Sehrwald. Ibid., 1889, xxvi, 447.

135. Calderone. Arch. di farmacologia e terapeutica, 1895, 53.

136. Strasburger. Zeit. f. klin. Med., 1903, xlviii, 491.

137. Rovighi. Zeit. f. physiol. Chem., 1892, xvi, 20.

138. Bourchard. Leçons sur les Auto-intoxications. Paris, 1886.

139. Loebisch. Wiener Med. Presse, 1901, xlii, 1273.

140. Stern. Zeit. f. Hyg., 1892, xii, 88.

141. Nuttall and Thierfelder. Zeit. f. physiol. Chem., 1895, xxi, 108; 1896, xxii, 62; 1897, xxiii, 231.

142. Levin. Ann. Inst Past., 1899, xiii, 558; Arch. f. physiol., 1904, xvi, 249.

143. Schottelius. Arch. f. Hyg., 1899, xxxiv, 210; 1902, xlii, 48.

144. Madame Metchnikoff. Ann. Inst. Past., 1901, xv, 631.

145. Moro. Verhandl. d. gesellsch. f. Kinderheilk. Sitz, Sept. 27, 1905. Jahrb. f. Kinderheilk, 1905, lxii, 589.

146. Lyons Arch. f. Hyg., 1897, xxviii, 30.

147. Cramer. Ibid, 1893, 151.

148. Dreyfuss. Zeit. f. physiol. Chem., 1894, xviii, 358.

149. Nashimura. Arch. f. Hyg., 1893, xviii, 318.

150. Duclaux. Ann. Inst. Past., 1889, iii, 413.

151. Nägeli and Löw. Journ. f. prakt. Chem., xxi, 97.

152. Theobald Smith. Trans. Am. Phys., 1895.

153. Brown. Ann. Report Mass. State Board of Health, 1909.

154. Seelig. Virchow's Arch., 1896, cxlvi, 53.

155. Kendall. Boston Med. and Sur. Journ., 1910, clxiii, 332.

156. Auerbach. Arch. f. Hyg., 1897, xxxi, 311.

157. Heilner. Zeit. f. Biol., 1906, xlviii, 145.

158. Cohnheim. Deut. Med. Woch., 1903, 248.

159. Cohnheim. Vorlesungen über Allgemeine Pathologie, 1882, ii, 140.

160. Hirschler. Zeit. f. physiol. Chem., 1886, x, 306.

161. Winternitz. Ibid., 1892, xvi, 460.

162. Gorini. Cent. f. Bakt., 1893, xiii, 790; ref. Maly's Jahresbr., 1893, xxiii, 633.

163. Schmitz. Zeit. f. physiol. Chem., 1894, xix, 378.

164. Iwanoff. Ibid., 1904, xlii, 464.

165. Rubner. Arch. f. Hyg., 1902, xliv, 451.

166. Rovighi. Zeit. f. physiol. Chem, 1892, xvi, 20.

167. Simnitzki. Ibid., 1903, xxxix, 99

168. Seelig. Virchow's Arch., 1896, cxlvi, 53.

169. Albu. Duet. med Woch., 1897, xxiii, 509.

170. Blumenthal. Virchow's Arch, 1896, cxlvi, 65.

171. Bienstock. Arch. f. Hyg., 1901, xxxvi, 390.

172. Tissier and Martelly. Ann. Inst. Past., 1902, xvi, 865.

173. Passini. Zeit. f. Hyg., 1905, xlix, 135.

174. Maly. Herman's Handbuch, v [2], 239.

175. Senator. Berl. klin. Woch, 1868, iv, 254.

176. Bourchard. Bull. de la Soc. Biol., 1884, 665.

177. Baumann. Pfluger's Arch., xiii, 285; Zeit. f. physiol. Chem., 1886, x, 123.

178. Wassiliew. Jeschendelnaja klinitscheska Gazeta, 1882, Nos. 12–14.

179. Ortweiler. Mitteil. a. d. med. Klinik zu Würzburg, ii, 153.

180. Fr. Müller. Ibid., ii, 342.

181. Nencki. Ber. d. chem. Gesell., viii, 336, 722.

182. Hoppe-Seyler. Zeit. f. physiol. Chem., 1888, xii, 1.

183. Albu. Berl. klin. Woch., 1895, xxxii, 959; Duet. med. Woch., 1897, xxiii, 509.

184. Jaffe. Virchow's Arch., 1877, lxx, 72.

185. Salkowski. Ibid., 1878, lxxiii, 408.

186. Brieger. Zeit. f. physiol. Chem., 1878, ii, 241.

187. Sucksdorff. Arch. f. Hyg., 1886, iv, 355.

188. Müller. See No. 180.

189. Ortweiler. See No. 179.

190. Krauss. Zeit. f. physiol. Chem., 1894, xviii, 167.

191. Biernacki. Deut. Arch. f. klin. Med., 1892, xlix, 87.

192. Eisenstadt. Arch. f. Verdauungskrankh., 1897, iii, 155; Diss. Berlin, 1897.

192a. Poehl. Maly's Jarnesbr., 1887, xvii, 277.

193. Bachmann. Zeit. f. klin. Med., 1902, xliv, 458.

193a. Biernacki. Deut. Arch. f. klin. Med., 1892, xlix, 87.

194. Winternitz. Zeit. f. physiol. Chem., 1892, xvi, 460.

195. Rennert. Diss. St. Petersburg, 1893.

196. Skorodumow. Diss. St. Petersburg, 1895.

197. Albu. Deut. med. Woch., 1897, xxiii, 509.

198. Schmitz. Zeit. f. physiol. Chem., 1894, xix, 378.

199. Gussarow. Diss. St. Petersburg, 1895.

200. Nasarow. Ibid., 1895.

201. Rovighi. Zeit f. physiol. Chem , 1892, xvi, 20.

202. Strauss. Zeit. f. klin Med , 1894, xxiv, 441.

203. Matteoda. Diss. Geneva, 1894.

204. Eisenstadt. Loc. cit., 192.

205. Solucha. Diss St. Petersburg, 1896.

206. Wereschtschagin Ibid., 1895.

207. Kopecki. Ibid., 1900.

208. Strauss. See 202.

209. Poehl. Maly's Jahresbr , 1887, 277.

210. Brudzinski. Jahrb. f. Kinderheilk, 1900, lii, 469.

211. Fischer. Vorlesungen über Bakterien. Zweite Aufl., 284.

212. Rovighi. Zeit f. physiol. Chem., 1892, xvi, 20.

213. Embden. Ibid., 1894, xviii, 304.

214. Krauss, Gottwald and Rothmann. Zeit. f physiol. Chem., xii, 16.

215. Ibid.

216. Ibid.

217. Salkowski. Deut. Gesell., 1876, iv, 408.

218. Salkowski and Jaffe Virchow's Arch., 1877, lxx, 70.

219. Albertoni. Arch. Italien de Biol., 1895, xv, 321.

220. Bey. Revue génerale de clin. et de thérapeutique, Mai 31, 1888.

221. Fischer. Zeit. f. Chir., xxii, 225.

222. Annual Report Mass. State Board of Health, 1890, 135.

223. Kendall. Boston Med. and Sur. Journ., 1910, clxiii, 322; Pediatrics, 1910, September.

224. Jehle and Charleton. Zeit. f. Heilk, 1905, viii, 402.

225. Southard and McGaffin. Boston Med. and Sur. Journ., 1909, clxi, 65.

226. Southard and Richards. Ibid., 108.

227. Conradi. Deut. Med. Woch., 1905, 25.

228. Gay. Penn. Med. Bulletin, 1902.

229. Lentz. Zeit. f. Hyg., 1902, xli, 559.

230. Todd. Journal of Hygiene, 1904, iv, 480.

231. Velagussa. Cent. f. Bakt., 1898, xxiv, 750.

232. Nepper. New York Med. Journ., 1908, 980.

233. Kendall. Boston Med. and Sur. Journ., 1910, clxiii, 398.

234. London. Zeit. f. physiol. Chem., 1906, xlix, 328.

.

TUBERCULOSIS AMONG GROUND SQUIRRELS (CITELLUS BEECHEYI, RICHARDSON).[*]

GEORGE W. McCOY.

(*Passed Assistant Surgeon, U.S. Public Health and Marine-Hospital Service.*)

AND

CHARLES W. CHAPIN.

(*Assistant Surgeon, U.S. Public Health and Marine-Hospital Service.*)

(*From the Federal Laboratory, San Francisco, Cal.*)

General remarks. — While examining ground squirrels for plague infection, we have observed among these rodents five examples of a disease due to an organism which so far as our observations have gone agrees with Bacillus tuberculosis. The infection is probably very rare, as we have encountered it only five times in the examination of about two hundred and twenty-five thousand squirrels. Cases may be more frequent than these figures would indicate, as unless there is some feature calling attention to a squirrel as probably tubercular the evidence of infection is only obtained when the guinea-pig inoculated from it dies or is chloroformed. We ordinarily use the cutaneous or "vaccination" method for the diagnosis of plague, and as we do not know how successful this procedure is in infecting guinea-pigs with tuberculosis some cases may have escaped detection. There is nothing to indicate that the disease is localized in distribution. Two cases were from one county. The others were from widely separated parts of California.

The fact that the organism isolated from these squirrels agrees rather closely in its essential characteristics with the bovine type of the tubercle bacillus leads us to suspect that the infection may have been derived from cattle, especially as pastures are favorite localities for squirrel colonies.

The lesions of tuberculosis in naturally infected ground squirrels. — We present here brief protocols taken from our laboratory records:

---

[*] Received for publication June 23, 1911.

Squirrel No. 464. — Shot by G. Received March 29, 1910. A large male. This rodent presented no lesions beyond a caseous axillary gland. The capsule was markedly thickened. Smears were not made. A guinea-pig inoculated subcutaneously with some of the caseous matter was chloroformed on the twenty-ninth day and found to have the following lesions : Bilateral caseous inguinal buboes ; splenic granules up to three millimeters in diameter, and smaller granules in the liver. Acid-fast bacilli were found in smears from the bubo, liver, and spleen. The infection was continued for about ten months by carrying it through a series of guinea-pigs by subcutaneous inoculation. The animals usually died between forty and fifty days after inoculation. In every case the lesions were those of tuberculosis. Organisms morphologically and tinctorially indistinguishable from tubercle bacilli were found in each animal.

Squirrel No. 483. — Shot by G. Received May 9, 1910. A large female. The lungs contained several caseous nodules up to .5 centimeter in diameter. Liquefaction had occurred in the center of these lesions. The omentum contained a large number of nodules varying in size from a mustard-seed up to two millimeters in diameter. Smears from the lung showed a few acid-fast bacilli. No other organisms were seen. Sections from the lungs showed a general infiltration and localized necrosis. A goodly number of giant cells were scattered throughout the affected tissue. Large numbers of acid-fast bacilli were found in sections stained by the method for demonstrating B. tuberculosis. A guinea-pig inoculated subcutaneously with a fragment of the lung died on the forty-fifth day. The usual lesions of tuberculosis were found. Smears and sections from the lesions showed acid-fast bacilli in considerable numbers. The microscopical appearances of the lesions were strongly suggestive of tuberculosis. The infection was carried through a series of animals as in the preceding case. One guinea-pig in the series died in twenty-nine days with characteristic lesions of tuberculosis. A rabbit inoculated subcutaneously with a piece of bubo from one of the guinea-pigs died on the sixty-second day with well advanced lesions of tuberculosis.

Squirrel No. 698. — Shot by M. Received Oct. 23, 1910. A grown male. Both lungs were converted into gray mottled masses adherent to the chest wall and to the diaphragm. Numerous grayish yellow bodies were found throughout the partially consolidated tissue. Microscopically there was found a general, though not uniform, infiltration of lymphocytes and polynuclear cells. Acid-fast bacilli were found in abundance in all parts of the tissue. A piece of the lung was placed in a subcutaneous pocket on a guinea-pig's belly. This rodent died on the twenty-fifth day with the usual lesions of tuberculosis, and numerous acid-fast bacilli were found in smears. The infection was continued as in the preceding cases. A guinea-pig vaccinated with the lung of this squirrel died on the forty-second day, and at autopsy presented the usual appearance of tuberculosis.

Squirrel No. 918. — Shot by K. Received Feb. 7, 1911. Full grown female. The only lesion was a purulent mass at the site of the left posterior inguinal gland. A guinea-pig which was vaccinated from this material was killed on the thirteenth day. The only lesions were bilateral caseous inguinal buboes. Smears from these buboes contained a few acid-fast bacilli.

Squirrel No. 1047. — Shot by P. Received May 11, 1911. The only lesion was a pea-sized abscess in the right lobe of the liver. A guinea-pig vaccinated with the pus was chloroformed on the eleventh day. The lesions were bilateral caseous inguinal buboes. Acid-fast bacilli were found in smears and in cultures. This case has not been studied beyond the examination of the primary cultures. The remainder of the work recorded in this paper is based upon the four preceding cases.

The nature of the infection was not suspected at the time the squirrels were dissected except in the two animals with lung lesions. In both of these acid-fast bacilli were found in smears

Morphological and cultural characteristics of bacilli. — No attempt will be made to discuss the differentiation of the human and bovine types of the tubercle bacillus further than to mention the better growth of the former upon artificial media, the greater virulence of the latter for rabbits, and Theobald Smith's reaction curves. For a general review of this subject the reader is referred to the recent exhaustive paper of Park and Krumweide. [1]

In size, shape, and staining reactions the organisms found in smears from the originally infected squirrels (two cases), from the artificially inoculated rodents, and from the cultures, were similar to tubercle bacilli obtained from other sources. In general the bacilli were shorter and thicker than those found in sputum studied for comparison, but these differences were not constant.

Plain egg (Dorset) and glycerine egg (Lubenau) were used in all of the primary cultures from animals. A guinea-pig inoculated with sputum from a case of pulmonary tuberculosis furnished a human strain which proved invaluable for purposes of comparison. A bovine culture that had been carried for a number of generations on artificial media was also used as a control, but on account of the long sojourn on laboratory media we do not attach much importance to its cultural characteristics for purpose of comparison. From the first our human control grew well, one might say with luxuriance, upon plain egg, and glycerine potato ; still better upon glycerine egg; and satisfactorily upon glycerine broth. Our squirrel strains carried at the same time and under the same conditions grew very feebly upon plain egg and glycerine egg and scarcely at all upon glycerine potato and glycerine broth. Subsequent cultures have not noticeably improved in their growth so far (fifth generation).

In order to judge the rapidity of growth, cultures were planted and each tube given a check number, the readings being recorded without the observer knowing the source of the culture. In each case the best growth in a series was rated at 10. Three tubes, numbered 1, 2, 3, of each strain were planted on each medium. The results are shown in the following tables. In each case, except the "Bovine," the generation under observation was the 4th, and the reading was made at the end of twenty-two days' incubation at 37° C. It will be noted that our control culture "Bovine," which had been carried long on artificial media, gave in one experiment a better growth than our recent human strain. There was this difference in the two plantings: one, Table "A," was made by transferring directly from an egg culture to the new media, plain egg, glycerine egg, and glycerine potato, while in the experiment represented by Table "B" the egg culture was emulsified in normal salt solution, and the inoculation made from the emulsion. This may account for some of the discrepancies.

TABLE A.

Culture.	Plain Egg.			Glycerine Egg.			Glycerine Potato.		
	1	2	3	1	2	3	1	2	3
Squirrel 483 . . .	1	Contaminated.	Contaminated.	1	O	O	O	O	O
Squirrel 698 . . .	1	1	1	1	1	1	O	O	O
Squirrel 464 . . .	2	1	Contaminated.	1	Contaminated.	Contaminated.	O	O	O
Squirrel 918 . . .	1	1	1	O	O	O	O	O	O
Human (recent) .	3	3	3	10	10	10	10	10	10
Bovine (old) . .	2	2	2	5	5	5	2	2	1

TABLE B.

Culture.	Plain Egg.			Glycerine Egg.			Glycerine Potato.		
	1	2	3	1	2	3	1	2	3
Squirrel 483 . . .	O	O	O	1	1	Contami-nated.	O	O	O
Squirrel 698 . . .	2	2	Contami-nated.	2	O	O	O	O	O
Squirrel 464 . . .	O	O	O	2	O	O	O	O	O
Squirrel 918 . . .	O	O	Contami-nated.	O	O	O	O	O	O
Human (recent) .	5	O	Contami-nated.	5	3	3	5	8	10
Bovine (old) . .	10	10	10	3	3	O	3	1	2

In another experiment, glycerine potato alone was used, the transfers being from plain egg media. Three of the squirrel cultures were used for this test. The strain "Bovine" (old) had a growth represented by 3, while the strain "Human" (recent) was marked 10. None of the three squirrel cultures, 483, 698, 918, gave any visible growth.

Smith's[2] change of reaction in glycerine broth has not been studied, chiefly because our squirrel strains have not grown well on the medium.

Pathogenicity. — The cultures have been tested for virulence, using rabbits, guinea-pigs, and ground squirrels. In the following experiment each rodent was given .01 of a loop of a twenty-day-old culture, the guinea-pigs and ground squirrels receiving the culture subcutaneously, while the rabbits were inoculated intravenously.

The term "loop" of a culture as employed here is a very vague quantity, as our squirrel strains grew so poorly that it was almost impossible to collect anything resembling a loopful of culture. The loop of culture of the human strain was certainly much the larger amount so that the error but emphasizes the difference in virulence. The second generation on artificial media was used except in the case of the culture "Bovine." The age of the latter is not known.

RABBITS.
(1/100 loop intravenously, 20-day-old culture.)

Culture.	Day of Death.	Weight at Inoculation (Grams).	Weight at Death (Grams).	Lesions.
Bovine (old) . . .	21	1885	Not recorded.	Macroscopic tubercles in lungs, liver doubtful. Microscopically confirmed in both and spleen and kidney.
Human (recent) .	Killed on 65th day.	1100	1953	No lesions.
Squirrel 483 . . .	17	1630	1245	Macroscopic tubercles in lungs and spleen.
Squirrel 698 . . .	54	1730	1610	Macroscopic tubercles in lungs, spleen, liver, and kidneys.
Squirrel 918 . . .	22	1800	Not recorded.	Macroscopic tubercles in lungs and spleen.

Guinea-pigs (including results of tuberculin injection). — One of the guinea-pigs inoculated subcutaneously with .01 of a loop of culture Squirrel 918 died on the sixty-seventh day. It presented at autopsy the usual lesions of tuberculosis. Seventy-three days after inoculation the others were given a dose of tuberculin subcutaneously.[3] The product was a dark syrupy fluid marked "Bovine Tuberculin." We do not know its source or the manner in which it was prepared. The test animals were each given one cubic centimeter of the agent while three small controls were given two cubic centimeters each. The results are shown here.

EFFECT OF TUBERCULIN ON INFECTED GUINEA-PIGS.

Culture used to Infect Guinea-pigs.	Dose of Tuberculin (Cubic Centimeters).	Weight at Original Inoculation with Tubercle Bacilli (Grams).	Weight at Death (Grams).	Result.	Lesions.
Bovine (old) . .	1	820	700	Died within 15 hours.	Tuberculosis.
Human (recent) .	1	630	520	Died within 15 hours.	Tuberculosis.
Squirrel 483 . . .	1	480	336	Died within 15 hours.	Tuberculosis.
Squirrel 698 . . .	1	410	480	Died within 15 hours.	Tuberculosis.
Control	2	. . . . . .	190	Died in 30 hours.	Extreme necrotic reaction at site.
Control	2	. . . . . .	190	Died in 48 hours.	Extreme necrotic reaction at site.
Control	2	. . . . . .	225(a)	Sickened but recovered.	

(a) weight at inoculation.

The control guinea-pigs received in proportion to weight an average of about five times the dose of tuberculin given to the test animals. The two that died presented no lesions beyond necrotic areas at the site of inoculation. The test animals all died during the night following the injection of the tuberculin, which was administered late in the afternoon.

Ground squirrels. — Each ground squirrel culture used to inoculate a rabbit and a guinea-pig was used to inoculate a ground squirrel. The dose was .01 of a loop subcutaneously. None of these rodents died. On the seventy-fourth day they were given one cubic centimeter of tuberculin subcutaneously. Three controls (ground squirrels) were inoculated with the same dose. The test animal infected with culture 483 died about forty hours after receiving the tuberculin. The other test squirrels sickened but recovered. They were chloroformed four days later. The controls remained well. The lesions found in these squirrels are shown below:

Culture used to Infect Squirrels.	Lesions.
Bovine (old)..........	Slough about 1 centimeter by ½ centimeter at site, caseous axillary gland.
Human (recent)	Slough about 1 centimeter by ½ centimeter at site, caseous axillary gland.
Squirrel 483..........	*Died after tuberculin injection:* slough at site, bilateral caseous axillary glands, 1 centimeter by ½ centimeter; lungs, liver, omentum full of small tubercles; spleen few tubercles. Lesions all confirmed by microscopical examination.
Squirrel 698..........	Slough at site. Small caseous mass in right axillary gland. Large caseous mass in mediastinum. Two small nodules in spleen; a few white granules in liver. A few nodules, all smaller than a pea, in lungs. Small caseous masses in the mediastinum.
Squirrel 918..........	Slough at site, discharging bubo in each axilla. Large caseous mass in mediastinum. Lungs compressed by the mediastinal mass but no tubercles found. Liver and spleen negative.

Acid-fast bacilli were found in smears and in sections from all of the squirrels. Microscopically the lesions in each case were consistent with tuberculosis. Cultures from each squirrel exhibited the characteristics of those used for inoculation. The culture "Human" grew much better than any of the others, showing that passage through a ground squirrel had not noticeably changed its characteristics in this respect.

One other series of rabbits was inoculated for the purpose of ascertaining the virulence of the cultures. Two animals were used for each culture, one being given .01 of a loop, the other .001 of a loop. In every case the inoculation was into an ear vein. Four squirrel strains were available for this experiment. A twenty-five-day-old plain egg growth was used in each case.

RABBITS.

(1/100 of a loop of culture intravenously.)

Culture used to Inoculate Rabbits.	Day of Death.	Weight at Inoculation (Grams).	Weight at Death (Grams).	Lesions.
Human (recent) .	Killed 60th Day.	1705	2250	Lungs full of translucent gray tubercles. No other lesions.
Squirrel 464 . . .	29th	1709	1760	Macroscopic tubercles in lungs, liver, spleen, and kidneys.
Squirrel 483 . . .	46th	1690	1575	Macroscopic tubercles in lungs, spleen, liver, kidneys.
Squirrel 698 . . .	28th	1585	1325	Macroscopic tubercles in lungs, liver, spleen, kidneys, heart muscle, and lymph glands.
Squirrel 918 . . .	29th	1615	1630	Macroscopic tubercles in lungs, liver, spleen, kidneys, lymph glands, and heart muscle.

RABBITS.

(1/1000 of a loop of culture intravenously.)

Culture used to Inoculate Rabbits.	Day of Death.	Weight at Inoculation (Grams).	Weight at Death (Grams).	Lesions.
Human (recent) .	5th	1180	Not recorded.	No lesions.
Squirrel 469 . . .	46th	1350	1230	Macroscopic tubercles in lungs, liver, spleen, kidneys, and lymph glands.
Squirrel 483 . . .	51st	1305.	1350	Macroscopic tubercles in lungs, liver, spleen, kidneys, and omentum.
Squirrel 698 . . .	37th	1415	1425	Macroscopic tubercles in lungs, liver, spleen, and kidneys.
Squirrel 918 . . .	Killed 60th Day.	1505	1700	Macroscopic tubercles in lungs, kidneys, and omentum. None in spleen.

It is clear that the squirrel strains are much more virulent for rabbits than the human (recent) control culture. The rabbits inoculated with squirrel cultures remained practically stationary, or lost in weight, while those infected with the human culture gained very materially. The lesions in the case of the rabbits given the squirrel cultures were all well marked and clearly progressive.

A ground squirrel weighing three hundred and eighty grams was inoculated subcutaneously with .01 of a loop of the same growth of culture, Strain 918, that was used to inoculate the rabbits in the last experiment. This squirrel lost fifty-five grams in weight and died on the sixty-fourth day, presenting the following lesions: A large caseous mass was found at the site of inoculation. The axillary, pelvic, and mediastinal glands were converted into caseous buboes. Miliary tubercles were found in the liver and in the spleen. There was a partial diffuse consolidation of the lungs and definite tubercles were present. Smears showed acid-fast bacilli. This is the only example we have had of a squirrel dying of the infection.

SUMMARY.

Five cases of natural tubercle infection in ground squirrels have been observed.

The gross pathology in these animals was as follows: Two had lesions of the lymph glands alone, one of the lungs alone, one of the lungs and the omentum, and one of the liver alone.

As compared with a human strain of the same age and number of generations the growth on artificial media was far less luxuriant than that of the culture derived from man.

The virulence of the squirrel cultures, judged by their effect on rabbits, agrees with that of the bovine type of B. tuberculosis.

REFERENCES.

1. Journal of Medical Research, xxiii, No. 2, October, 1910, 207–368.

2. Journal of Medical Research, xxiii, No. 2, October, 1910, 186–204.

3. Anderson. Bulletin No. 56 of the Hygienic Laboratory, U.S. Public Health and Marine-Hospital Service, Washington, D.C., 1909, 183.

PRECIPITATION TESTS FOR SYPHILIS.[*]

LAWRENCE W. STRONG.

(*From the Laboratory of the Manhattan Eye, Ear, and Throat Hospital.*)

The refinement of antigen for the Wasserman reaction has been from the aqueous extract of a syphilitic liver through the guinea-pig or beef heart to the lipoids.

In this progress its specificity from the Ehrlich antigen-antibody concept has been entirely abandoned.

Of late certain other reactions, namely precipitates, have been obtained by the use of cholesterin and sodium glyco-cholate as " antigens " and here even the phenomenon of complement fixation has been dispensed with.

In this precipitation reaction confusion seems to have arisen from the use of the word antigen, which implies a relationship to the Wasserman reaction, with which indeed it may be associated though it is due to an entirely different cause.

Recently Herman and Peritz[4] have simplified the precipitation test of Elias,[5] Neubauer and Salomon and claim for it a high degree of accuracy. They note, however, that sera of many conditions other than syphilis may precipitate their suspensions of cholesterin and sodium glycocholate, depending upon their concentration, which must be accurately graded for syphilitic serum.

The writer examined a series of eighty-two cases by the Herman and Peritz method, comparing the results with the Noguchi[7] test for syphilis. In this series there was an agreement in twelve positive and forty-three negative cases, a total of fifty-five, and a discrepancy in the remaining twenty-seven. Of these discrepancies the Noguchi method was negative and the cholesterin positive in eighteen, of which ten were syphilis, under treatment, and the remaining eight were fracture, osteomyelitis, cardiac, tuberculosis, lupus erythematosis and an old serum. The other nine

---

[*] Received for publication May 27, 1911.

discrepancies where the Noguchi method was positive and the cholesterin negative comprised six cases of syphilis under treatment, one case each of Pott's disease, osteomyelitis, and cardio-nephritis.

In approximately ten per cent of this series the cholesterin test was positive in cases not syphilis. It being evident from this experience that the test as given was unsatisfactory, an attempt was made to obtain the same precipitation with the antigen used for the Noguchi test, but without the intervention of complement.

Thirty-five tests were made in a series of thirty-two cases, using equal parts of serum and of freshly prepared emulsion of antigen. This was a three per cent solution of the acetone insoluble fraction of beef heart, dissolved in methyl alcohol and diluted one to five; thus twice as strong as for the Noguchi reaction. It was found that this antigen-serum mixture caused a precipitation similar to the sodium glycocholate, and with fair regularity. Of the series twenty tests were negative agreements and seven positive agreements. Of the discrepancies three cases of syphilis under treatment were positive to Noguchi and negative to the antigen, and an equal number were positive to the antigen and negative to Noguchi. One case without syphilis gave a positive precipitate to the antigen. It will be seen that in this small series the results are fully as satisfactory with the sodium glycocholate, but the presence of a precipitate in a negative case destroys its value as a test.

Many more tests were made without tabulation, in which it was demonstrated that while this precipitation is fairly constant it may be entirely absent in positive cases giving a strong Noguchi reaction, depending apparently on the strength and freshness of the antigen emulsion.

Ternuchi and Toyada,[10] using a similar acetone insoluble residue which they term cuorin, in strength similar to the above, also obtained precipitates with syphilitic serum.

The nature of the precipitate. — Noguchi has shown that the globulin content of syphilitic serum is increased and may be precipitated by butyric acid. The appearance of the flocculation is similar to that obtained by glycogen and by the antigen or cuorin. Furthermore, sera from patients suffering from other diseases than syphilis produce the precipitates and the heightened globulin content.

Tests for globulin were therefore made as follows: To a serum giving a strong Noguchi reaction an equal volume of freshly prepared antigen emulsion was added. In eighteen hours there was a distinct precipitate, which was further separated by strong centrifugalization. The residue was redissolved in normal salt solution and became perfectly clear on the addition of one drop of one-tenth normal sodium hydrate. Upon passing a stream of carbon dioxide through the solution a distinct precipitation occurred which was again redissolved on the addition of alkali. Half saturation of the clear solution with ammonium sulphate also causes precipitation. This demonstrates that the precipitate is globulin.

Noguchi also states that the syphilitic antibody is contained in or precipitated with the globulin.

To test this, the precipitates formed by the use of antigen were separated from the clear supernatant fluid by centrifugalizing and both were incubated separately with complement. The precipitate was found to give the Noguchi reaction while the supernatant fluid did not.

Bauer [1] also has shown the same phenomenon. Jacobsthal [5] has shown by the ultramicroscope that there is a precipitation in the Wasserman reaction from the union of antigen and serum. By centrifugalizing he was also able to obtain the Wasserman reaction in the sediment while the upper part gave none. From this he concludes that the Wasserman reaction itself is a precipitation, while the fact that either may be obtained without the other proves that though they may be associated they are distinct.

Ternuchi and Toyada state that heating serum to 56° C. for one-half hour destroys the power of forming precipitates with cuorin.

This is contrary to the observations of other writers on

precipitins, although these have employed other agents as cholesterin and not cuorin. In a series of tubes, using both my antigen which corresponds to cuorin, and cholesterin and sodium glycocholate, I found precipitates both in the active and inactive sera, and I conclude that heating to 56° C. for one-half hour has little effect on the precipitation. They also state that dialysis brings out precipitates even in normal sera. This affords very strong proof that the precipitation is of globulin, for dialysis quickly removes the sodium chloride which restrains precipitation of the globulin. I was able to confirm this observation on two sera which I dialyzed.

Calcar[3] on dialysis states that the flocculation of albumins is a colloid phenomenon. If a colloid solution, a '' sol.'' goes into the '' gel.,'' we have flocculation. If one removes the salts of blood serum by dialysis globulin is precipitated, because it is insoluble in water.

Consequently, flocculation or precipitation tests of serum cannot be expected to have value in the diagnosis of syphilis, since they are based upon an increased globulin content of the serum, which may be present in other conditions than syphilis.

To study the relationship of globulin to the Wasserman test the following technic was employed :

Fresh sera were dialyzed in quantities of from five to ten cubic centimeters in one-half liter of sterile distilled water for eighteen to twenty-four hours.

Dialysis was performed through gold beater's skin fastened to glass cylinders of one to one and one-half inches in diameter. These were sterilized by steam.

At the end of this time a current of CO_2 was blown down into the serum from a capillary pipette. This caused precipitation of the globulin, which was then centrifugalized and weighed.

The globulin was then redissolved in two cubic centimeters of .9 salt solution, and one cubic centimeter was used for the test and the other for the control, the other components being added directly to this globulin solution.

Twenty-four sera with their corresponding globulins and

with the albumin residue were tested by the Noguchi method.

> Serum, globulin, and albumin positive,   5 cases.
> Serum and globulin positive, albumin negative, 6 cases.
> Globulin positive, serum and albumin negative, 4 cases.
>   (15 positive cases.)
> All three negative,   9 cases.
>   (9 negative cases)

Thus we find the globulin giving a positive reaction in four cases where syphilis was suspected clinically but where the reaction with the whole serum was negative. In all other cases the globulin and serum correspond. The albumin after the removal of the globulin gave a positive reaction only in cases where the serum was very strongly positive.

These observations point to a removal of the syphilitic antibody from the serum albumin and its condensation in the globulin.

Entirely at variance with these results are those published by Friedmann.[11]

His technic consisted in precipitating the globulin with a saturated solution of ammonium sulphate and then dialyzing for two days in fish bladders in running water.

The objections to this method are the addition of a foreign body, the ammonium sulphate, which even forty-eight hours dialyzing will not completely remove, and, more important, the length of time of dialysis which allows decomposition changes to occur in the serum.

The same sera were tested by Friedmann's technic and by my own, and the results were found to be at variance.

Serum " A " dialyzed for twenty-four hours in running water and twenty-four hours in distilled water still showed a trace of sulphates when tested with barium chloride. It was cloudy, and showed presence of spore-bearing bacilli.

The globulin gave a positive reaction by Friedmann's technic, while the whole serum and the globulin were negative by my technic. This was a normal serum.

When globulin was dialyzed for forty-eight hours by my

method also, it became anticomplementary.   These findings occurred in several sera thus tested.

Friedmann states that he did not obtain constant results from his technic, and that frequently also the globulin without extract (antigen) showed a marked hindering of hemolysis, but not a true inhibition.   He gives a series of cases where the globulin plainly was anticomplementary, and another where there was no inhibition with or without antigen.

His assumption, therefore, of an inherent inhibitory power in the globulin of normal sera is not justifiable from his own technic.

No evidence of an inhibitory activity appears in the globulin, using the technic I have described.

This study suggests that the meiostagmin reaction which is dependent upon differences in surface tension, may also be due to increased globulin content of the serum.

REFERENCES.

1.   Bauer.   Biochemische Zeitschr., ix, 301, 1901.
2.   Calcar.   Dialyse, Eiweisschemie und Immunität.
3.   Elias.   Wien. Klin. Woch., 1908, xxiii.
4.   Herman and Peritz.   Med. Klin., vii, 2, 1910.
5.   Jacobsthal.   Mün. Med. Woch., 1910, lvii, 215, 1036.
6.   Lowenberg.   Münch. Med. Woch., x, 35.
7.   Noguchi.   Serum diagnosis of syphilis.
8.   Porges.   Berl. Klin. Woch., 1907, 7.
9.   Rosenfeld.   Deutsch. Med. Woch., x, 4.
10.   Ternuchi and Toyada.   Wien. Klin. Woch., 1910, xxv, 1910.
11.   Friedmann.   Zeitschrift für Hygiene und Infections-krankheiten., 67, 1910.

NOTES ON TWENTY–TWO SPONTANEOUS TUMORS IN WILD RATS (M. NORVEGICUS).[*]

PAUL G. WOOLLEY, M.D., AND WM. B. WHERRY, M.D.

(*From the Laboratory of the Cincinnati Hospital, Cincinnati, O.*)

We must express our regret that the following report does not deal with the inoculability of spontaneous tumors in wild rats. This apparent lack of energy is due to the fact that the tumors were found during the systematic examinations of rats captured or killed in San Francisco during the campaign for the eradication of plague (1907–08). The routine of bacteriological examinations left no time for the experiments on implantation that we should have carried out under other circumstances.

In the literature that we have had at our disposal we have been able to glean but little information as to the incidence of tumors in wild rats. The general fact has been elicited, however, that sarcomas have been most frequently reported, and Opolant makes the statement that whereas in mice ninety-five per cent of the tumors are adenocarcinomas of mammary origin, in rats ninety-five per cent are sarcomas, — a statement that does not correspond with Tyzzer's results in the case of mice nor with McCoy's or ours in the case of rats. It is true, however, that, except for McCoy's statistics, few epithelial tumors have been reported, and this is at the bottom of the current belief in the predominance of the connective tissue group of tumors. In McCoy's series of ninety-nine there were forty-eight epithelial tumors, a percentage of 48.4 per cent. In our series of twenty-two there were fourteen or 63.64 per cent. In McCoy's series there were thirty sarcomas and eighteen fibromas; a total of forty-eight tumors of fibroblastic origin. The other tumors were one lipoma, one endothelioma, and one angioma. In our series there were seven sarcomas and one fibroma.

Our cases, like those of McCoy, were all in Mus norvegicus

* Received for publication June 29, 1911.

(decumanus), a fact that is undoubtedly explained by the relatively small numbers of Mus rattus and Mus alexandrinus in the rat population of San Francisco.

The twenty-two tumors which we report were found in the course of the examination of about twenty-three thousand rats, and since but one of our rats exhibited more than one growth it is apparent that our results compare well with those of McCoy, who found that, on an average, one rat in a thousand was affected with a tumor of one sort or another. Of McCoy's ninety-nine reported cases, thirty were sarcomas, twelve carcinomas, and one endothelioma, a total of forty-three malignant growths. In our series were seven sarcomas, one epithelioma, one adenoma, and three renal adenomas, a total of eleven malignant growths. In McCoy's series sixteen tumors showed metastasis; in ours but four. The tumors that gave rise to metastasis in both series were with one exception (McCoy's case of renal carcinoma) sarcomas.

The cases were as follows:

Case No. 1 (3837). Adult. Sex ? — A tumor about the size of a walnut was observed in the right axilla. It was not adherent to the skin and was not closely bound to the surrounding tissues. It was rather soft and on section appeared whitish and had a lobulated appearance. It was thought that it had a definite relation to a mammary gland.

Sections showed that it was a typical fibro-adenoma with no evidence of malignancy. The sections showed a grossly lobulated appearance. The various lobules were composed of central collections of parenchymatous cells surrounding, in a single row or occasionally several rows, central duct-like spaces which were filled with coagulated proteid material. In some instances the coagulated material had an inspissated appearance. Occasionally no lumen was present in the centers of the parenchymatous masses so that the lobules were composed of solid masses of cells. The connective tissue was abundant and well formed and contained considerable numbers of mast cells.

Case No. 2 (5386). Adult female. — The tumor in this case was in the left groin and measured 3.5 x 8.5 centimeters. Upon its surface was an ulcerated, punched-out area five millimeters in diameter. The tumor was not adherent, was soft, and on section white and lobulated.

Sections showed that this tumor was similar to the preceding (3837), but differed in that the parenchyma was somewhat better developed, and that the connective tissue was more mature, and showed a tendency to become hyaline. The epithelial cells showed a more constant increase in the number of layers and less coagulated proteid in the spaces. There was some tendency to intercanalicular growth. At a single point there was evidence of invasion of the supporting tissue by the epithelial cells. Mitotic figures were few. Occasional direct divisions were observed.

Case No. 3 (2). Adult. Sex? — A very large tumor 7.5 x 13 centimeters in the subcutaneous tissue of thorax, not adherent, white, lobulated. It differed from 5386 only in absence of tendency to infiltration and in absence of mitosis.

Case No. 4 (5755). Adult female. — The tumor was situated beneath the skin on the left side of the thorax. It was soft, lobulated, whitish in color and not adherent to the skin or surrounding tissues.

Microscopically this growth showed a more preëminently adenomatous structure than the foregoing tumors. The whole mass was generally lobulated. It also showed a beautiful alveolar arrangement. The adenomatous parts were divided and subdivided into large and small masses by well developed, ripe connective tissue. The alveoli were composed chiefly of solid masses of cells, with only occasional evidence of lumens which were filled with inspissated coagulated proteid material. In the preceding tumors the cells had vesicular nuclei, and a homogeneous granular protoplasm that frequently showed a basophilic tendency; in this the nuclei were small, though still vesicular, while the protoplasm was spongy and clear. There was no evidence of reaction on the part of the supporting tissue; no sign of infiltration of the epithelial parts; no evidence of rapid growth at any place.

Case No. 5 (16646). Adult female. — In this case there were two tumors, one the size of a hen's egg under the skin on the left side of the thorax; one — a smaller one — in the right groin. Neither was adherent. Both showed the same structure as the growth in Case No. 5755.

Case No. 6 (5298). Adult female. — The tumor was situated on the left side of the thorax and measured 5 x 2.5 x 2.5 centimeters. It was hard, cut with difficulty, and was not adherent. This tumor was an example of fibro-adenoma in which the cell masses showed, as a rule, central lumens surrounded by but a single row of cells (Fig. 1). The whole growth was formed of alveoli each of which was divided and subdivided into small acini by a scanty fibrous tissue. The gross masses were limited by a well-formed — in some places hyaline — connective tissue. The lumens were filled with coagulated material. There was no evidence of malignancy.

Case No. 7 (17142). Adult female. — The animal was pregnant. In the right inguinal region there was a globular, soft, lobulated, white tumor measuring about 2.5 centimeters in diameter. It was not adherent. The tissue of this growth was composed chiefly of long, free and interlacing processes, which were constructed of a narrow, central, vascular, connective tissue covered with one to several layers of epithelial cells (Fig. 2). These papillary processes originated from the thin fibrous capsule that surrounded the whole mass. In this capsule a few eosinophilic cells were observed. No mitotic figures were found.

Remarks. — The eight tumors described in the preceding paragraphs represent growths of an adenomatous type, the variations within the group being determined by the relative development of the glandular tissue and the stroma. In all but one case (5) the tumors were single. In the one case two similar tumors were present in different parts of the body. All the animals in which the sex was recorded were females, a fact that indicates the greater tendency of females in rats, as in human beings, to exhibit mammary growths. This sexual difference is shown in McCoy's figures in which thirty out of thirty-four mammary tumors were found in female rats. In none of the seven animals from which these growths were taken was there any evidence of a causative factor.

Case No. 8 (6712). Adult female. — In this animal a slightly adherent, soft ovoid tumor the size of a hazel nut was discovered in the right inguinal region, and upon the upper lip among the large whisker hairs was a small epithelial growth. The mass in the inguinal region proved to be a typical, pure, soft fibroma. The growth on the lip was due to a localized hyperplasia of the epithelium — a hyperkeratosis — with no evidence of down growth and no sign of malignant change. It was not apparently due to Sarcoptes alepsis.

Case 8 furnishes the only example of a pure fibroma in our series. In McCoy's series there were sixteen subcutaneous fibromata.

Case 9 (13487). Adult. Sex? — The tumor in this case was a papilloma of the bladder associated with calculi. The calculi were sent to Prof. H. B. Ward for examination for ova of parasitic worms. Up to the present time no ova have been demonstrated.

The growth was composed of elongated branching and interlacing papillary projections, each with a central connective tissue framework carrying blood vessels, and covered with approximately normal, though occasionally hyperplastic, epithelium. The spaces between the interlacing columns were sometimes cystic, sometimes completely filled with cells of a polygonal squamous type, but without evidence of keratinization. In such cellular masses the central cells not infrequently showed degenerative changes. There was no evidence of malignant change.

Interspersed between the cells at various places in the growth were ovoid bodies, composed of a central round protoplasmic mass with one or two rounded chromatin masses. These central structures were surrounded by a clear space limited externally by a thin sharply demarkated capsule. These structures are apparently the result of retrogressive metamorphosis of epithelial cells, though at first it was suggested, because of the fact that the bladder contained calculi, and that in M. norwegicus bladder worms (Trichosoma Sp?) were not infrequently found, that they might be ova.

Remarks. — Case No. 9 represents the only example of bladder tumor which we have seen.

Among McCoy's cases there was no example of such a tumor, which indicates that, in spite of the frequency of bladder worms and calculi, new growths of the bladder are very uncommon.

Case No. 10. Adult female. — Above the labia was an ulcerated surface 2.5 millimeters in diameter, where the skin was thickened, hard, and ulcerated. Attached to the border of this area was a nodular mass 10 x 5 x 6 millimeters, which was firm in consistence and pinkish on section.

The labial growth showed hyperplasia of the epithelium with distinct invasion of the subjacent tissues. The surface of this hyperplastic growth was ulcerated, and the tumor itself was infiltrated with considerable numbers of polymorphonuclear leucocytes. The epithelial cells in the invading columns showed numerous mitotic figures, symmetrical and asymmetrical, and some epithelial giant cells.

The nodular mass connected with the tumor was composed entirely of granulation tissue.

Remarks. — Epitheliomatous growths are not frequent in rats so far as our records show, and in McCoy's series there is no example.

Borrel, Gastinel and Gorescu [3] believe that acarids, particularly Demodex folliculorum, play an important rôle in the production of epitheliomas in man. While the analogy is not complete, it may be worth mentioning that whereas hyperkeratosis of the ears, lips, and nose, due to Sarcoptes alepis, was extremely common among the Norway rats on the Pacific Coast, yet no tumors were found at these sites.

Case No. 11 (8741). Adult female. — A flattened ovoid tumor 3 x 2.5 centimeters was found in the right groin. It was not adherent, was readily peeled out, was soft and had a lobulated appearance.

An emulsion of this tumor was made in physiological saline and injected subcutaneously into a small white rat. No growth of parasitic or tumor origin appeared, and two months later the rat was chloroformed and examined. Nothing abnormal was discovered.

The bulk of the tumor was composed of groups of cells of squamous epithelial character limited by a well developed, partially hyaline, connective tissue. The minor portion was a glandular tissue with a delicate supporting connective tissue (Fig. 3). The epithelial islands of the major part varied in size and shape. The small islands, those which

showed for the most part little or no keratinization, were generally round.
The larger ones, those formed of an external zone of more or less normal,
or flattened epithelial cells and a central area of cells showing extreme
keratinization, or merely masses of keratohyalin, were polymorphous,
lobulated, trefoil-shaped, or round.   About the smaller islands there was
a well marked small round-cell infiltration with occasional polymorphonuclear leucocytes.   About the larger there was merely a well formed connective tissue.

In many instances in both the smaller and the larger islands the epithelial
cell boundaries could not be distinguished, so that the central, cellular
or keratin masses seemed to be surrounded by a more or less complete
syncytial layer.   In various places the cells contained " inclusions," usually acidophilic, each surrounded by a clear achromatic zone.   At no
place could prickle cells be observed.

The glandular portion of the tumor showed longitudinal and cross sections of ducts and acini, some with a central lumen; some completely
filled with cells.   Among the typical epithelial cells were others that had
a yellow granular appearance, the result of the presence in the protoplasm
of numbers of fine yellow granules of lipoid (?).

The lining epithelium of the ducts and acini were composed of one or
several layers of cells of cylindrical or cuboid form.   In the supporting
tissue there was very little evidence of reaction, except for occasional
basophilic polymorphous cells, and a few eosinophiles.   No mitotic figures
were found.   The blood vessels were slightly congested.

In a few sections small colonies or clumps of organisms were found.
The individuals of these colonies were rounded or spindle-shaped with
granular and vacuolated bodies.   These were arranged tip to tip in chains
within the clumps.   Associated with them there were a few structures
resembling mycelial threads.   The identity of this organism which was
found only in the adenomatous part of the tumor cannot be more than
guessed.   Its relation to the tumor can only be conjectured.

Remarks. — We believe that there are but four similar
tumors on record — one reported by Tyzzer, and three by
Murray; all in mice.   These tumors are interesting examples
of metaplasia due most likely to continued irritation of one
sort or another, and are apparently comparable in their
metaplastic changes to the interesting lung tumors described
by Tyzzer in mice.   It is difficult, however, to discover what
the cause of the irritation was in our case.   There were
microbic parasites present, but only in the purely adenomatous parts.   It is possible that the absorbed secretions from
these organisms was responsible, in part, for the changes in
the tumor, but to us this possibility seems remote.

Case No. 12 (10375). Adult. Sex? — This rat showed a soft tumor the size of a filbert in the submaxillary region, which was adherent to an adjacent lymph gland.

Sections showed that the tumor was composed of small polymorphous cells with relatively large vesicular nuclei and a slightly granular protoplasm, and with no evidence of orderly arrangement, except that they are more compactly related to the blood vessels. Mitotic figures were numerous and in some instances asymmetrical. It was a small polymorphous cell sarcoma (Fig. 4).

Case No. 13 (9). Adult female. — This rat showed a large tumor of the right humerus, ovoid in shape and measuring 4 x 3 x 3 centimeters. It surrounded the upper part of the humerus and involved the shoulder joint. It was soft and whitish and apparently very vascular. It was not adherent to the skin. The humerus itself was not involved. It was apparently a periosteal tumor.

The sections showed a partially alveolar structure due to the perivascular arrangement of the tumor cells which were of a short spindle shape (Fig. 7). There was no evidence of endothelial origin of the tumor. Occasional giant cells were present. No chondroblasts were found. Mitoses were frequent.

Case No. 14 (20804). Adult male. — The animal showed rather marked post-mortem changes. Just above and attached to the right adrenal was a soft tumor mass the size of a small walnut. It was smooth and glistening and the surface was mottled with reddish white. Between it and the stomach were similar smaller nodules, and extending anteriorly across the abdominal cavity was a large lobulated mass 3 x 2.5 x 1 centimeter and in the mesentery another measuring 2 x 2 x 1.5 centimeters. Beneath the liver were other smaller rounded masses, and a single one in the left kidney. Throughout the liver were other tumor growths, and the peribronchial lymph glands were fused together into a tumor-like mass. The lungs were nodulated and on section the areas of consolidation had the same appearance as the other tumor masses.

This tumor was a typical lympho-sarcoma, the origin of which it was impossible to determine. The character of the growths in all the situations was identical.

Case No. 15 (7). Adult. Sex? — In the liver region a tumor measuring 2 x 3 cubic millimeters, irregularly ovoid, and apparently closely associated with the gall-bladder was found. It was adherent to the edge of the right lobe of the liver, and was composed of a soft grayish or whitish tissue with areas of congestion. In a cleft in the anterior of the tumor mass was a free Cysticercus fasciolaris. The vesicle or bladder which usually encloses the cysticercus was not found. A cysticercus within its cyst was found projecting from the lower border of the right lobe of the liver. Part of the tumor was fed to a white rat with no result.

The tumor proved to be a polymorphous cell sarcoma composed of round and spindle cells, with numerous giant cells scattered throughout (Fig. 5). Mitoses were frequent.

Case No. 16 (8). Adult female. — Attached to the under surface of the lower lobe of the liver was a pinkish, lobulated tumor the size of a large hen's egg. Throughout the omentum and mesentery were hundreds of metastatic nodules varying in size from that of a large pea to .5 millimeters in diameter. As in the previous case free Cysticerci fasciolaris were found lying in smooth-walled channels within the tumor mass. In the upper part of the duodenum was a small ulcer near which an intestinal worm was attached. In the non-secreting portion of the stomach were two small crater-like elevations of a pale color. The stomach contained dipterous larvæ and parasites resembling tricocephalus. The intestine contained tape-worms. In the cecum were some anchylostoma-like worms.

The tumor in this case was composed of polymorphous and giant cells in close association with thousands of ova (Fig. 6) It had apparently originated in the liver. The metastatic nodules were composed of cells similar to those of the primary growth and contained no ova.

The gastric lesions were due to the presence of small subepithelial cystic spaces containing ova that resemble those of Anchylostoma. About these there is no evidence of malignant change.

Case No. 17 (22058). Adult female. — In the gastric omentum there was a large lobulated pinkish white tumor the size of a large hen's egg. Scattered throughout the mesentery were similar, but smaller, rounded and ovoid nodules, 3–6–8 millimeters in diameter. Other similar masses were found in the subperitoneal tissues Microscopic study showed that the tumor was a large spindle-cell sarcoma with metastases of the same type. Giant cells were present in small numbers and mitoses were scanty.

Case No. 18 (4). Adult male. — In the abdomen attached to the diaphragm and to one lobe of the liver was a tumor mass, 4 x 3 x 3.5 centimeters, irregularly ovoid in shape and composed of a soft almost fatty tissue. Attached to the diaphragm were three or four smaller nodules two to six millimeters in diameter; flattened, rounded, and apparently composed of the same sort of tissue as the larger growth. Cysticerci were present in the liver substance and two were present in the tumor. The tumor was a very vascular polymorphous cell sarcoma and occasionally presented a perfect perivascular, alveolar appearance. There were no ova present (Fig. 7).

Remarks. — This group of cases is extremely interesting to us for the reason that several of them (15, 16, and 18) were associated with parasitic worms, and because several of them (14, 16, 17, 18) produced metastases.

In one of them (16) the sarcomatous changes were evidently associated with the presence of enormous numbers of ova (Fig. 6). In another (17) tumors were present in the mesentery, but in the absence of parasites or ova outside the intestine. In still another (16) merely cystic spaces were present in the stomach walls, and in these spaces ova of another sort of parasite were present. In all the cases directly associated with the presence of immature tape worms, Cysticercus fasciolaris, the malignant tumors affected the liver. These facts bring up the questions, whether it is the worms themselves or their excretions that are to blame for the tumor, or whether it is the ova that are chiefly to blame, as in biharziosis of the bladder and intestine; whether the teniæ alone, either through their presence and secretion and ova are active in producing these tumors; and whether the liver is more prone to undergo malignant change as the result of these various influences. Saul did some implantation experiments using various portions of the Cysticercus fasciolaris. Rats inoculated with the head end and middle portions died of evident toxemia. One inoculated with the tail end developed a fibro-sarcoma at the site of implantation. Reimplantation with a portion of his tumor resulted in infection.

It is interesting that so many tumors of rats are associated with parasitic worms. In twelve of McCoy's cases parasites were associated with tumors. One of these growths was a fibroma. The other eleven were sarcomas. All were of hepatic origin.

Case No. 19 (8994). Adult. Sex ? — All of the left kidney except the upper pole was replaced by a tumor mass, oval in form, the size of a hen's egg, smooth, and encapsulated. On section it appeared partially fatty and necrotic, partially hemorrhagic. The tumor mass appeared on microscopic examination to be composed of three parts; one a mass of recent hemorrhage; one a laminated mass of partially organized clotted blood; one, the larger part, composed of renal and tumor tissue. The kidney substance itself showed cloudy swelling, edema, and interstitial accumulation of small round cells. The interstitial changes were more marked in the immediate vicinity of the tumor mass, and in this region there were also occasional giant cells of renal epithelial origin. Though the tumor itself was as a rule well demarcated from the kidney substance, there were points at which an apparent transition could be made out, so that it seemed evident that the papillo-adenomatous tumor mass was of renal origin. In occasional spots there seemed to be some evidence of malignant reversion in the connective tissue.

Case No. 20 (3). Adult male. — In this animal the central part of one of the kidneys was occupied by a grayish yellow and pinkish white mass which was bounded by approximately normal renal tissue.

Microscopically this tumor resembled the preceding one. There was less hemorrhage in the tumor, and more evidence of chronic interstitial changes in the kidney substance. The tumor showed adenomatous, cystic, and intracystic arrangement, with transition from renal substance to tumor. There was also some evidence of sarcomatous change in areas where the cells had a localized tendency to assume a spindle form, especially in those parts of the tumor where, with a rich vascular supply, the cells had a more or less perfect radial perivascular arrangement.

Case No. 21 (5). Adult female. — The organs seemed generally healthy except the left kidney, which was enlarged, and in its anterior half had a whitish appearance. The left adrenal was immediately in juxtaposition with the tumor area but separate from it.

The tumor was composed of cells smaller than those of the two preceding cases, of a more solid type and with fewer giant cells. Within the tumor itself were well-preserved glomeruli which had been surrounded in the progressive growth of the neoplasm. In this case as in the others there was evidence of renal origin in the gradual transition of renal cells at the tumor borders.

Remarks. — In spite of certain likenesses to adrenal tissue we believe that these tumors have originated in renal tissue and not in misplaced adrenal rests; that they are malignant renal adenomas, therefore, and not hypernephromas, in the sense of Grawitz. This seems reasonable because of the evidences of malignant transformation in the cells immediately about the tumor proper. We realize that this transformation may be explained as being the effect of the presence of the tumor rather than the cause of it, but we think that there are greater similarities between the kidney proper and the tumors than between adrenal tissue and the tumors.

In McCoy's series of eleven renal tumors, one produced metastasis. Of our cases none metastasized.

SUMMARY.

Total rats examined,	23,000	
Total rats with tumors,	21	
Total tumors,	22	
Total epithelial tumors,	14	63.64%
Total connective tissue,	8	36.36%
Tumors of breast,	9	40.9%
" " kidney,	3	13.64%
" " bladder,	1	4.54%
" " skin,		4.54%
" " connective tissue,		4.54%
" " liver,	3	13.64%
" " mesentery,	1	4.54%
" " submaxillary gland,		4.54%
" " periosteum,		4.54%
" " lymph glands,	1	4.54%
Malignant,	11	50.50%
Metastasis,	4	18.18%
Associated with parasites,	3	13.64%

BIBLIOGRAPHY.

McCoy. The rat in its relation to public health. Public Health and Marine Hosp. Service, Wash., 1910, 64.

Wherry, Walker and Howell. Jour. Amer. Med. Assoc., 1908, l, 1165.
Borrel, Gestinel and Gorescu. Ann. d. l'Inst. Past., 1909, xxiii, 97.
Saul. Centr. f. Bakt. I. Abth. Orig., 1908, xlvii, 444.

DESCRIPTION OF PLATE II.

FIG. 1. — Mammary adenoma from Case No. 6.
FIG. 2. — Papillary adenoma from Case No. 7.
FIG. 3. — Metaplastic mammary tumors from Case No. 11.
FIG. 4. — Sarcoma from Case No. 12.
FIG. 5. — Sarcoma with ova from Case 16.
FIG. 6. — Sarcoma, Case No. 18.

(Printed without author's corrections.)

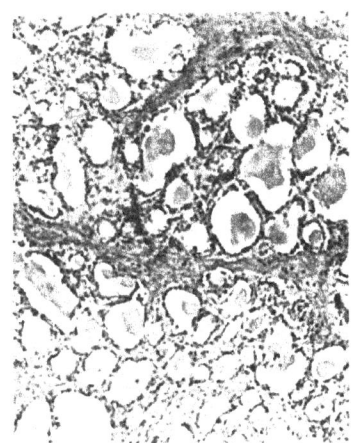

FIG. 1.

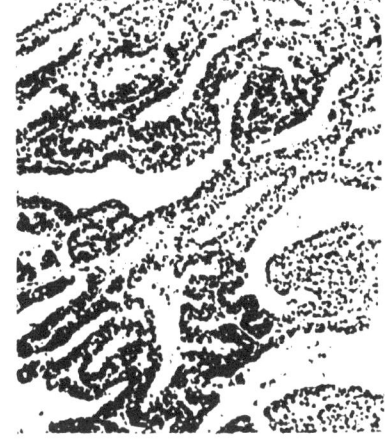

FIG. 2

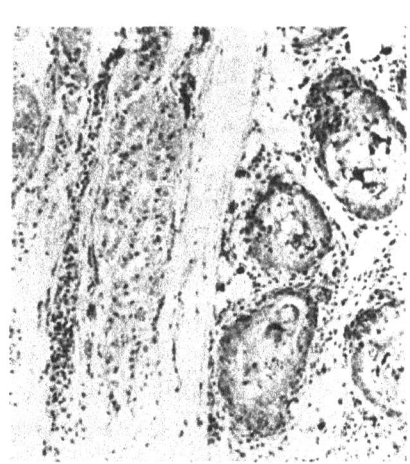

FIG. 3.

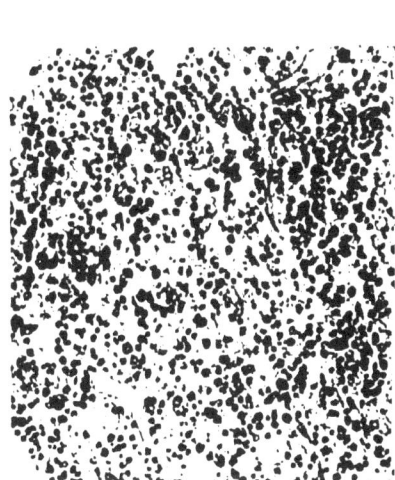

FIG. 4.

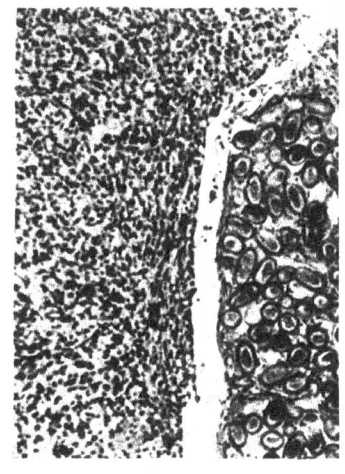

FIG. 5.

FIG. 6.

THE ISOLATION OF TYPHOID BACILLI FROM URINE AND FECES WITH THE DESCRIPTION OF A NEW DOUBLE SUGAR TUBE MEDIUM.[*]

F. F. RUSSELL, M.D., WASHINGTON, D. C.

(*From the Laboratory of the Army Medical School.*)

It is probably but a matter of a short time before public opinion will demand that health authorities exercise the same supervision over typhoid convalescents and carriers that they now practise with respect to diphtheria convalescents and carriers. It is almost impossible to find an article on the epidemiology of typhoid, whether written for the medical profession or laity, which does not refer to the dangers of infection from typhoid carriers. As the education of the public progresses, its present attitude of indifference may be succeeded by a typho-phobia and ultimately by a desire for reasonable prophylaxis based on a better understanding of the manner in which the disease is spread.

At the present time we are passing through a similar transition period in the history of tuberculosis. It is not long since, at least so far as the general public is concerned, that tubercular disease was looked upon as an act of Providence, with no special danger of contagion through contact with the tuberculous. As the result, however, of a very active anti-tuberculosis campaign, the public is being rapidly educated in the epidemiology of the disease and instances of phthiso-phobia are not uncommon. In America it is only since the meeting of the International Congress on Tuberculosis in 1908 that the public has been able to overcome its fears and to appreciate the results to be obtained by a widespread aggressive campaign against tuberculosis based on rational principles. Public opinion now demands special hospitals, or at least particular wards in general hospitals for the consumptive and, too, that health authorities afford adequate facilities for the examination of tuberculosis material.

[*] Published by permission of the Surgeon General, U.S. Army. Received for publication July 12, 1911.

We should be prepared to expect during the next few years a somewhat similar attitude on the part of the public regarding typhoid fever. At the present time the public press is filled with articles, both good and bad, on the dangers of the "typhoid fly," and from time to time water-borne epidemics furnish startling news. The health authorities of most of the larger cities and of many states to-day maintain laboratories in which the serum of suspected typhoid cases may be examined for the Widal reaction; but with this test their activity ceases except for the collection of statistics of morbidity and mortality. Nothing further toward the control of typhoid is attempted anywhere in civil laboratories in the United States or in any country excepting in certain parts of Germany, where an anti-typhoid campaign aiming at the ultimate extermination of the disease, or at least the reduction of the number of cases to an irreducible and insignificant minimum, was instituted by Prof. Robert Koch in about 1890.

The researches of Reed, Vaughan and Shakespeare[1] on the origin and spread of typhoid fever in the military service during the Spanish-American War established for all time the importance of contact in the extension of the disease. The German investigations have resulted in confirming this work and in discovering the typhoid bacillus carrier; so that in the course of the next few years we may reasonably expect to learn the relative importance of the chronic and temporary carriers and of contact in general as compared with water-borne infections.

Great explosive outbreaks due to infected water supplies such as occurred at Plymouth, Pa., Ithaca, N.Y., and Butler, Pa., attract so much attention by the appalling numbers suddenly attacked that we are apt to lose sight of the fact that the total amount of sickness and number of deaths due to the contamination of drinking water is small compared to the steady, unbroken chain of infections continually occurring through more or less direct contact.

For such reasons as these we believe health authorities will soon feel compelled to recommend and adopt regulations governing the degree of isolation and disinfection to be exercised in hospitals and private houses; to recognize that typhoid convalescents are an important source of new infection and consequently should be controlled by some regular and systematic examination of the urine and feces. As Lentz[2] has shown, about four per cent of all cases may be expected to become chronic carriers, and it is advisable for health authorities to have a list of the names of such persons.

At the present time the isolation of the typhoid bacillus from urine and feces is looked upon as difficult and uncertain, and in the absence of a good system it is doubtless attended with difficulty. During the past two years we have examined a large number of specimens of urine, feces or blood and have found the following routine simple and reliable; in fact the search for typhoid bacilli has become as easy as that for colon bacilli in water supplies.⟩

The specimens of urine and feces are obtained in as fresh condition as possible and are plated on Endo's fuchsin agar. At different times we have used nearly all of the special media designed for this purpose including the media of Hiss,[3] Conradi-Drigalski,[4] Hesse,[5] Loeffler's[6] malachite green and rein blue, and the picric acid, brilliant green medium of Conradi,[7] but nothing has given better results than Endo.[8] Its superiority lies in the simplicity of its preparation and the delicacy of fuchsin as an indicator for the acid produced by colon bacilli. A quantity of Endo may be prepared ready for use in about two hours or it may be made up at leisure, tubed, sterilized and stored in the dark for use within two or three weeks.

It is prepared as follows: Into a saucepan pour one liter of tap water marking the level of the fluid on the inside of pan; add thirty grams of shred agar, ten grams of peptone, five of salt and five of Leibig's beef extract, cook until dissolved (three-quarters hour), and filter through sterile gauze and cotton. If the agar is clean the use of eggs for clearing is not necessary, although with the average quality of shred agar on the market a more satisfactory medium is obtained by clearing with egg white. The preparation of the filter is important if rapid filtration without loss of medium is to be obtained; we proceed as follows: a square of absorbent cotton is split into two sheets of equal thickness and the upper is rotated through an angle of 180° until the fibers come to lie at right angles to those of the lower sheet. A somewhat larger square of gauze is placed over the cotton and the whole inverted in a glass funnel with the gauze layer out. The whole is heated in a steam sterilizer for a half hour or more to sterilize the filter and to improve the filtering power of the absorbent cotton.

After the greater part of the agar has run through the cotton filter, the four corners of the square of gauze are gathered up, twisted, and the remaining fluid squeezed out of the cotton. By following this technic the quantity of medium lost in filtration is insignificant. The agar, as a result of the prolonged boiling, is practically sterile before filtration, and the use of a sterile filter keeps it free from contaminations. This stock agar may be kept on hand in quarter and half liter flasks ready for making the finished Endo which cannot be held in stock for any length of time. Since the acidity of agar is apt to increase with storage, no attempt is made to standardize it until just before use, when the reaction is adjusted to .2 per cent acid to phenolphthalein. In our experience the proper reaction has proved the most important single element in obtaining uniform and successful results. Both the amount of fuchsin and sodium sulphite may be varied considerably and good results still be obtained if the medium has a suitable reaction. To the .2 per cent acid agar is now added 1.8 cubic centimeters of a filtered saturated solution of basic fuchsin in ninety-five per cent alcohol. This solution apparently keeps well but has not been made up in large quantities; one bottle, however, has been in use for as long as six months. The flask is well shaken to diffuse the dye. Next is added enough of a freshly prepared ten per cent solution of sodium sulphite crystals in sterile water to almost decolorize the mixture, it usually requiring from twenty-five to thirty cubic centimeters of the solution to obtain the proper tint. It should be remembered that the end point of the decolorization is the restoration of the original color of the medium, and that this will vary according to the amount of color in different lots of agar. The medium while still warm has a pale rose color, but that fades on cooling. The last step is the addition of ten grams of lactose dissolved by the use of moderate heat in about fifty cubic centimeters of sterile water. This gives approximately a liter of three per cent agar .2 per cent acid to phenolphthalein. The finished product is poured into large Petri dishes, fifteen to twenty centimeters in diameter, in a layer about two millimeters deep. Any agar remaining is stored in large test-tubes (two hundred and fifty millimeters by twenty-eight millimeters), each of which contains enough medium for one plate. Large test-tubes are used in preference to flasks or bottles, because sterilization in the Arnold can be accomplished with a shorter exposure to heat and consequent danger of breaking down the lactose. If the large test-tubes are packed loosely in a basket, ten to fifteen minutes in the Arnold on two successive days is sufficient, if reasonable care has been taken in the preparation of the medium. No great quantity of medium is tubed since it does not keep well, and its preparation is simple and not time consuming.

The freshly poured plates stand open with the covers completely removed until the surface is perfectly dry, when each plate is inoculated with the material to be examined, urine, feces, bile media, or the fermentation tubes used for the presumptive test of colon bacilli in water analysis.

For this purpose the well-known right angled glass rod is used; its tip is immersed in the fluid to be examined and a drop or two deposited in the center of the plate; the spreader is carried through the drop to the periphery of the plate and then carried completely round its circumference after the manner of the hands of a clock; the spreader is not carried over any portion of the plate twice. In this way only one plate is necessary for each sample, since isolated colonies can practically always be found on some portion of the surface. After inoculation the plate remains completely uncovered until the surface is again perfectly dry. This requires ten to fifteen minutes, after which the plates are covered and placed bottom up in the incubator. While the plates stand open they are of course contaminated by air organisms, but as these are readily distinguished from typhoid their presence has not been troublesome.

After fifteen to twenty-four hours incubation the typhoid colonies appear as clear, colorless dew-drops of varying size on a colorless background. The typical colony, when the media has the optimum reaction, is about one millimeter in diameter; the margin is irregular, being idented and the surface veined, both outline and surface suggesting a grape leaf. Quite often, although different strains vary considerably, the appearance of the colony under the low magnification of a hand lens reminds one of the contours of a relief map of a mountain, showing ridges and valleys sloping away from the peak. On some plates the resemblance of the typhoid colonies to a barnacle is striking. In the examination of these large plates for suspicious colonies the microscope is of little use; it is much more convenient to hold the plate upright in the left hand and go over it rapidly with a hand lens, altering the position of the head and plate rapidly until just the right amount of light is obtained and, if possible, using the edge of some dark object as a background.

This method so distinctly reveals the structure and absence of color of the typhoid colony that its identification is easy, though unfortunately it is true that the typical morphology is not always obtained and the worker is often compelled to rely mainly on the color reaction. The structure of the colony seems to depend on the particular strain of the organism, the number of colonies to a given area, on the dryness of the surfaces of the medium and its reaction, and on other unknown conditions, all of which are so rarely favorable at the same time that the conclusion has been forced upon us that the morphology of typhoid colonies is less reliable for purposes of identification than the fermentation reactions.

Typical colon colonies are colored with the fuchsin of the medium and are readily differentiated from typhoid though there are in some stools many " slow " colonies, which in the first twenty-four hours are practically indistinguishable from typhoid, and as such colonies may give rise to considerable confusion the plates must be incubated for forty-eight hours or some special method used to bring out the colon characteristics.

In Germany, where these methods have been developed and used most extensively, it is customary to identify the suspected typhoid colonies by immediate microscopic agglutination reactions on slides; this naturally means the preparation of innumerable emulsions in immune serum and their subsequent examination with the microscope. If the organisms from suspected colonies are properly agglutinated the case is regarded as proven and completed. In our own work we have found that when the plates show large numbers of more or less typical colonies of bacilli agglutinable in immune serum, the problem is easy and there is little chance of error, although we know from our own work and that of Frost[9] that organisms other than typhoid can be agglutinated in typhoid immune serum.

When the typical typhoid colonies are few, or the plates negative, we have found the search for typhoid bacilli by immediate agglutination laborious and time-consuming and now use it only in occasional instances as in Endo plates made from blood cultures or from bile media inoculated with the small clot from blood furnished for the Widal reaction.]

While this method of immediate agglutination from Endo plates is apparently quite direct and simple, it really requires considerable time and more than the average skill and judgment, so has been abandoned by us in favor of a simpler method. We have felt that we could work with greater certainty and exactness and as quickly arrive at a correct diagnosis by making use of a tube cultural test between the plates and the final agglutination. The concluding agglutinative test is of supreme importance and is carried out by the macroscopic method in a uniform manner which admits of ready control against error.

At first we transferred suspected typhoid colonies to litmus milk tubes, thus enabling us to exclude both the colon bacilli and all alkali formers; but as this method was slow and failed to leave the cultures in a suitable medium for the final agglutination test it was early abandoned. We next transferred our suspicious colonies to glucose broth in Durham fermentation tubes. This procedure had so great an advantage over litmus milk that it was possible to exclude the colon bacillus after six or eight hours' incubation. The alkali formers were easy to exclude by testing for acidity with litmus. The delicacy of litmus papers varies, but with Squibb's blue paper, laid upon moistened filter paper, the test is sufficiently delicate, as then the moisture is evenly distributed and the colors show well against the white background. It was also ascertained that our entire stock of typhoid cultures produced in eighteen hours sufficient acid in glucose broth to give a distinct reaction with litmus. This test was somewhat simplified by adding one per cent of a five per cent aqueous solution of litmus to the Durham fermentation tubes; this modification of the medium appeared at the time to be adequate if the sterilization in the Arnold was watched to guard against breaking down the litmus by long exposure to heat. In this stage all tubes showing gas and all which failed to show well marked acidity, were discarded. As this intermediate tube culture narrowed the field considerably, comparatively few cultures remained to be agglutinated. The macroscopic agglutination test was carried out with all remaining cultures by pipetting one cubic centimeter of the growth into ordinary test-tubes already containing one cubic centimeter of rabbit typhoid immune serum diluted to one in one thousand (liter, one in ten thousand to twenty thousand). If the tubes did not show an immediate clumping they were placed in the incubator for one to two hours when the readings were made.

While glucose litmus broth tubes were as a rule quite

satisfactory, they were not faultless; the litmus not infrequently broke down during sterilization and they failed to distinguish between paratyphoid and colon, and for our purpose this was equally important. Our problem was to find a medium which in one tube would enable us to distinguish between the alkali formers, typhoid, paratyphoid and colon and also leave the culture in a suitable condition for macroscopic agglutination. Typhoid in glucose media produces sufficient acid to differentiate it from the alkali formers, but fermentation tubes furnished no clue for the separation of paratyphoid and colon. It would be useless to substitute lactose for glucose, as we would then have no way of distinguishing between typhoid and the alkali formers (B. fecalis alkaligenes, etc.). Our problem was to find some carbohydrate in which typhoid produced enough acid to distinguish it from the alkali producers and which was fermented with the production of gas by colon and not by paratyphoid. A search through our records of fermentation experiments failed to show the existence of any such convenient carbohydrate. A mixture of the qualities of both glucose and lactose was needed and we set ourselves to find out the most suitable way of combining them. Neisser of Frankfort uses a glucose-lactose agar stab culture medium to distinguish between colon and paratyphoid; but in the manner used by him it does not permit the exclusion of alkali producers or leave the culture in a suitable form for immediate macroscopic agglutination in typhoid or paratyphoid immune serum.

We at first thought of using one-tenth of one per cent of glucose and one per cent of lactose in fermentation tubes and differentiating between colon and paracolon by the amount of gas; but as the volume of gas produced during the first twenty-four hours is subject to marked variation the idea was relinquished without trial. In the meantime it occurred to me that if solid media were used, two sugars might be used in one tube without mixing them, so for a time we used media prepared in this way. About five cubic centimeters of glucose litmus agar were put into each tube and after sterilization and cooling enough sterile lactose

litmus agar was poured in to make a good slant when the
tubes were incubated over night to develop any contamination.   The tube was inoculated by stroking the surface of
the slanted lactose agar and stabbing through the overlying
lactose into the glucose agar forming the butt of the tube.
When so inoculated with typhoid the tube has an appearance
after eight to eighteen hours incubation which is almost
specific for typhoid.   The sloped surface shows the usual
non-spreading, colorless growth of typhoid on a blue background of unchanged medium.   In the butt of the tube,
however, the medium is changed to a bright uniform red
color.   Only rarely do we find an organism from feces or
urine which gives rise to a similar appearance, and in these
instances the growth has failed to clump in typhoid immune
serum.   Further observations have usually shown such
organisms to be pigment producers or other non-pathogenic
organisms.

The practical use of this medium simplified the isolation
of typhoid and paratyphoid bacilli from the urine and feces.
Endo plates are made on the day the material is received;
these are fished on the second day and inoculated to double
sugar tubes, and on the third day all those showing the characteristic picture are agglutinated, after which the report can
be made.   Experience has shown that we rarely have pure
cultures to deal with when working from surface colonies on
Endo plates; consequently a series of agar plates are poured
from the double sugar tubes and pure cultures obtained,
which are proved out on all the usual cultural media.

After this double sugar tube had been in use some time it
was learned that it was not necessary to keep the two sugars
separated and at the present time the medium is prepared
as follows: enough five per cent aqueous solution of litmus
(three to five per cent) is added to plain agar (two or three
per cent), which usually has a reaction of about .8 per cent
acid to phenolphthalein, to give it a distinct purple violet
color, the amount of litmus depending on the original color
of the agar; dark requiring more than light, and the reaction
is then adjusted by adding sodium hydrate until the mixture

is neutral to litmus.   Next, and last, one per cent of lactose and one-tenth of one per cent of glucose dissolved in a small amount of hot water is added and the medium tubed for slants. The sterilization is done in the Arnold and because of the danger of breaking down the lactose must not be carried too far; if the tubes are packed loosely in the sterilizer basket to allow good circulation of the steam, ten minutes on the first and fifteen on the second day has been time enough. The tubes are then slanted and stored in small quantities in a dark place.

On this double sugar tube the typhoid bacillus gives, after an incubation period of from eight to eighteen hours, an extremely characteristic appearance; the surface growth is filiform and colorless on a blue background, the upper part of the tube is unchanged in color but the lower part, the butt, is a brilliant uniform red. ⎩The entire point of the medium rests upon the difference in the changes produced by the growth of the typhoid bacillus under aërobic and under the imperfect anaërobic conditions found in the butt of the tube, where the bacillus obtains its oxygen by breaking down the glucose with the liberation of considerable acid; on the surface, however, in the presence of free oxygen, no acid is formed.

The colon bacillus, which is often slow in producing acid on the Endo plate, shows abundant gas and acid formation on this medium. The tube is reddened throughout, both above and below, and since the abundant lactose is attacked equally with the glucose there is exuberant gas formation.

The bacillus fecalis alkaligenes and other alkali formers leave the medium unchanged or slightly bluer. The staphylococcus reddens the tube above but leaves it blue below; the streptococcus intestinalis, when it grows well, gives a beaded growth and reddens the tube slightly throughout. B. subtilis, which is commonly found in feces, usually leaves the medium unchanged but may redden it below without producing gas, yet the heavy, rough surface growth suffices for its differentiation. B. pyocyaneus gives a greenish blue surface growth and leaves the color of the medium unchanged. B. proteus produces small gas bubbles in the depth and reddens and then decolorizes the butt very early, while the upper part of the tube is unchanged except for the spreading surface growth.

All dysentery bacilli alter the medium in the same manner as typhoid, yet the quantity of acid produced is small and the reddening is usually confined to the line of inoculation. This reaction is so characteristic that we use this medium regularly in isolating dysentery bacilli; in fact, the same media and technic are used for both typhoid and dysentery.

The paratyphoids leave the upper part of the medium unchanged, the surface growth is like typhoid but in the butt of the tube in addition to the reddening are found a few small gas bubbles. The only organisms which may simulate the paratyphoids are slow colons and these must be thrown out by agglutination tests and further observation of the sugar tube.

We have, then, in this double sugar tube a medium on which typhoid, paratyphoid and dysentery give a characteristic appearance; a medium which enables one to throw out of consideration almost all colons and many other organisms which bear some resemblance to typhoid on the Endo plates, and which leaves the culture in a suitable condition for macroscopic agglutination on the second day following the receipt of the material. It has been in almost daily use for two years and in connection with our simplification of the Endo medium has made the examination of a large number of specimens of urine and feces a comparatively simple procedure.

One routine method, in short, is to make surface inoculations on Endo plates as the material is received; these are incubated over night and all suspicious colonies inoculated into both the butt and on the surface of double sugar tubes. On the second day macroscopic agglutinations are made from those showing the characteristic fermentation reactions, and the results reported.

BIBLIOGRAPHY.

1. Report on the Origin and Spread of Typhoid Fever in U.S. Military Camps, during the Spanish War of 1898. Government Printing Office, 1904.

2. Lentz, O. Med. Klinik. Berl., 1907, iii, 253.

3. Hiss and Zinsser. Text-book of Bacteriology, Appleton, N.Y., 1910.

4. Von Drigalski und Conradi. Ztschr. f. Hyg. u. Inf. Krankh., 1902, xxxix, 283.

5. Hesse. Ztschr. f. Hyg. u. Inf. Krankh., 1908, lviii, 441.

6. Loeffler, F. Deüt. Med. Wochsch., 1907, xxxiii, 1581.

7. Conradi, H. Munch. Med. Wochschr., 1906, December 4, 2386.

8. Centralbl. f. Bakt., 1904, xxxv, 109.

9. An organism (*pseudomonas protea*) isolated from water agglutinated by the serum of typhoid fever patients. W. H. Frost. Bull. No. 66, Hyg. Lab. U.S. Mar. Hospl. Serv., Washington.

THE ISOLATION OF BACILLUS TYPHOSUS FROM BUTTER.[*]

D. H. BERGEY, M.D.

(*Assistant Professor of Bacteriology, University of Pennsylvania.*)

(*From the Laboratory of Hygiene, University of Pennsylvania.*)

The isolation of Bacillus typhosus from food materials is frequently a more or less hopeless task, because of the presence of large numbers of other bacteria that thrive much more readily on the ordinary culture media than does the typhoid organism.

A variety of special media have been suggested to assist in the isolation of Bacillus typhosus, and amongst those that are regarded as helpful are von Drigalski-Conradi's litmus-nutrose-lactose agar,[1] Endo's fuchsin-sulphite agar,[2] and Loeffler's malachite green agar.[3]

In addition to these special media for the cultivation of Bacillus typhosus, several enriching methods have been suggested for the purpose of favoring the development of the typhoid organism and the exclusion of the other organisms, with subsequent cultivation on the special media just enumerated. The enriching media which have been found most helpful are Hoffmann and Ficker's,[4] and the use of bile as first suggested by Conradi.[5]

Recently some samples of butter were submitted for examination, because it seemed evident that it was the cause of a small epidemic of typhoid fever occurring in an institution. The method employed for the isolation of the typhoid bacillus consisted in transferring about five grams of butter by means of a sterile scalpel to a test-tube containing ordinary nutrient bouillon. After incubation for several days some of the bouillon was removed from the tubes and transferred to a plate into which had been poured some of the special agar media mentioned above, namely, the Drigalski-Conradi, the Endo, and the Loeffler agar media. The bacteria were distributed over the surface of the agar in the first plate by means of a sterile glass spreader, and from

[*] Received for publication July 5, 1911.

the first plate the bacteria were carried over on the spreader to a second and to a third plate. In this way each sample of butter was inoculated on a series of three plates of each of the three special agar media.

After incubation for from twenty-four to forty-eight hours suspicious colonies were fished out of these plates and transferred to ordinary agar slants and, at the same time, a tube of dextrose agar, liquefied and kept at 40° C., was also inoculated. From each colony two transplants were made, the one on the ordinary agar medium, and the other into the glucose agar. After incubation for twenty-four hours all the cultures that showed fermentation in the dextrose agar medium were discarded and only those that failed to show fermentation were kept for further study.

The first observation on the non-fermenting cultures was made to establish the general morphology of the organisms, and where this resembled that of the typhoid bacillus they were subjected to further study. Each of the cultures resembling the typhoid bacillus was inoculated into the Hiss serum-water media containing the following carbohydrates: lactose, sorbite, raffinose, dextrin, saccharose, dulcite, adonite, and inulin. Observation has shown that Bacillus typhosus produces a slight acidity in the lactose medium and a more definite acidity with coagulation in the sorbite medium, and leaves the other media unchanged. All the cultures isolated from the plates that gave a reaction in the Hiss serum-water media corresponding to that of Bacillus typhosus were then tested as to their agglutinability with the sera of rabbits immunized with Bacillus typhosus.

Altogether, eight samples of butter obtained from the institution were studied according to the method outlined, and typhoid bacilli were found in one of the samples.

The special agar media that seemed to be most helpful, in that it was possible to pick out the colonies of Bacillus typhosus from amongst the colonies of other bacteria in the butter, were the Drigalski-Conradi agar and the malachite green agar. No typhoid organisms were recovered from the plates made with the Endo medium. This result is probably due to the fact that the Endo medium may not have been properly prepared, though four different batches of it were tried. These special agar media are rather difficult to make up, as the original descriptions are somewhat involved and the process is rather long.

From the results obtained it is evident that these special media are of great value, under certain circumstances, for the recovery of Bacillus typhosus from infected food materials.

After the sample of butter which contained the typhoid bacilli had been in the ice-chest for several weeks it was impossible to isolate any additional typhoid bacilli from it. Whether this was due to the fact that the organisms had all died out or whether they were not uniformly distributed cannot be stated. Either of these possibilities may explain the negative results obtained with the other samples of butter.

REFERENCES.

1. Zeitsch. fur Hyg., xxxix, 283.
2. Centralblatt fur Bakter., orig., xxxv, 109.
3. Deutsche Med. Woch., 1906, 289.
4. Hygiene Rundschau, xiv, 1904, 1.
5. Deutsche Med. Woch., 1906, 58.

NOTE ON A PEPTID–SPLITTING ENZYME IN WOMAN'S MILK. [*]

Louis M. Warfield, A.B., M.D., Milwaukee, Wis.
(From the Pathological Laboratory of the Milwaukee County Hospital, Wauwatosa, Wis.)

During the course of some experiments with saliva and glycyltryptophan reported recently[1] the question arose whether or not any of the other secretions of body glands possessed a ferment capable of splitting the di-peptid. As it had been found that even the faintest acidity of the saliva was sufficient to inhibit or destroy the ferment, the next secretion to be examined was woman's milk, which is known to have an alkaline reaction.

Ten cubic centimeters of fresh breast milk, alkaline in reaction, were placed in a small bottle with two cubic centimeters of glycyltryptophan (prepared by Kalle & Co., Biebrich a/Rhein, and purchased through Arthur H. Thomas Company, Philadelphia, Pa.), one cubic centimeter of toluol floated on top, and the mixture set in a thermostat at 37° C. for twelve hours. It was then tested for tryptophan by adding a few drops of three per cent acetic acid to two to three cubic centimeters of the incubated mixture, then adding, drop by drop, a solution of bromine water. The presence of free tryptophan, indicating the splitting of the di-peptid, is shown by a rose or rose-lilac color which deepens on standing and turns yellow upon the addition of an excess of bromine water.

Tryptophan was present in the mixture of milk and glycyltryptophan. This ferment is destroyed by heating the milk to 75°–80° C. or above. When the specimen of milk is rendered faintly acid by the addition of either HNO_3 or HCl, there is no splitting of glycyltryptophan. Controls of untreated, normal milk and part of specimen treated, were always incubated for the same length of time. The controls were always positive for tryptophan.

It was thought that the addition of formaldehyde sufficient

---

[*] Received for publication July 5, 1911.

to inhibit bacterial growth might destroy the ferment. This was found not to be the case. The glycyltryptophan was split with as much readiness apparently as that with the control specimen.

It was thought that a Pasteurizing temperature, 165° F. (74.5° C.), would probably destroy the ferment. A specimen was divided into two parts, one part Pasteurized, the other used as a control. At the end of fourteen hours in the thermostat, the Pasteurized portion showed a faintly positive reaction with bromine water, while the control was strongly positive. Repeated trials gave always the same result.

Heating milk to 60° C. for one hour in a thermostat does not destroy the ferment.

The specimens of milk were obtained from several women, one of whom had a high degree of myocardial incompetency with dilatation, and who died three weeks after birth of her child. Her milk was scanty. The other women were healthy.

Apparently it made no difference, so far as the presence of this ferment was concerned, whether the milk was rich or poor in fats or solids, whether it was the first flow, the middle, or the last strippings. It appeared that as long as there was any milk secreted, the ferment was present.

A search through the literature has not revealed the record of any ferment like the one here described.

So far as we know, polypeptids are split only by proteolytic enzymes of which trypsin in the pancreatic juice is the best known example.

Austin[2] in a very careful piece of work on the proteolytic enzyme of woman's milk comes to the conclusion that there is no evidence of auto-digestion of human milk. It would appear that this enzyme, here described, has no action on the proteids of the milk for, as Austin shows, the rest nitrogen (that is the difference between the total nitrogen in milk and the proteid nitrogen) is not materially influenced by heating to 98° C., a temperature sufficient to destroy any known enzyme.

What the significance of this peptid-splitting ferment is, or what part it can possibly play in metabolism, are questions which, I frankly admit, I do not know, nor have I any theory to advance.

This brief report is merely for the purpose of calling attention to such a ferment in the hope that, in the hands of others, its place in the body economy may be found.

REFERENCES.
1. Bull. Johns Hopkins Hosp., 1911, xxii, 150–152.
2. Jour. Med. Research, Boston, 1908, xix (n. s. xiv), 309.

CARCINOMA INVOLVING THE ENTIRE KIDNEY.*

LINDSAY S. MILNE, M.D.

(From the Russell Sage Institute of Pathology.)

Renal tumors have for long presented a wide field of discussion and still are amongst the most difficult to understand.

Their numerous histological variations and the latitude of possibility of their embryological relationships has naturally created very varied and confusing classifications. In practically all works on the subject, sarcomata, carcinomata of the kidney, mixed tumors, hypernephromata, endotheliomata, cancers of the renal pelvis, etc., are described. Extreme differences in their relative frequency, however, are quoted, and only very rarely have the tumors in question been illustrated.

The present case is interesting as it varies from the more common types of renal tumors and presents more than usual difficulty in its classification.

It occurred in a somewhat emaciated woman, aged sixty-eight, about whom very little was known except that she had suffered for some time from considerable dyspnea and other symptoms of cardiac insufficiency. The urine had not been examined systematically but was reported on one examination as containing a trace of albumin and no blood. She died very soon after coming under observation. No abdominal growth was suspected.

At the autopsy the right kidney was found to be almost completely involved in tumor growth (Fig. 1). It was twelve centimeters long and seven centimeters wide and was regular in outline and very hard in consistence. The general shape and markings of the kidney were fairly well preserved and on section the organ presented a uniform white appearance. The capsule was considerably thickened and adherent to the surface. It was particularly thick towards the upper pole and the right suprarenal appeared to be involved and destroyed in this thickening.

The pelvis of the kidney was not dilated but was completely filled with firm white tumor tissue. The differentiation between the tissue in the pelvis and the renal structure was very ill defined although in part the outline of the calyces could be determined. There was no papillomatous appearance to this tissue occupying the pelvis of the kidney and no stones

---

* Received for publication July 2, 1911.

were found embedded in the tumor or in any part of the genito-urinary tract. With the exception of the upper inch the right ureter was not involved.

The tissues along the renal vessels were infiltrated, and the right renal artery and vein seemed to be almost completely obstructed by tumor growth. There was also considerable extension of the growth along the aorta and round the inferior vena cava. The lumen of the inferior vena cava between the entrance of the renal veins and almost to the hilus of the liver was very much reduced. There were several small, firm, whitish metastases in the liver, particularly towards its under surface and near the hilus. These secondary growths in the liver were mostly about one centimeter in diameter. No other metastatic foci were found in the bones or in any other organ.

The left kidney was ten centimeters long and six centimeters wide, and showed some acute degenerative change but only a very slight degree of chronic nephritis. The left suprarenal showed no special abnormality. There was no pyelitis or cystitis.

The heart was slightly dilated, but the myocardium, particularly of the left ventricle, was considerably atrophic. The tricuspid and mitral orifices were both slightly dilated. The aortic and mitral valve segments were somewhat thickened and the coronary arteries and aorta were markedly atheromatous.

Microscopically the entire kidney was involved in tumor growth, although in many parts glomeruli, and occasionally other portions of the kidney structure, persisted amongst the tumor and inflammatory tissue (Fig. 2). The glomeruli indeed seemed specially resistant, some appearing well preserved and others in all stages of fibrous occlusion. The tumor cells were chiefly of medium size, cubical in shape, and their nuclei relatively large and brightly staining. At first sight they seemed to be arranged in acini, yet on closer examination were found to be aggregated in small masses enclosed in narrow elongated spaces (Fig. 3). The younger tumor cells existed in small groups and were mostly cubical although many were somewhat spindle shaped. They seemed to be very actively multiplying and karyokinesis was common. They extended along the lymphatics and also apparently by the uriniferous tubules.

In the cortex, particularly, the tumor cells commonly assumed a very large size (Fig. 4). These also were sometimes multinucleated and often contained several small vacuoles closely resembling the bodies described by Russell.

Numerous evidences of phagocytosis were also found. These large cells were often irregular in shape and often very attenuated. Like the smaller cells they were chiefly in the lymphatics or secreting tubules.

Indeed in every direction the growth seemed to be advancing very rapidly. The tumor cells extending along a lymphatic or uriniferous tubule destroyed the lining cells and rapidly proliferated till they filled the space. Not uncommonly they became attached to the wall of the space and so produced a somewhat acinous arrangement.

Besides evidences of marked inflammatory reaction round the invading cells a curious reaction of the lining endothelium of the lymphatics was also common and accounted for the most perplexing appearances. In the advance of the tumor cells the endothelial cells seemed to become swollen and acinous like. Throughout the kidney no single focus could be found which was definitely suggestive of the original growth.

The tissue occupying the renal pelvis was composed, as was the kidney, of small masses of flattened somewhat irregular epithelial looking cells enclosed in a dense stroma of connective tissue. The walls of the renal vessels and vena cava were similarly involved (Fig. 5).

In the liver the metastases showed only a very slight degree of inflammatory reaction. The tumor cells were aggregated in long narrow masses which corresponded to the spaces once occupied by the liver cells. At the margins of the metastasis the tumor cells could be seen extending along the line of the trabeculæ and destroying the liver cells.

At the present time practically every tumor of the kidney is classed as a hypernephroma. Their relative frequency is certainly overestimated, but a large group of the renal tumors do conform to the type originally described by Grawitz. They are not diffusely infiltrating growths, although they may assume a large size and completely obliterate the kidney structure. Usually they are encapsulated and have a very destructive yellowish color. They are extremely subject to hemorrhage and degenerative changes and their extension is very commonly in the blood vessels. The histological type, also, is more or less characteristic, although it varies between tissue closely resembling the adrenal (Fig. 6) and spindle cells (Fig. 7). They have, however, as a rule, some sort of acinous arrangement. Their etiology has been very much disputed, as is the case with all renal growths. The majority of authors follow Grawitz in his belief that they arise from adrenal tissue included in the kidney. Others again, such as Wilson and Wills,[1] deny absolutely on histological and embryological grounds any such origin and consider them as nephromas. Sudeck[2] believed these Grawitzian tumors to be an atypical proliferation of renal tubules which was initiated by a process of repair. In only two of eight cases which occurred amongst the last seven hundred autopsies in this institute were there any chronic inflammatory changes in the kidney. A number of authors also have claimed important etiological relationships to included portions of the Wolffian body in the cortex or capsule of the kidney. It has also been shown by a number of authors that hypernephromata have no special chemical or histological peculiarities to adrenal tissue which might not equally be derived from kidney epithelium. However this may be, adrenal rests have been described in almost every portion of the genito-urinary tract, and in spite of what is said to the contrary there is at least a group which have both grossly and histologically very close resemblances to adenomata of the suprarenals themselves.

In two cases which came to autopsy in the Institute there were small hypernephromata in the renal cortex of the kidney and adenomatous enlargement of the suprarenals. In one case the suprarenal corresponding to the side on which a small hypernephroma was found in the kidney was enlarged to the size of a walnut. In the other case both suprarenals were greatly enlarged, left measuring six centimeters long, 3.5 centimeters broad, and 1.5 centimeters thick, and the right 5.5 centimeters long, 3.5 centimeters broad, and 1.5 centimeters thick. In the upper pole of the left kidney in this case were several small tumors of apparently the same structure as the suprarenal.

Even the smallest of these hypernephromata do not resemble renal tissue or the adenomata which so commonly are found in the cortex (Fig. 8).

Even admitting the marked variations found in hypernephromata the present case did not seem to fall into this category. The tumor was too diffusely infiltrating in the kidney and in the lymphatics along the renal vessels, its absence of primary focus, and its composition of large flattened irregularly arranged cells were all against its inclusion as a hypernephroma.

Its lack of any acinous arrangement to some extent was also against any growth from embryonic renal adenomata or even from adult renal tubules.

There was nothing in the growth to suggest a mixed embryonic tumor as described by Wilms.

In relation to endothelioma, however, there was some considerable difficulty. The tumor cells were very extensively invading along the lymphatics and the reactive processes of their endothelium produced what might at first sight be taken as positive evidences of the development of the tumor cells. (This reaction in the lymphatics has also frequently been described in the lymphatic glands in situations where no cancer existed,[3] and is considered due to a toxic process. In connection with cancer this appearance in the lymphatic glands is liable to be mistaken for a metastases.) These appearances were also to some extent reproduced by the tumor cells themselves, as they commonly seemed to become spread along the wall of the lymphatics or uriniferous tubules in which they happened to be. From this situation they proliferated rapidly and sometimes they formed very definite looking acini. Both the reaction of the lymphatic endothelium and the arrangement the tumor cells assumed in their extension produced an artificial acinous arrangement. Appearances at first sight, therefore, were very suggestive of endothelioma. The tumor cells, however, were much more irregular, more vesiculated and more obviously cancerous in type than is commonly noted in the endotheliomata.

The pelvis of the kidney was filled with tumor tissue in this case. There was no ulceration or papillomatous growth which might have been suggestive of this being the primary focus. Also no stones and no signs of pyelitis were found which might have proved an etiological factor. The kidney, however, was uniformly infiltrated by tumor cells which were rapidly extending in its lymphatics system and uriniferous tubules. The kidney was involved in an older inflammatory change than the invasion of the tumor cells, suggesting that the obstruction of the renal vessels had preceded the invasion of the kidney. The flattened irregular type of tumor cells as well as their arrangement and method of extension were more suggestive of pelvic epithelium than of any other possible source. If the pelvis there be considered as the most probable source of this growth two methods of origin have to be considered, from the pelvis proper or from some embryonic inclusion of a portion of the pelvis in the kidney.

Remnants of pelvis epithelium are described, included in the medulla of the kidney. Lubarsch,[4] for instance, described a nodule of pelvis epithelium in the medulla of a rabbit's kidney, and Ruckert[5] found small cysts lined by squamous epithelium in human hypoplastic kidneys. There was, however, no such focus discovered in this kidney from which the tumor might have disseminated or extended into the renal pelvis and in this way caused the widespread infiltration of the kidney.

Recently Beneke[6] has described a case which is of considerable interest in relation to tumors of the kidney. In this case the kidney was uniformly infiltrated by tumor cells without being markedly enlarged. The tumor cells in many places were also very large and irregular. There were metastases along the renal vessels and in the liver. In his case there was a history of trauma and at autopsy a cancerous ulceration of the pelvis and pyelitis was found. He also showed what has also been noted by Kaufmann and others, that tumors of the renal pelvis and also of the urinary bladder, although commonly papillomatous, may also be diffusely infiltrating.   He concludes that the tumor in his case was derived from the renal pelvis and that it was the consequence of regenerative processes.

The etiology of the present case also seems most likely to be related to the renal pelvis, and it certainly illustrates well the difficulty of classification of the renal tumors.

REFERENCES.

1. Journal of Med. Research, 1901, xxiv, 73.
2. Virch. Arch., cxxiii and cxxxvi.
3. Falkner. Cent. f. Gynäc., 1903, 50.
   Meyer. Zeit. f. Gynäc., 1903, 49.
   Brunet. Zeit. f Gynäc., 1906, 54 and 56.
   Setzenfrey. Zeit. f. Gynäc., 1906, 57.
   Kaufmann. Spec. Path., s. 985.
4. Lubarsch. Zent. f. Path. Anat., 1905, xvi, 344.
5. Ruckert. Verh. d Path. Ges., 1903, vi, 70.
6. Beneke. Virchow's Archiv., 1911, cciii, 463.

(Printed without author's corrections.)

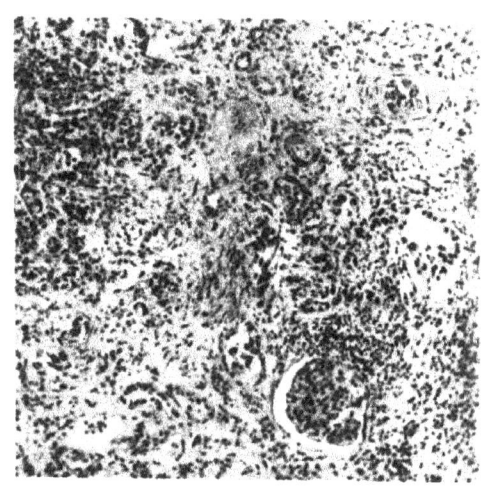

2

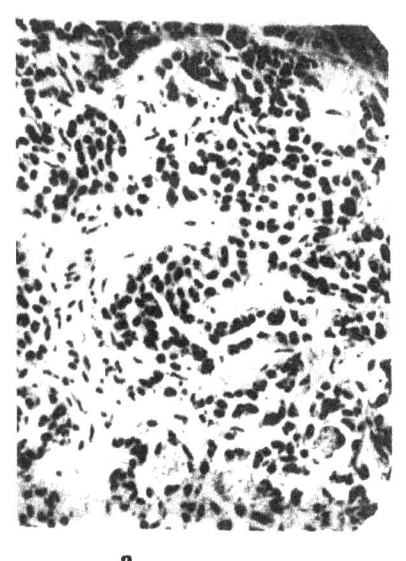

3

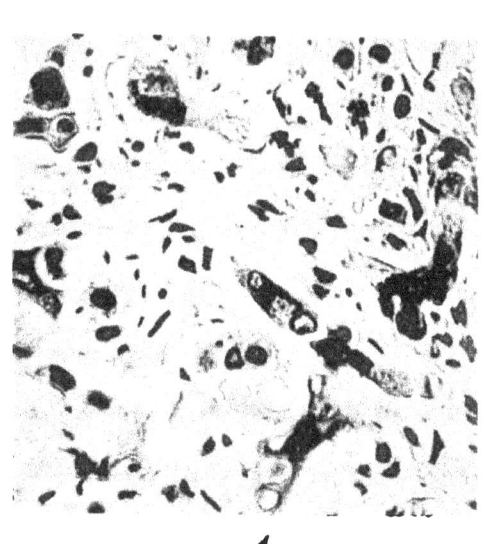

4

Carcinoma

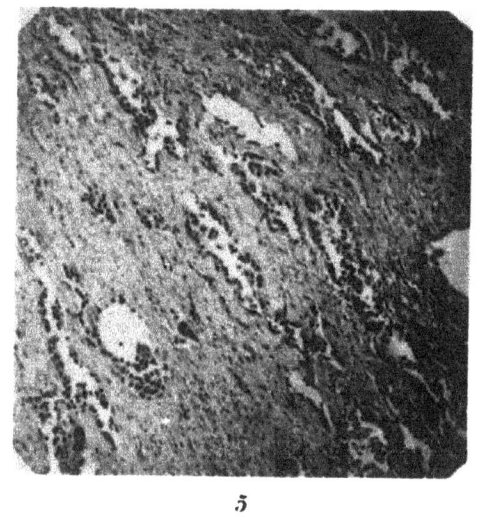

5

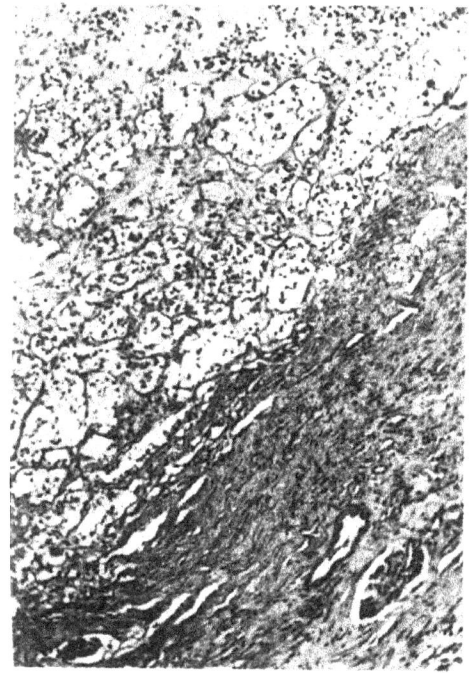

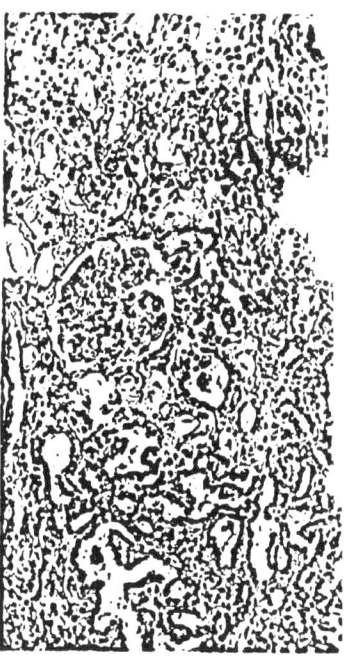

6

Milne

A STUDY OF A CASE OF THROMBO–ANGITIS OBLITERANS.*

HARLOW BROOKS, M.D.

(From the Medical Service of Montefiore Home, New York City.)

Notwithstanding the considerable number of cases of thrombo-angitis obliterans which have been collected, by Buerger (70) and others, there still remain many points for consideration in the study of the disease and especially concerning the manner in which the thrombosis is brought about. There can no longer exist any question as to the primary essential pathologic anatomy of the disease, nor of its definite differentiation from Reynaud's disease or from erythromelalgia, although the clinical distinction in many cases is by no means so clear as the pathological.

The important subject of ultimate etiology, however, still remains unsolved. Not only is this so, but also a logical explanation of the fact that the disease occurs with striking frequency among the Jewish people. This is, however, no invariable rule and two of my clinical cases have been respectively in a Bavarian and in a lady of Irish-American stock.

For this reason it still seems justifiable to report individual instances of the disease when accompanied by adequate pathological studies in the hope that, from a larger mass of more diversified observations, facts of direct importance regarding these mooted points may be obtained.

After having several of these cases in my services I have been impressed with the idea that the vascular lesions alone do not satisfactorily explain all of the manifestations present and I have felt, largely as a matter of clinical observation, that Buerger has perhaps underestimated the part which neuritic alterations take in the disease whether they be primary, secondary, causative or independent in origin. The degree and persistence of pain in the diseased extremities and the marked devitalization of the tissues have led the

* Received for publication July 19, 1911.

writer to question if in this disease where the pain is usually so excruciating — so much more so than in most instances of mere vascular obliteration as in phlebitis or endarteritis obliterans, other factors of importance might not enter, especially since the disorder develops almost exclusively in patients of a distinctly neurotic type.

This is further suggested by the striking character of the symptoms produced by the lesions. These are, in most instances, of a much more general nature than would be expected in a disease solely caused by cutting off of the circulation in a single vessel. The fact that anastomoses do not follow which relieve the obstructed circulation and especially the definitely inflammatory nature of the changes which are so different from simple thrombosis, are also points of great importance. Why should the disease be usually symmetrical on both sides if it be due to a purely local condition, and why should it spread above the line of amputation in so strikingly large a number of cases? If essentially a thrombotic process, why is this definite tendency to extend and spread so much more certainly exhibited than in phlebitis or simple thrombosis?

The writer is therefore anxious to report the following case of the disease from his service at the Montefiore Home, together with a study of the lesions found in the amputated extremity, some of which appear to have thus far escaped observation, and also to suggest an additional theory as possibly explanatory of some points of the question. Concerning the problem of why a primary thrombosis of these peripheral vessels takes place, the writer has as yet nothing to suggest, although he fully realizes that this is the crux of the problem.

It may not be out of place to review very briefly the three most probable theories explanatory of the pathological anatomy of the thrombo-angitis obliterans.

Von Winiwarter and Friedlander looked upon the disease as due to a primary endarteritis obliterans. This theory is disproved by the fact that many of the involved vessels do not show, at least to any marked degree, an obliterative endarteritis. Furthermore, most instances of thrombo-angitis differ quite materially clinically from the course of endarteritis obliterans. Again, the factor of age serves to partly differentiate in as much as thrombo-angitis is strikingly a disease of young and adult life, while endarteritis obliterans is commonly seen only in actual or comparative senility.

The theory of Weiss and von Manteufell is that the disease is caused by a primary endarteritis with the formation of a white descending thrombus with subsequent organization, arteritis and so on. This theory is overthrown by the fact that Buerger has shown that the disease does not extend from above, but that beginning below it progresses upward — a fact that is apparent to any close clinical observer quite independent of pathological study. Furthermore, Buerger has demonstrated that red and not white thrombus is the precursor of the process and that the lesion may be set up in trunks entirely devoid of endarteritic alterations.

Briefly expressed Buerger's well substantiated theory is that the process primarily originates with a thrombosis the tendency of which is to extend upward into the larger trunks, though it may at the same time also progress outward toward the periphery or to adjacent trunks. Secondary to this, organization of the thrombus occurs, perhaps canalization, and an arteritis, often followed by a periarteritis then develops.

Thus far the work of Buerger appears to be unassailable, and in so far as my study of Buerger's and my own specimens and cases goes nothing is to be added as to the arterial lesions.

Patient, M. P., age forty-seven years, Russian Hebrew, carpenter.

Family history: Father died of pneumonia, mother of asthma. One brother died of typhoid and a sister of asthma. No history of tuberculosis, cancer, diabetes, vascular or other similar disorders in family.

Previous history: Denies all children's diseases. Had pneumonia at twenty. No history of traumatism.

Habits: Used alcoholic beverages and tobacco in moderation. Sexually moderate; denies gonorrhea and syphilis. Appetite fair; bowels irregular. Lived under fair hygienic conditions.

Marital: Married at nineteen, wife nineteen. Had twelve children (two miscarriages). Nine children living and well, the others died from various children's diseases.

Present illness: Dates back eight years. First had sticking pains in right chest while at work, worse on bending forward and taking a deep breath. Consulted a physician who declared he had pleurisy. Was admitted to Mt. Sinai Hospital where his chest was tapped and sixteen ounces of clear fluid aspirated; at this time he was told that he had pulmonary tuberculosis and he was advised to go to the country. The second day after leaving the hospital (after two weeks' residence) he suddenly had an attack of hemoptysis, losing about four ounces of blood. Cough and expectoration were very severe, he suffered from night sweats and had fever in afternoons. Had a few attacks of hemoptysis after this, the last one occurring six years ago, since which time he has not coughed and his pulmonary symptoms have practically disappeared without treatment.

Went to work and gained in weight. About four years ago, patient noticed that both ankles (on dorsal surface) became swollen simultaneously, and he complained of burning pains and frigid sensations in the soles of feet. After resting, swelling and pain would disappear, but as soon as he would return to work they reappeared. Had no fever, no redness, and no discoloration of feet at the time. A few months later had pain in muscles of both legs and could not walk more than two blocks or so on account of this pain. Shortly after this he noticed small red blotches on both feet and hands, appearing first in one limb, and going to the others; was told he had phlebitis at the Presbyterian Hospital. At this time had pains in hands and feet, but no fever. The spots and pain left after a few weeks.

Two years ago he noticed that the big toe of right foot had become reddened and he had very severe pain in this member, preventing him from sleeping. Other toes normal. A couple of weeks later the toe cracked in center on plantar surface. The fissure became dark gradually and the pain was excruciating. The other toes were normal. He was admitted to N.Y. Hospital, where an amputation was performed at the middle of the right leg. Wound healed, after six days. At this time the left foot looked normal, but patient could not use his artificial leg owing to the pain and to a hernia.

About four months ago patient noticed that he began to have pain in the left foot, even while at rest; pain was continuous and very severe. Two months ago redness and swelling appeared over the inner border and tip of big toe. The adjoining toe also seemed affected and later the other three. The last cause no pain, but the inner two toes are very painful. The small ulceration noticed two weeks ago at the tip of big toe has become darker lately. Pain in " both feet " is very bad (patient still has a feeling as if the right foot were with him and he often wants to grasp same to stop his pain). When patient holds foot in a downward position, the entire dorsal portion feels numb and dead.

Upper extremities: Occasionally has numbness and slight pains in them; and also at times has hemicranial pains.

No polyemia, emaciation, excessive pallor, or other symptoms of diabetes.

About eight months ago he received an artificial limb, but cannot walk well (even with crutches) owing to pain.

Chief complaint: Severe pains in both lower extremities, especially in left big toe.

Examination: Adult male patient of good general nutrition, cannot walk without assistance of crutches owing to amputation of right leg and pain in left toe. Musculature very good. Adipose tissue, moderate. Mucous membranes, fair color, healthy. Lymph nodes, negative. Head of normal size and shape. Temporal arteries only slightly visible and normally tortuous. Pupils, equal, react to light and accommodation sluggishly. No occular palsies; nystagmus (lateral and vertical) present. Tongue, moist, coated, protrudes straight. Teeth in poor condition. Chest, well developed, symmetrical. Lungs, apparently normal. Heart, boundaries normal, no murmurs heard anywhere; the sounds are short and snappy, regular. Pulses, equal, fair size; tension +, some arterio-sclerosis; occasional drop-beat noticed — seventy-five per minute. Abdomen prominent, no abnormal masses: no tenderness, no rigidity.

Extremities: Upper. Hands warmer than arms or forearms; upon squeezing hands there is severe pain.

Lower, Right. Has been amputated at the middle of the leg; stump looks normal, but is very tender to touch. The fibula is at a slightly lower level than the tibia. On touching the stump or extending the leg the muscles go into spasmodic contractions which last a few seconds.

Left. Thigh and leg appear normal. The toes are all dusky red, especially the great one. The redness gradually diminishes but extends to about the middle of the foot. The tip of the big toe looks whitish, as though pus were present. At the inner end of the nail of big toe is a small ulcerated area one-half inch by one-eighth inch, dark brown in appearance and exceedingly tender to touch; this last is also true of the entire toe and to a lesser extent of the other toes as well. The toes are cold to touch but patient says they often feel very warm. No pulsation can be made out in the dorsalis pedis artery. Upon raising the extremity the toes become blanched in a short while, and if patient keeps them slightly raised he feels perhaps a little easier. If limb is extended pain is worse.

On inner portion of foot there is a small reddened and very painful area about one centimeter in diameter.

Blood pressure: Syst. one hundred and fifteen millimeters Hg.°.

Weight: One hundred and forty-five pounds.

Blood examination:

Hemoglobin,	90 per cent.
R.B.C.,	5,280,000.
W.B.C.,	14,400.

Differential count :

Baso.,	1.
Neutro.,	64.
Eosino.,	6.
Small lympho.,	5.
Large lympho.,	19.
Large mono.,	3.
Transitionals,	2.
	Sl. poikilocytosis.

Urine examination :

Quantity in 24 hours,	ss.
How passed,	Normal.
Color,	Clear amber.
Odor,	Foul.
Reaction,	Acid.
Sp. gr.,	1026.
Alb.,	None.
Sugar,	None.
Acetone,	None.
Diacetic acid,	None.
Indican,	Excess.
Amorphous,	Cylindroids.
Organic constituents,	Few.
Epithelia W.B.C.	

Sputum examination : T.b. negative.

Bed-side notes : March 4, 1911. Patient receives hot air baking of the left foot for twelve minutes, temperature up to 220°.

March 7, 1911. His pain is markedly lessened, but the appearance of the foot is about the same.

March 12, 1911. Dr. Beer removed some of the gangrene tissue at the inner tip of the big toe. Pain very great so that patient has to be kept under morphine.

The patient suffered severely for some two weeks after this history was taken. There was slow progress of the gangrenous area and he was finally persuaded to submit to surgical treatment. The operation was performed by visiting surgeon Goodman, assisted by the house staff. Amputation was done in the lower third of the thigh. It was noted at the time that the tissues were suspiciously dry and not well vitalized. The patient subsequently suffered great pain in the stump adjacent to the line of incision. Proper agglutination and union did not occur and the tissues, although handled aseptically and with the greatest surgical care, soon showed suppuration and necrosis. For a time it seemed certain that general infection had taken place and that death must occur, but the patient gradually rallied and finally made an excellent recovery. He is at the present time (July, 1911) in excellent general condition, is gaining weight, and the wound now appears healthy and is granulating up in a normal way.   The pains in the right lower and the upper extremities has entirely disappeared and the patient is grateful, cheerful, and happy.   In the light of what was found microscopically, I feel justified in assuming that infection of the wound probably took place from the severed arteries.

Immediately after the amputation, the leg was turned over to me through the courtesy of Dr. Goodman, although the case had been transferred to his service.   The following gross examination was then dictated, and the tissues were placed at once in suitable fixing solutions :

Gross examination of leg. — The end of great toe presents an area of dry gangrene measuring 3 x 2.5 centimeters.   The skin of remaining toes is glossy and white.   Over the os calcis, a little laterally and externally placed, is a cyanosed area.   The skin of the sole and that of the leg shows no gross alteration.

The general subcutaneous adipose is fairly abundant and shows nothing obviously abnormal.   That of the great toe is edematous and the gangrene is demonstrable for about one centimeter into the tissues.

The muscle is dry, looks like corned beef, and is of a singular brown color like cardiac muscle showing brown atrophy.

The entire anterior tibial arteries and their accompanying veins are blended into a hard cord-like mass from the middle third downward. The artery apparently contains a thrombus, but in the upper third the trunk appears to be at least partially open.   The anterior tibial nerve is about four times the usual size, is firm, hard, and apparently shows an interstitial neuritis.

The small arterioles throughout leading from deeper structures to the skin stand up prominently and are mostly obviously thickened.   The dorsalis pedis artery shows marked thickening and is apparently bloodless.   The muscles of the foot show the same peculiar color and texture noted above.

Microscopic examination. — Soleus muscle.   There is no macroscopic increase in the connective tissue of this muscle, but microscopically the connective tissue cells of the endomesium show proliferation.   The cross striation is well marked throughout and most of the fibers show longitudinal striation in addition.   In a few areas this longitudinal striation is extreme and each fiber is broken up into definite fasciculi.   This is especially well shown in transverse sections, where it is also demonstrated that swelling as well as atrophy of the fibers is sometimes present.   Many of the fibers are distinctly atrophied and a fine brown intra-cellular pigment is evident in places.   The most striking change shown, however, is an active proliferation of the nuclei of the muscle cells and some fibers show an almost continuous row of newly developed nuclei. Occasional fibers show a hyaline degeneration of very marked degree, and in the diffuse hyalin material resulting from the breaking down of the fiber are found masses or foci of proliferating nuclei.

The arterioles in the endomesium though generally somewhat thickened are not markedly so, and in so far as these vessels are concerned a microscopic arterio-sclerosis is not demonstrable, neither do the veins or capillaries show changes of allied nature.

Sections through the muscles of the anterior tibial group and of the small muscles of the foot. distant from the area of gangrene, show precisely similar changes in both degree and quantity.

Sections taken through the gangrenous area of the great toe show a sharp line of demarkation from the relatively uninvolved tissue. The necrotic areas show a remarkable degree of complete necrosis of all structures extending down to the periostium. Along the border, infiltration with poly-nuclear leucocytes is evident and a beginning necrosis is taking place, even of the elastic connective tissue associated with a hyperplasia of the endothelial cells lining the lymph spaces and channels. The cells of the sweat glands imbedded in the subcutaneous tissue show in many cases active hyper-plasia and some, but not all, of the blood vessels and nerve fibers show thickening of their supporting stroma. There is, however, no evidence of an extreme generalized arterio-sclerosis. The skin shows only the changes incident to inflammation and necrosis in the neighborhood of the necrosed areas.

Sections of the skin of the leg and foot, wide distant from the site of the gangrene, show no apparent change in the epidermal layers. The dermis shows in a few places circum-vascular and circumneural hyperplasia, but no general hyperplastic dermatitis. The subcutaneous fat is apparently not altered. The blood vessels of the true skin show universal though not extreme arterio-sclerosis and the veins are, for the greater part, relatively normal. Some of the nerve trunks, however, show an interstitial neuritis, very marked in some trunks and entirely absent in others. Paccinian bodies and touch corpuscles which chance to be included in sections from the sole of the foot show little apparent change, although possibly hyperplasia is indicated in the nuclei of some paccinian bodies by karyokinetic-like chromatic figures. Some of the sweat glands show proliferation of their cell nuclei.

Sections through the anterior and posterior tibial nerves and their larger branches show a marked degree of long standing hyperplasia of their supporting connective tissue. The sheath immediately surrounding each individual fasciculus shows evident thickening, apparently of more recent character, while extending into the bundle is a fine overgrowth of very delicate fibrils. Many of the nerve fibers show degeneration of the axis cylinder process and of the myeline sheath (Marchii method). The blood vessels throughout the nerve trunks show moderate arterio-sclerosis, which approaches the obliterating stage only in rare instances. In a few places tiny arterioles are found filled with blood clot and surrounded by a zone of tissue infiltrated by lymphocytes. The degree of nerve degeneration seems altogether too great to be accounted for by the changes in the endoneural blood vessels.

Sections through the obliterated arteries and their accompanying structures show the lumen of the artery to be filled by an organized mass of highly vascularized connective tissue containing well formed veins, arteries, lymph spaces and apparently in some instances minute nerve fibers. The internal elastic membrane is in many areas intact and in some the lumen of the vessel has obviously been filled primarily by a blood clot which has, however, in most instances been completely replaced by fibrous invasion which has in some patches passed on to adult cicatrization. In the more recently diseased trunks the walls are invaded by a dense mass of lymphocytes and hyperplastic connective tissue cells which has completely destroyed in some levels every resemblance to an artery. A few stretches show a tunnelization of the old lumen apparently by a new-formed large-sized blood channel.

At some levels an active perivascular inflammation has taken place with necrosis both of the arterial walls and of the surrounding tissues and a diffuse suppurative necrosis has followed. In yet other areas, associated with the intra-vascular changes and the resulting fibrosis are found ovoid collections of giant cells mostly with centrally situated nuclei.

In all of the sections, in so far as the anatomical structure goes, the appearance is that of an inflammatory and hyper-plastic process which has originated within the arteries and has passed from them outwards and not changes such as would have originated without and proceeded inward. The veins appear to be involved secondarily only and no evi-dences of long standing phlebitis are present. In the group of small arterioles and venules surrounding the central dis-eased vessel are many which show no pathological changes whatever. Most show arterio-sclerosis to a greater or less degree, and occasionally tiny trunks are seen in which changes analogous to those described in the larger trunks are taking place. In some, though rarely, a chronic endarte-ritis approximating obliterans is demonstrable. Most of the nerve trunks included in these groups show a thickening of the epineurium. A few present fibroid invasion of the fiber and many show axis cylinder degeneration.

Remarks. — This process is evidently not arterio-sclerosis and especially not of the types commonly known as arterio-capillary fibrosis or endarteritis obliterans. The production of connective tissue in and about the walls of the original blood vessels is insufficient for this condition. The process does not appear to have originated in the walls of even the arteries most diseased, but rather from their lumen. This conclusion is further justified by the fact that some branches of the main thrombosed vessel, even such as are of consider-able size, show no changes of moment.

Notwithstanding the fact that gross examination indicated a marked degree of interstitial neuritis, this process is not so marked as one would expect from gross inspection, and the neuritis appears to be essentially parenchymatous in nature.

The number of degenerated fibers in the anterior tibial nerve is far greater than could be accounted for by the mere upward extension of fibers immediately diseased about the thrombosed vessels; it must therefore be assumed that a distinctive neuritis is also present.

The lack of marked alterations, aside from those of moderate thickening in the walls of the veins, suggest that the disease is primarily one of the arteries rather than of the veins.

The atrophic changes in the voluntary muscles, apparently hitherto undescribed, are probably secondarily due to the neuritis.

To recapitulate briefly. The lesion found is a thrombo-angitis obliterans, a process extending upward and corresponding pathologically to the lesions described by Buerger. In addition, in this instance, there appear an interstitial and parenchymatous neuritis of more marked degree than reported by him. Alterations, probably secondary in origin, also exist in the voluntary muscles of the amputated extremity. These changes are definitely of an atrophic nature, and judging from analogy are probably secondary to the changes in the nerve trunks and not to vascular disease.

In the last two particulars, at least in apparent grade, this case definitely differs from those reported by Buerger, but that these changes might have been present in his cases does not seem to be excluded by his as yet published reports.

As a result of my study of this and other cases, especially in comparison with specimens and cases of Reynaud's disease and intermittent claudication, for all instances of claudication are by no means to be included under Buerger's disease, I have been led to believe that the change, notwithstanding its distinctly inflammatory and thrombotic lesions, is also largely trophic in origin. This theory is borne out by the points already suggested in the introduction to this study and by the very striking and general neuritic and atrophic alterations found in this particular case and probably also present in many other examples of this disease. The character of the lesions and the subjects in which the disease occurs and the symptomology are all strongly corroborative of this theory.

[I wish to express my thanks to Dr. Felberbaum of the Pathological Laboratory for technical assistance in this study.]

THE VALUE OF THE " HORMONE " THEORY OF THE CAUSATION OF NEW GROWTH.[*]

I. LEVIN AND M. J. SITTENFIELD.

(From the Department of Pathology of Columbia University, College of Physicians and Surgeons, New York.)

Pawlow and more recently Starling have shown that the secretion of the juice of a digestive gland is not always caused by a nervous reflex, but by a complex chemical mechanism. Starling, for instance, assumes the following steps in gastric digestion: The first products of. digestion act on the pyloric mucous membrane, and produce in this membrane a substance which is absorbed in the blood stream, and carried to all the glands of the stomach, in which it acts as a specific of their secretory activity. Such a substance Starling calls a gastric secretin, or gastric " hormone." Such chemical messengers or hormones, *i.e.*, chemical agents exciting certain specific activities, Starling considers of great importance in functions of the organism.

There is a tendency in modern writings on experimental cancer problems to apply Starling's theories to the explanation of the qualitative and quantitative difference in the power of growth of the cells of malignant tumors. Ehrlich[1] thinks that Starling's experiments may indicate that there are substances circulating in the organism which may stimulate the body cells to resist the athreptic influence of cancer cells. Bashford, Murray and Haaland[2] think that the biological difference between normal cells of the mamma and those of an adeno-carcinoma of the same organ may be due to the qualitative differences in the hormones. Askanazy[3] believes in accordance with Starling's conceptions that certain hyperplasias developing in genital organs subsequently to the formation of tumors in the ovary, testis or pineal gland may be due to the influence of embryonal tissue formed by the tumor. R. Frank[4] indicated in a

[*]Conducted at the expense of the George Crocker Special Research Fund, Received for publication July 25, 1911.

recent study that certain experimental proof adduced by Starling in support of his theory may not have been correctly interpreted.

The hormone theory is apparently applicable to many phenomena of the functions of glands of external and internal secretions. It would seem, however, a priori that its value for the interpretation of the phenomena of growth of tumor cells could be only very indirect and general.

There appeared recently a short publication by A. S. Grünbaum and H. G. Grünbaum[b] in which the authors attempt to prove experimentally the direct interdependence between growth of a tumor and abnormalities in the function of internal secretion of glands. They presume a priori the possibility that an excess of hormone in the organism, together with a lesion or irritation of the tissue complementary to the hormone, might cause unlimited growth. Their experiments consisted in a simultaneous inoculation of tissue of normal parotid gland with tumor tissue into an animal which was previously shown to be immune against this tumor. Another set of experiments consisted in the removal of the parotid gland of a tumor-bearing animal. The first experiment was done on eleven immune animals. In every animal there appeared a small nodule, of which the largest was .3 centimeter in diameter. This remained for about twenty days and was then absorbed. Their conclusion is, that the parotid gland is able to assist growth of sarcoma in immune rats. The removal of the parotid was done on three animals. In two, the tumor did not diminish in size, but fatty and fibrous changes were noticed. In the third the tumor regressed from dimensions 2 x 1.5 centimeters to 1 x .6 centimeter. They conclude that the removal of the gland caused some change in the growth.

The tumors employed by the Grünbaums were Bashford's and Ehrlich's sarcoma of the rat. The writers had opportunity to inoculate Ehrlich's rat sarcoma for various purposes in hundreds of animals. Judging from experience thus gained, the conclusion must be drawn that nodules of which the largest is .3 centimeter in diameter, and which subsequently absorb, are not true tumor growths but occur frequently after an inoculation of tumor tissue in an immune animal.   For a while there may even be observed in such a nodule proliferation of the tumor cells, but they ultimately become necrotic and absorb.   Fatty degeneration and fibrous and necrotic areas are observed frequently in large tumors of animals which received no previous treatment.   Recession of a large sarcoma within a month to half the former size is not very common.   Still the occurrence of this phenomenon in one animal is not sufficient to show the influence of the removal of the thyroid gland.

In order to gain a personal impression of the possible influence of parotid tissue on the growth of sarcoma in a resistant animal, the writers repeated the first series of experiments of the Grünbaums.   A simultaneous inoculation of parotid gland tissue and sarcoma was made in twenty immune rats.   No animal developed an actual growth of tumor.   The small nodules observed in seven of the animals did not appear different either on gross or microscopic examination from the nodules which develop occasionally in an immune animal after inoculation of sarcoma tissue.   It would appear as a result of these experiments that the Grünbaums did not give the correct interpretation to their results.

BIBLIOGRAPHY.

1. P. Ehrlich. Beiträge z. Exp. Pathol. und Chemoth., Leipzig, 1909.
2. Bashford, Murray and Haaland.   Third Scientif. Rep. Imperial Cancer Research Fund, London, 1908, 356.
3. Askanazy. Zeitschr. f. Krebsf, 1910, xi, 397.
4. Frank and Unger.   Arch. f. Internal Medic., 1911, vii, 812.
5. A. S. Grünbaum and H. G. Grünbaum.   Journ. of Pathol. and Bact., 1911, xv, 289.

THE

Journal of Medical Research.

(NEW SERIES, VOLUME XX.)

VOL. XXV., No. 2.   DECEMBER, 1911.   Whole No. 128.

PURE CULTURES OF AMEBÆ PARASITIC IN MAMMALS.[*]

ANNA W. WILLIAMS.

(*From the Research Laboratory, Department of Health, New York City. Dr. Wm. H. Park, Director.*)

Introduction. — In February of this year the writer published the results of growing on brain-streaked agar a pure strain of an ameba isolated in pure mixed culture from a case of human amebic dysentery.[1]   The present paper is a report of the continuation of the study of pure cultures of amebæ.   It presents the results of growing in pure cultures several strains of amebæ obtained from the intestines of different mammals.   The culture media used were crushed brain, liver, kidney, and other tissues of rabbit or guinea-pig, streaked on nutrient agar.

It seems desirable, since there is some ambiguity in the statements of several of the writers on the subject of growing amebæ in pure cultures, to give first a short critical review of the previous reports.

Critical review. — We find in the literature three meanings to the term " pure culture " as applied to amebæ.   First, a culture of amebæ is called pure when it is grown free from other protozoa, though any other living microörganisms, such as bacteria and yeasts, may be present.   This meaning is given more frequently in the older reports (Cunningham,[2] 1881 ; Celli and Fiocca,[3] 1894, and others) ; still, even at present, some writers continue to use the term in this, to say the least, non-bacteriologic sense (Saul,[4] 1904 ; Thomas,[5]

[*] Received for publication Aug. 8, 1911.

1906, etc.). Second, " pure mixed " cultures, that is, cultures of a single strain of ameba with a single species of
bacterium, are frequently and wrongly called pure cultures.
Third, a single species of ameba growing in successive culture generations without the presence of other living organisms is rightly called a pure culture ; and, as it is in this sense
that the term pure culture is used in this report, the work of
only those investigators who have given this meaning to the
term will be reviewed.

Amebæ have been cultivated in mixed cultures more or
less successfully since 1856 (Auerbach [6]). Much later
(1898) successful " pure mixed " cultures were made by
Frosch [7] and by many following him. Full bibliographies
have been given by Musgrave and Clegg,[8] Walker [9] and
others on parasitic amebæ and their saprophytic relatives,
and very large bibliographies they make ; but in the whole
list are found only four writers who claim to have obtained
amebæ growing in strictly pure cultures.

1.   Kartulis [10] in 1893 said that he obtained by growing
in sterile hay infusion a pure culture of ameba from a bacteria-
free liver abscess ; but he did not mention culture generations, and in 1900 in an article on tropical dysentery [11] he
ignores his earlier statement and says, " Parasitic amebæ are
seldom cultivated and then in impure cultures." He is still
quoted, however, as obtaining pure cultures of amebæ.

2.   Casagrandi and Barbagallo [12] in 1897 said that they
obtained once a strictly pure culture of ameba (Ameba
spinosa) a saprophyte, on alkaline, five per cent fulcus cris-
pus medium ; but these authors also say nothing about
culture generations, and nothing further about technic.

3.   Tsugitani [13] in 1898 tried to grow three strains of
saprophytic amebæ with several species of killed cultures
of bacteria and he found that one strain grew with one
species of dead bacteria. Nobody has been able to corrobo-
rate his work, though many others have tried to grow
amebæ with killed cultures of bacteria. He also says nothing
about culture generations. It seems, however, in the light
of some work Miss J. T. Chase and the writer have been

doing with killed cultures as food that Tsugitani probably obtained in this way a continued growth of his ameba. We have grown one strain of ameba for ten culture generations on agar cultures of dysentery bacilli killed by heating in normal salt solution for two hours at 60° C. Though we obtained undoubted growth, it had to be helped along, and was, at best, a very slow and scanty one.

We see from the above review that no one has reported a continued growth in successive culture generations of pure cultures of amebæ. While, on the other hand, Musgrave and Clegg,[14] 1906; Walker,[9] 1908; Craig,[15] 1909, and the majority of investigators who have tried to grow pure cultures of amebæ all make emphatic statements as to the impossibility of obtaining such cultures by any methods tried.

The conclusions of Musgrave and Clegg[14] are interesting as showing how near these authors came to the using of tissue cells as a medium. They say: "Pure cultures of amebæ which will continue to propagate in media free from other living microörganisms have not been obtained. Amebæ from liver abscesses and the intestines would not thrive in various filtered extracts from human and animal livers although these proved excellent culture media for bacteria. Liver and other animal tissue extracts obtained in a sterile manner furnished no better results than the filtered products. The work of all recent authors as well as our own seems to point to the impossibility of such a procedure."

Walker[9] says: "These cultures of bacteria-free amebæ . . . grew when transplanted on fresh media only when supplied with a living culture of bacteria." Again: "Amebæ will not grow apart from bacteria or other microörganisms upon which they probably feed instead of obtaining their nourishment directly from the culture medium. Therefore, pure cultures in the bacteriological sense are not possible."

In looking carefully over the literature the writer has found that no one apparently has tried crushed tissue cells as a medium; though, considering the number of bacteria-free liver, brain, and lung abscesses that have been reported it seems rather strange that the use of such a medium has not occurred before to some worker.

It is true that a number of writers have recognized the probability of amebæ using tissue cells in some way as food. Thus, Councilman and Lafleur [16] (1891) say: " No microorganisms other than amebæ were found in the smallest and most recent abscesses . . . and even in the large the bacteria which were found did not seem to have any causal connection with the absccss.   None of the organisms which we are accustomed to consider pus organisms were found. The lesions were of a different character from those produced by bacteria."

Kruse and Pasquale [17] (1894) say: " We are not of the opinion of Kartulis that the amebæ are simply carriers of the pus forming bacteria into the liver, but we believe that, as in the ulcerated intestinal wall, the amebæ have a direct action in the degenerative process.   On the other hand, we do not believe with Councilman and Lafleur that the existence of bacteria in liver abscesses and intestinal ulcers is a non-specific complication or an entirely unrelated phenomenon; because the presence of bacteria is far too constant. . . . Even in ' sterile abscesses,' there are probably a few bacteria."

Musgrave and Clegg [14] (1906) say: " In all intestinal lesions and in many liver and lung abscesses there is present, in addition to amebæ, one or more varieties of bacteria, and cultural experiments demonstrate that under these circumstances a definite bacterial symbiosis often exists between the amebæ and some of the organisms which are present, whereas at the same time the ameba exists in partial tissue symbiosis.   In other instances, even in mixed amebic and bacterial lesions, the association of the bacteria seems no longer to be necessary for the life and action of the amebæ, for no symbiosis with any of the bacteria which are present can be found by cultural methods.   It is probable that, in these cases, the amebæ have taken on a symbiosis with the animal tissues; in other words, have become true parasites and that the presence of the bacteria is without significance."

Lesage[18] also came close to the point of using tissue cells as a culture medium. He recommends his medium especially for the isolation and growth of " Entameba histolytica " from abscesses of liver or intestines in amebic dysentery. He obtained a leucocytic exudate (that from the guinea-pig best), placed it in the ice-box for one day and then centrifugalized. The supernatant liquid was the culture medium. In this medium the whole contents of a liver abscess were placed. If intestinal contents were used they were first put into the peritoneal cavity of the guinea-pig to lessen the number of bacteria. The amebæ developing in this culture medium, the author says, have all the characteristics of Entameba histolytica as described by Schaudinn. The cysts in pus from old liver abscesses opened in this medium after two to three days at room temperature. The liberated amebæ developed into adult amebæ which after a short period produced new cysts. Lesage says nothing about culture generations, neither does he describe his amebæ minutely, so we can only infer that he obtained the usual one cycle of development which is often obtained with encysted amebæ on many kinds of culture media. His medium was not bacteria free.

Original work. — In a problem of this kind where the claim put forward has been disallowed so strongly by so many good workers and where error undetected may so readily creep in, it seems necessary to give a more minute description than usual of the whole work.

First in regard to the strains employed — 1. Ameba coli. This culture was obtained through Dr. G. N. Calkins in 1906, who received it the year before directly from Musgrave's Laboratory. It is the one Musgrave and Clegg labelled 11524 and with which they performed many of their experiments. At the time it came into the writer's possession it had been in a sealed agar plate for a year after growth with cholera bacillus. After moistening this dried culture with sterile normal sodium chloride solution, transplants on protozan agar were made from it on which after four days a certain number of cysts opened and some motile amebæ appeared. The culture used in this work is the one hundredth culture generation from a single ameba isolated from one of the early transplants. The original culture was obtained by Musgrave and Clegg from the feces of a case of amebic dysentery in man. Whether it is Entameba histolytica or whether it was pathogenic is not definitely proved, though Musgrave and Clegg's experiments are certainly strong evidence in favor of its pathogenicity.

2. Ameba limax. This ameba was isolated from potato parings and tap water, in 1906, by M. A. Wilson of this laboratory. It was identified as one of the limax group by Dr. Calkins, who studied some phases of its life cycle with the writer.[19] It had been carried on with a yellow air bacillus and with B. dysenteriæ (Shiga type) since its isolation and was in its one hundred and tenth culture generation at the time of these experiments.

3. Two strains of tiny amebæ, size one to twelve microns, cysts two microns, isolated several times in succession in 1907 from feces of two cats respectively. They grow very rapidly, use up the bacteria (coli-typhoid group), and encyst quickly. They have to be transplanted every two days at room temperature if motile amebæ are required. These organisms are interesting because their cysts are formed exogeneously, as Entameba histolytica is said to form its cysts.

4. Ameba intestinalis sent the writer by Walker in 1906; described by him as a new species[9]; carried on by writer with B. dysenteriæ; in its ninety-sixth cultural generation at the time of these experiments.

5. Ameba cobayæ sent by Walker in 1906; described by him as a new species;[9] in ninetieth cultural generation at beginning of this work.

6. One strain of ameba from dog similar to Ameba intestinalis.

All of these were grown, as routine, with the dysentery

bacillus (Shiga) on nutrient agar one per cent acid to phenolphthalein, and were transplanted every seven to thirty
days.   All of the above strains with the exception of Ameba
limax were presumably parasitic, in a certain sense at least,
in the animals from which they were isolated.   Whether
any of them were capable of becoming tissue parasites is a
question which the present work may help to clear up.
This will be discussed later.

Obtaining amebæ free from other living organisms. —
Two methods have been employed, differing only in detail
from those successfully used by other investigators.
The first method, that of transferring motile amebæ from the
zone beyond the growth of the bacteria, was reported in the   ✓
first paper.   The second method was used with those strains
of ameba that digested the bacteria quickly, that is, with
strains numbered 3, 4, 5, and 6.   These strains under certain conditions in from one to two weeks grown on nutrient
agar apparently digest all of the accompanying bacteria
(B. dysenteriæ) and remain finally a mass of cysts.   These
cysts, transplanted on nutrient agar without bacteria, remain
in large part encysted.   A few cysts may open, but the
freed ameba soon encysts again unless suitable food is within
reach.   In order to be sure that no bacteria have been
carried over with the amebæ the new transplants are allowed
to remain, some at 36° C. and others at room temperature
for twenty-four to forty-eight hours.   If no bacteria show
themselves in this time it is reasonably certain that none
have been carried over.   The control plates are allowed to
remain longer.   The nutrient agar (one per cent acid to
phenolphthalein) used in most of the work allows contaminations readily to show themselves.   The first method, that of
transferring motile amebæ, is to be recommended in most
instances.

Obtaining sterile tissues for food. — Several different
organs of guinea-pigs and rabbits and the brains of dogs
have been used.   The animals are killed by chloroform.

Their bodies are soaked for ten minutes to half an hour in five per cent carbolic acid. The organs are removed as quickly as possible in a room practically bacteria-free. Different sterile instruments, well sharpened, are used for each slip. The abdomen is opened, the periphery of the liver is cut off and placed on a freshly made agar plate; then the spleen, the kidneys, and any other tissue needed are rapidly cut out and put each on a separate plate, and lastly, the cerebrum is removed in the same manner. The tissues are now cut into tiny pieces with sharp sterile scissors, and are taken up with sterile forceps and spread over several freshly-made agar plates. These plates are then placed in the thermostat (36° C.) for twenty-four hours. With all these precautions for insuring sterility there is no need of washing the freshly-removed tissues first in five per cent carbolic acid, as was recommended in the first paper. Emulsions of liver and brain in sterile neutral glycerine have given a moderately good result any time up to three weeks. They have not been tested later, neither have they been tried for more than three successive culture generations. These glycerinated emulsions of tissue were always allowed to remain for twenty-four hours in the thermostat spread on freshly-made agar plates, before being planted with the amebæ. This allowed the excess of glycerine to be absorbed into the agar.

Growth on sterile tissue media. — The broken up tissue, fresh or glycerinated, after being in the thermostat for twenty-four hours is ready to add to the bacteria-free amebæ. A small quantity of the tissue is gently mixed up with the amebæ on each agar plate, and the plates are placed some at 36° C. and others at 20–24° C. (room temperature). If the thermostat cultures show slight or no growth in twenty-four hours, a loop or two of sterile 0.8 per cent sodium chloride solution may be added.

Usually in twenty-four hours at 36° C. or in three to five days at room temperature, with certain strains of amebæ, there is an abundant growth on brain, liver, and kidney tissue prepared in this way.

In order to rule out the presence of contaminating organisms, especially of those difficult to demonstrate or of those growing very slowly, many control plates were made, at first with each transfer, later at less frequent intervals.   The controls spoken of in the first paper — blood, broth, normal salt solution, plain agar — have been used, and others also, such as various crushed tissues which gave no growth with the amebæ.   Then, as an added control — and a very important one — three sets of spreads have been made at practically each transfer, one stained by Giemsa's method, one by Heiderheim's, and the other with fuchsin-methylene blue mixture.

"Ameba coli," Ameba intestinalis, Ameba cobayæ, and the strain from the dog all grow on normal brain tissue with the greatest ease.   Ameba coli is now in its fifty-fourth culture generation on brain tissue, transplanted every two to four days at 36° C., since December 31, 1910, and in the twentieth culture generation at room temperature transplanted every one to four weeks.   Ameba intestinalis, started later, is in its thirtieth culture generation at 36° C. Ameba cobayæ was carried on for fifteen culture generations, and the strain from dog's intestines for ten culture generations.   Ameba coli is in its thirty-fifth culture generation on kidney and in its thirtieth culture generation on liver; Ameba intestinalis is in its sixteenth culture generation on liver and on kidney, both at 36° C.   The attempts to grow these two strains with lymph nodes, suprarenals and spleen have so far given negative results, but the attempts have not been many.   All of the many efforts to grow Ameba limax and the two tiny amebæ on crushed brain tissue have given negative results.   No attempt has been made to grow them on other tissue.   Ameba coli has been grown successfully on dog's brains for five culture generations.

The points which have been considered in the study of pure cultures of amebæ on tissue media are the following:

(1) Temperature, (2) rate of growth, (3) morphology and life cycle, (4) pathogenesis, (5) isolation from cases

of human dysentery, (6) action of amebæ on diseased tissues in which no specific microörganisms have been positively recognized.

1. Temperature. — Most students of parasitic amebæ state that these organisms grow best at room temperature. The truth of this statement depends upon the definition of the word best. Walker [9] says: "Cultures of parasitic and of free-living amebæ will grow on artificial media at room temperature or at 37° C., but they maintain their growth and vitality longer at room temperature." He does not call attention to the fact, however, that they grow much more slowly at room temperature than at 37° C. At the latter temperature, according to the writer's experience, all amebæ with proper food grow very rapidly and abundantly, therefore, if rapidity and abundance are the qualities which constitute goodness in a culture, growth for these strains is best at thermostat temperature.

The pure tissue cultures grow even more quickly and abundantly than the pure-mixed cultures at 36° C. Cultures at this temperature must, of course, be transferred more frequently than at room temperature, usually every two to four days, while the room temperature cultures need not be transferred under a month. The thermostat cultures, it is true, encyst sooner than the room cultures and the individual transplant loses its vitality sooner, but frequent new transplants at 36° C. show no loss of vigor, on the contrary, they seem to have increased ability to multiply rapidly. Lasting cysts are better produced at the lower temperatures (20–25° C.) on thinly poured, slightly nutritious culture media that quickly dries after cysts are formed.

Only the two temperatures, 36° C. and about 22° C., have been tried by the writer. It is possible that 30° C., or thereabouts, might prove better for maintaining most characteristics of amebæ.

2. Rate of growth in cultures. — This varies with temperature, condition and kind of tissue media, age of

transplant, stage of development, species, and adaptation. The effect of temperature has just been noted. The age of transplant and stage of development which usually go hand in hand have a good deal to do with rate of growth at the beginning. Thus, an encysted culture will develop in a period increasing with the age of the cysts, while transplants of motile amebæ on fresh tissue media are checked but slightly in their growth, especially if many be transplanted, *i.e.*, a small platinum loopful. Occasionally, however, for unknown reasons, the new transplants encyst, if motile, or remain encysted unless they are stirred or after twenty-four hours a loop of normal sodium chloride is added to the culture.

Growth is somewhat more rapid and abundant on tissue freshly isolated, or kept in ice-box for a short time than when kept in the thermostat first for twenty-four hours. If perfectly fresh tissue is used it must be most carefully controlled for the presence of extraneous organisms. On the whole, these amebæ grow more quickly and abundantly on brain media than on kidney or liver media, and they grow better on kidney than on liver media. Strains freshly isolated from pure-mixed cultures with bacteria do not grow as well as after several transplants.

3. Morphology and life cycle. — So far only a few interesting points concerning this topic have been noted; interesting, because they contradict several statements given by various authors in regard to the differential diagnosis of pathogenic and non-pathogenic amebæ in human beings.

The observations on only two of the strains of amebæ — Ameba coli from human amebic dysentery and Ameba intestinalis from normal dog's intestines — will be reported at present.

Before beginning the report of the personal study, the question of the effect on morphology and life cycle of amebæ when grown on artificial culture media should be considered. The general criticism is that "artificial

cultures" grow atypically, but just the nature and importance of this "atypical" growth in any instance has not
been made clear. Granted that in an occasional individual
case fundamental changes may occur on artificial media, in
general, changes under these conditions must be quite superficial or we should have to change our ideas in regard to the
frequency of mutation. Such changes must be only transitory variations, and species characteristics should be looked
for in those appearances which are similar under several
conditions of environment. For instance, in amebæ the
pseudopods are known to be decidedly variable within
limits, therefore until we know by sufficient controls what
these variations are we should not use size and shape of
pseudopods as species characteristics. The same is true for
motility of organism as a whole, as well as for its size, shape,
color, number of food vacuoles, position of nucleus, distinction between cytoplasm and endoplasm, contained bodies,
such as red blood cells, and lastly, for its staining reactions.
Peculiarities in all of these characteristics have been emphasized as aids in differential diagnosis. Thus Craig [15] says:
"If, in a freshly voided specimen of feces we observe large,
motile amebæ, showing a clear hyaline ectoplasm, the distinction between the ectoplasm and endoplasm being
marked, an absence of a nucleus, or a nucleus situated
near the periphery of the endoplasm, small in size, poor in
chromatin, and having a dimly-defined nuclear membrane
with two or more vacuoles, and with or without red blood
corpuscles in the endoplasm, we may be sure that the organism observed is Entameba histolytica;" also, "In Entameba
coli the ecto- and endo- plasm are both grayish in color and
there is never observed the greenish color which is not
uncommon in Entameba histolytica. In Entameba histolytica the endoplasm is very often observed to contain one or
several red blood corpuscles, while in Entameba coli red
blood corpuscles are very, very rarely observed in the endoplasm." Further, "In well-stained preparations the two
species of ameba can be distinguished by the difference in
the staining reactions of the ecto- and endo- plasm. In

Entameba histolytica the ectoplasm stains more intensely with Wright's stain than does endoplasm, while in Entameba coli the opposite is true.

Not one of these characteristics taken separately, nor all of them together, can be considered by the writer, in the light of the present study, as sound differential points, since amebæ from two such widely separated habitats as Ameba coli and Ameba intestinalis, and showing certain constant morphological differences,* vary so much on tissue media in regard to these characteristics.

Size. — On the whole the size of the amebæ varies more than it does in cultures growing with bacteria, particularly at 36° C. Ameba intestinalis is on the average slightly larger than Ameba coli. The largest forms measured are about fifty microns, the average of the majority about twenty-five microns, while of the many small forms (produced by budding) the smallest measure about two microns. We see then that these two strains, which are apparently different from each other, and one of which, at least, is probably not Entameba histolytica, both may show in artificial culture media many forms similar in size to those described for Entameba histolytica and therefore could and presumably would show such forms when grown in the intestines. Hence, this point, at present, seems not to be of worth in differential diagnosis.

Shape. — Both forms are round or slightly oval when at rest. Their pseudopods are quite variable in shape, depending upon the consistency of the medium and the age of the culture. Those of Ameba intestinalis are markedly lobose in any partially soft media which allow the amebæ to penetrate. The pseudopods under these circumstances are often long and narrow or finger-shaped, corresponding to those described for Entameba histolytica in feces. This is probably due to the fact that the conditions of growth as to consistency of media are somewhat similar. On solid media where the

---

* To be published in later report.

organisms remain on the surface or in more fluid media, under both of which conditions the organisms meet with less resistance in their progress, the pseudopods are apt to have spinous processes. In Ameba coli the spines are more pronounced than in Ameba intestinalis. Several spines usually pass out from a broad basal pseudopod. The pseudopods of Ameba intestinalis are more refractive than those of Ameba coli. Both of these species vary in motility according to age, culture media, and temperature. Young cultures (twenty-four hours old) in fluid culture media, at higher temperature (36° C.) are most motile. On the whole, Ameba coli is less motile than Ameba intestinalis. In cultures which have been constantly kept at 36° C., motility ceases in a short time when put at room temperature, to be resumed after a varying period according to the age of the organisms. This is true for all cultures kept at 36° C. Some species resume movements more quickly than others, for instance Ameba limax and the tiny amebæ which are all very motile in young cultures on practically any medium.

Cytoplasm. — The cytoplasm in both Ameba coli and Ameba intestinalis shows distinct separation into ectoplasm and endoplasm in the large forms when the organisms are in motion in a more or less fluid medium or on moist surfaces, but in media which they can penetrate this distinction is not so apparent. Thus Ameba coli in brain media often sends out spinous pseudo-pods into the granular brain food, in which the spines seem to become lost, so even in well fixed and stained specimens the border line between amebæ and food medium is not well defined. The reticular cytoplasm of such an organism seems simply to merge into the granules of the brain medium. The reticulum and cell contents show distinctly about the nucleus.

With both Giemsa and fuchsin-methylene blue stains the cytoplasmic reticulum is clearly demonstrated in these pure growths, since there are no bacteria to help false interpretations. With the former stain the cytoplasm of Ameba intestinalis takes a dark blue or purple, the ectoplasm staining more intensely than endoplasm, which, Craig claims, is true for Entameba histolytica; while with this stain the ectoplasm of Ameba coli takes the stain more faintly than the ectoplasm, that is, as Craig says, Entameba coli stains.

The more refractive pseudopods of Ameba intestinalis appear finely granular under the high power (x 1,500), while the pseudopods of Ameba coli are more hyaline.

At 36° C. both of these species generally show one to very many vacuoles, some large, others very small, sometimes one or two very large ones with none or several small ones. The largest amebæ contain the most vacuoles. Ameba limax at 36° C. may also show many vacuoles, therefore the presence of many vacuoles and large size seems not to be of value as a diagnostic point.

In regard to the importance in differential diagnosis of the presence of red blood cells within the amebæ, the present study proves conclusively that it also is not great. For all of the strains studied in this connection (Ameba limax, Ameba coli, and Ameba intestinalis) are markedly phagocytic for erythrocytes when adult motile amebæ are planted in fluid or moist fresh culture media containing fresh blood cells and examined in about twenty-four hours at room temperature. At this time a variable number of amebæ are found containing one to many red blood cells. Of the three species studied Ameba intestinalis is the most phagocytic. Some individuals may be so packed that the number of contained erythrocytes cannot be determined. Ameba limax is the least phagocytic of these three; usually one to four blood cells are seen in occasional amebæ, but a few times as many as eight have been counted.

After twenty-four hours a rapid digestion of the contained erythrocytes is evident, hence the cultures must be examined in about that time, or somewhat earlier if kept at 36° C. After forty-eight hours there is a marked diminution of the number of free red blood cells, and none, or only remnants are found in the amebæ, a number of which have encysted; and both motile amebæ and cysts may have a decidedly greenish color. This process occurs both in pure and in pure-mixed cultures. Though the amebæ apparently digest the red blood cells they do not seem to find in the latter sufficient if any food for continued life, since amebæ planted in blood alone or in blood added in varying quantities to broth or agar do not propagate after the first or second culture generation. There seems to be some increase in the phagocytic power of amebæ for erythrocytes after growing for several successive culture generations in suitable media containing fresh red blood cells.

Reproduction. — The nuclear changes and the methods of reproduction observed do not in general differ from those seen by the writer in pure-mixed cultures. Budding is a little more frequent. The nucleus is more varied in chromatin content and in position, but divisions occur by definite mitoses, and cysts develop essentially as they do in the latter cultures. Budding is comparatively more frequent than mitoses at thermostat temperature, while the opposite is true at room temperature.

The process of the separation of small portions or " buds " from the parent ameba has been repeatedly observed under the microscope and corroborated by spreads stained with Giemsa. The separated buds show a delicate blue reticulum dotted with small chromatin-staining granules which show distinctly only in well fixed well stained spreads. Many different sized buds may be seen in some young cultures but the small ones predominate. Under favorable conditions some of the larger of the separate buds encyst, which process somewhat resembles that described for encystment of Entameba histolytica.

No definite evidence of endogenous budding or spore formation has been obtained. The large rounded chromatin staining masses so often seen within the cytoplasm of the amebæ in bacteria-mixed cultures do not appear in these pure cultures. In their places are morphologically similar masses taking a light blue stain with Giemsa, or a reddish stain with fuchsin-methylene blue. The probable interpretation is that these masses are more or less digested food particles — bacteria in the pure-mixed cultures, and broken up tissue in the pure cultures.

Mitotic division is most frequently observed in young vigorously growing cultures.   Though in living individuals of both species the nucleus is often indistinct, in well fixed and stained specimens (spreads and impression slips fixed while moist) the mitotic changes may be clearly seen. Probably no amitotic division occurs unless the larger " budding " may be so considered.

The precystic, cystic, and postcystic stages require such prolonged, minute, and control study in order correctly to interpret the changes seen, that their description is left for a future report.   We may state, however, that from our morphological study we think the assumption fair that amebæ parasitic for mammals show essentially the same morphological changes in tissue culture media as they do in the intestines.

It is certainly true that the worth in differential diagnosis of the various characteristics described above, singly or taken together, is much less than others would have us believe. Indeed the recent demonstration of another species of ameba, presumably pathogenic for man, Ameba tetragena, and having many of the characteristics of Entameba coli, has already thrown grave doubt upon the worth of these points in diagnosis.

It seems much more probable that Musgrave and Clegg and those agreeing are right in thinking that several different species of amebæ produce or help produce human dysentery. And this brings us to the next point to be discussed.

4.   Pathogenesis. — The fact that certain amebæ may be nourished by tissue cells alone, as well as by bacteria alone, is strong evidence in favor of the view that amebæ are of an importance equal to, if not greater than, bacteria in the production of amebic dysentery and amebic abscesses.   It may be that in the living host the tissue needs first to be injured in some special way.   The fact that pure cultures of Ameba coli have been inoculated into the liver and brain of two

guinea-pigs and two rabbits with a negative result favors this idea. But the injury need not necessarily be directly from living bacteria.

These inoculations were made only after the fourth culture generation on brain tissue, so the amebæ may not have become sufficiently accustomed to the brain media to produce positive results, or there may have been other controllable causes. That certain varieties of amebæ, however, become strict tissue parasites seems from this study to be true. The question as to the nature of saprophytism and parasitism in the group of amebæ needs much further study.

Pure cultures of the two strains, Ameba coli and Ameba intestinalis, were every day fed to two kittens, respectively, for four weeks. The growths on agar plates planted at varying times, thus allowing the mixture to contain cysts of different ages as well as motile amebæ, were mixed in milk, and the dose was increased from one to several plates (4–8). The amebæ were found in the stools after the sixth day. The kittens were first found to be ameba-free. The animal receiving Ameba intestinalis had several short attacks of diarrhea without blood during the feeding. The one receiving Ameba coli had one short attack of bloody diarrhea three weeks after the feedings were stopped. It apparently recovered in two days and both kittens are now well. The amebæ in these stools seem to be quite similar to those in the pure cultures inoculated.

5. Isolation of cultures of amebæ from cases of human amebiasis. — Unfortunately, since the beginning of the present study, the writer has had no opportunity to work directly with human amebic dysentery and therefore can only suggest that crushed-tissue media promises to be an aid in obtaining cultures of amebæ from this disease, especially from the bacteria-free liver and other tissue abscesses which are so frequently a complication and in which the amebæ have become accustomed to the tissue food. From these abscesses workers have so far had difficulty in obtaining growth of amebæ. Even Musgrave and Clegg, who have

reported more positive results than any one else, only obtained occasional growths in one or two plates out of many. In attempting growths from any one of these sources, especially from the intestinal contents, we must remember that it is not necessary at first to obtain bacteria-free cultures; but, since the bacteria present may easily overgrow the amebæ, an attempt should be made to keep the former down. Probably the best results would be obtained with alkaline water-agar plates spread with small quantities of brain, liver or kidney tissue to which the ameba mixture should be added in small amounts. Many plates should be made, some kept at room temperature and others at 36° C. If no growth is seen in forty-eight hours a loop or two of sterile normal sodium chloride solution should be added and the plates should be observed for some days further.

6.   Action of amebæ on diseased tissue in which no specific microörganisms have been positively demonstrated. — The possibility of using amebæ in pure cultures to aid in detecting microörganisms that have so far eluded us led to the present work. It was called forth by Clegg's work on the lepra bacillus. The pure cultures have been tried so far only with rabies brains, though plans have been made to use them with material from measles, scarlet fever, and acute poliomyelitis. With rabies brains these amebæ have grown well from the start; there seems to be no drawback to their growth either at room temperature or at blood heat, which is not always the case with normal brains used as food. Very little study has been made yet of these rabies brain cultures. They are virulent after a week at room temperature. Morphologically nothing definite has so far been demonstrated.

SUMMARY AND CONCLUSIONS.

1.   Successive pure cultures of certain strains of parasitic amebæ may be easily obtained by using as food, sterile brain, liver or kidney tissue freshly removed from the normal guinea-pig, rabbit or dog.

2. Morphologically the individual amebæ in these cultures do not differ essentially from those of the same species growing in the intestines of mammals.

3. Certain characteristics — namely, large size, marked motility, clearly differentiated and highly refractive ectoplasm, nucleus poor in chromatin and situated near periphery of endoplasm, endoplasm showing two or more vacuoles and containing two to many red blood cells — which as a whole have been pronounced by others to be of use in differentiating pathogenic from non-pathogenic forms in the human intestines, have been shown by this study to be possessed by two species of amebæ in pure culture, one at least, in all probability, non-pathogenic for man. Hence the diagnostic worth of these points is rendered doubtful.

4. Freshly removed tissue as food for amebæ promises to be of aid in obtaining pure cultures of these organisms from cases of amebiasis, especially from bacteria-free amebic abscesses.

5. Pure cultures of amebæ may be of use in helping to detect specific microörganisms in infectious diseases of unknown origin.

REFERENCES.

1. Williams, A. W. Pure cultures of parasitic amebæ in brain streaked agar. Proceedings Soc. Exp. Biol. and Med., 1911, vii, 56.

2. Cunningham. On the development of certain microscopic organisms occurring in the intestinal canal. Quar. J. A. Mic. Sci., 1881, xxi.

3. Colli and Fiocca. Ueber die Aetologie der Dysenterie. Centralbl. f. Bakt., 1 Abt., 1895, xvii, 309.

4. Saul. Ueber Reinkulturen von Protozoen. Arch. f. Anat. u. physiol., Physiol. Abt., 1904, 374.

5. Thomas, J. B. Report on the action of various substances on pure cultures of the ameba dysenteriæ. Am. J. of Med. Sci., 1906, cxxxi, 108.

6. Auerbach, L. Ueber die Einzellegkeit der Amöben. Ztschr. f. wiss zool., 1856.

7. Frosch, P. Zur Frage der reinzuchtung der amöben. Centralbl. f. Bakt., etc., 1 Abt., 1897, xxi, 926.

8. Musgrave and Clegg. Amebæ: Their cultivation and etiologic significance. Bureau of Gov. Lab. Biol. Lab., No. 18, 1904.

9. Walker, E. L. The parasitic amebæ of the intestinal tract of man and other animals. J. Med. Res., 1908, xvii, 445.

10. Kartulis. Einiges uber die Pathogenese der dysenterie Amöben. Centralbl. f. Bakt., etc., 1 Abt., 1895, ix, 365.

11. Kartulis. Dysenterie. Nothnagel's specielle Pathologie und Therapie, 1900, v, 1, 1.

12. Casagrandi and Barbagallo. Uber die Kultur von Amöben. Centralbl. f. Bakt., etc., Abt. 1, 1897, xxi, 579.

13. Tsugitani. Ueber die Reinkultur der Amöben. Centralbl. f. Bakt., etc., Abt. 1, 1898, xxiv, 666.

- 14. Musgrave and Clegg. The cultivation and pathogenesis of amebæ. The Philip. j. of Sci., 1906, 1, 909.

15. Craig, Chas. F. Studies upon the amebæ in the intestine of man. J. Inf. Dis., 1908, v, 324.

16. Councilman and Lafleur. Amebic dysentery. Johns Hopkins Hospital Reports, 1891, 11, 395.

17. Kruse and Pasquale. Untersuchungen uber Dysenterie und Leber Abscessen. Ztschr. f. Hyg., 1894, xvi, 1.

- 18. Lesage, A. Cultur du parasite de l'amibe, etc. C. R. Soc. Biol., 1907, lxii, 1157.

19. Calkins, G. N. Protozoölogy, 1909, Lea and Febiger, N.Y. and Phil., 1st Ed.

20. Clegg, M. T. Some experiments on the cultivation of *Bacillus lepræ*. The Philip. j. of Sci., 1909, iv, 77.

THE IDENTITY IN DOG AND MAN OF THE SEQUENCE OF CHANGES PRODUCED BY FUNCTIONAL ACTIVITY IN THE PURKINJE CELL OF THE CEREBELLUM.[*]

DAVID H. DOLLEY.

(From the Pathological Laboratory of the University of Missouri.)

SYNOPSIS.

Introduction.

The history of the human case.

Microscopic technic.

General considerations regarding the nature of the process as exhibited by the dog.

The comparison of the process in the two species.

An analysis of the literature as regards its bearing upon the sequence of changes.

The universal unity of the anatomical process which underlies function in nerve cells.

Introduction. — The anatomical course of functional activity and of the recovery from activity in the Purkinje cell of the cerebellum has formed the subject of several previous publications ('09c, '11). Dogs which had been exercised in a treadmill furnished the material. In addition, the identical sequence of changes induced in the same animal in the abnormal states of traumatic shock ('09a, '10) and acute anemia ('09b) have been brought in correlation and have afforded a large part of the data.

Though the concrete extension of the findings and conclusions to man is based essentially on the study and comparison of a single human case, it will be apparent that if the significance of the process be correctly interpreted, this is fully sufficient. However, it may be said that corroboration has been afforded through the survey of other human material from various sources.

The history of the human case. — The writer is indebted to Dr. C. W. Chastain of Plattsburg, Mo., at the time of the

[*] Presented in abstract before the American Association of Pathologists and Bacteriologists, Chicago, April 14, 1911. Received for publication Oct. 9, 1911.

(285)

occurrence attending physician to the State Penitentiary, for the account which is here in part abbreviated. The man, a negro, aged thirty years, was a prisoner in the above institution. He had begun to show symptoms of mental aberration which consisted " of imaginary whisperings concerning him by convicts and a supernatural control of his thoughts by certain officers of the institution." On several occasions he had been detained in the hospital for observation. On the night of May 26, 1906, word reached the authorities that he threatened trouble and the next morning the order was issued to detain him in his cell. However, he had already taken his place in the breakfast line and refused to return. Persuasion failing, the guard laid hand upon him, though not forcibly, whereupon the negro stabbed him in the neck and jumped upon him. Two convicts came to the guard's assistance and both received flesh wounds in the rough and tumble fight which ensued. Finally, the murderer made his escape and secreted himself in a cell. As soon as possible efforts to capture him were begun, but " he was defiant, going from cell to cell, climbing the uprights from walk to walk like a squirrel, hurling down heavy iron-bound oak buckets at any one attempting to enter, turning on the water faucets, etc." He evidently acted on the offensive as well as the defensive and this strenuous activity must have lasted over an hour. At last it was decided to shoot him and he was killed instantly in the act of hurling a bucket, falling from the top walk to the flagging below. The whole performance lasted at least two hours. After the inquest the body was sent to the anatomical laboratory of this university where it arrived within ten hours after his death.

The case then offers a unique and exceptional opportunity for the study of the most intense and concentrated activity possible to man, one calculated to bring out the reserve of power as no ordinary circumstances would. Furthermore, it is uncomplicated by factors of extrinsic origin, such as shock and hemorrhage, which it would be most uncommon to find unassociated with the traumatic death of an active individual. On account of its bearing on the comparison of this case with the experimental activity of dogs, the suggestion is made that the immediate mental excitement is certainly not without parallel in its influence on the particular cell under discussion in the experiments upon the lower animal.   This is of importance in both species in the consideration of the severity and extent of the changes which have been found to result.   No neurologist nor physiologist probably would deny that the effect of such a factor by itself would only be expected to be contributory in kind and augmentative in degree.   A possible question as to whether the changes to be described have any direct relation to the preëxisting mental aberration would seem to be sufficiently answered by the comparison with enough dogs of all ages to be assured of a normal standard.   Based on this standard, the changes which are set forth for the human cerebellum are those of immediate and purely normal activity.

Microscopic technic. — The technical experience in this case is considered to offer additional argument in support of the applicability of simple ordinary staining methods to the nerve cell.   The only preparation the subject received was for gross dissection and it consisted of the injection of the body as soon as it was received with a one-half of the full strength formalin solution, presumably about a twenty per cent formaldehyde.   The body was then placed in cold storage.   Some unknown time thereafter the brain was removed and left permanently in an eighty per cent alcohol. The material under consideration, therefore, received only this crude fixation and was between four and five years old, certainly an unfavorable combination.   And furthermore, there was no agreeable disappointment in the results of numerous preliminary attempts at staining with different reagents.

After these failures, application was made of a suggestion obtained from Lee ('06).   This is to the effect that certain primary fixing agents, notably for the present purpose, mercuric chloride, picric acid, and chromic acid, really play the same part as mordants and are efficient because their combination with the tissue has an affinity for the given stain. Accordingly, mercuric chloride and picric acid in saturated solution and chromic acid in one per cent solution were tested as mordants. Five micra paraffin sections, fastened to slides by the water method, were freed of paraffin by xylol, passed through the alcohols, and then soaked for different times in these solutions. The sections were then stained in a concentrated aqueous Thionin, following v. Lenhossek (quoted by Lee) who employed it for formol material. No noteworthy difference was apparent among the three mordants and the stains so obtained were sufficiently tenacious to admit of proper differentiation. A further improvement next resulted from the adaptation of Ohlmacher's statement (quoted by Lee) that formaldehyde is a powerful mordant for coal tar colors. Forty per cent formaldehyde was added to the Thionin solution in quantity necessary to make a four per cent solution and other sections were stained in this after the preliminary mordanting. The results were highly satisfactory, the effect of the formalin in the Thionin being to limit the stain to the basic chromatic element, both intranuclear and extranuclear. Further, as it stains slowly it becomes a progressive rather than a regressive stain, requiring no decoloration if stopped at the proper moment. However, it is easier to control by overstaining somewhat and following the subsequent differentiation under the microscope. Formalin by itself was not found to be a sufficient mordant. Fresher formalin material from other sources was also tested and it only appears to differ in a more intense and quicker effect.

For the sake of constancy in the routine work following, mercuric chloride was used as the initial mordant. The sections in this particular case were treated about three hours, washed thoroughly in running water, stained an hour in fresh formalin-thionin mixture, differentiated in alcohol, counterstained in alcoholic eosin, passed through absolute alcohol to xylol and mounted in balsam. The exact times evidently would depend upon the individual case.

With nothing but the practical end in view of getting a

satisfactory working stain, the experiments were not extended to other dyes, but from the striking results obtained it seems reasonable that it is only a question of a suitable mordant for the particular dye.   There is nothing peculiar, nothing round-about in the method, the technic is the most ordinary. The impression is not intended that the results are as good after formalin as they would have been after a sublimate fixation in the first instance, yet the differences between them in the successful preparations in no way interfere with an interpretation.

In the progressively more successful attempts it was most noticeable how uniformly and evenly the intranuclear basic chromatic material, both when found in the karyosome alone and in hyperchromatic nuclei, and the extranuclear (the Nissl substance) coincided in the degree and rate of their staining reaction.   Indeed, the nuclei of the granular layer are also to be included in this statement.   The basic substance within the nuclear membrane is conventionally chromatin.   For the nerve cell, that without reacts in just the same way.   In itself this is not held to be any decisive proof, but taken in connection with points to be considered later, it further substantiates for the nerve cell the identity between the basic substance without the nucleus and that within.   This only means that the substance we call chromatin is identical in nature wherever distributed in the nerve cell.   It .does not necessarily imply that all chromatin of somatic nuclei is one and the same substance.

General considerations regarding the nature of the process as exhibited by the dog. — Repetition of previously described details will be avoided so far as seems consistent with clearness.   The opportunity will be used to consider more broadly and from different points of view the inherent character of the changes and the rationality of their correlation with functional states.

The final interpretation of the sequence of the changes, grouped irregularly and varied as they are in any animal, has resulted from the weighing of evidence from various

sources. No small part of the evidence is purely morphological — the structural evidences of transition from stage
to stage are definite and distinct. There is no break in the
continuity of the process. Again the staining reaction supports the conception. To this phase the results obtained
for the present case constitute an added feature. Based on
the experience of others, as well as the personal one, the
statement seems justified that any ordinary nuclear stain may
be used successfully to demonstrate the Nissl substance, provided the fixation be proper therefor. This limitation is by
no means confined to nerve cells. The strongest indication
is derived from both the morphological and technical sides,
and is to be found in the orderly shifts in the chromatin
distribution throughout the process. Third, a large amount
of data, both published and unpublished, is now in hand to
show the actual correspondence with outward functional
states — the practical application is satisfactory. Finally,
the interpretation serves to harmonize the results obtained
by others and to show that what might appear to be discrepancies and differences are not real in large part at least.

Primarily, however, the interpretation rests upon the
application as a working hypothesis of the theories emanat-
ing from Richard Hertwig and Richard Goldschmidt of
the Munich Zoölogical Institute. It would have been
impossible except in the light which they shed and whatever
it possesses of a logical and rational nature is due to them.
Nevertheless, it is pardonable to point out that a sufficient
part of the confirmatory data is so outside of relation or connection with the theoretical side as to give the interpretation
a certain independent standing. On a purely morphological
and empirical basis, supported by an analysis of the consensus of opinion, the changes would fall most satisfactorily
in the same definite, constant, and orderly sequence.

The direct and indirect evidence has invariably served to
strengthen the belief that the basic staining element of the
body of the nerve cell, generally designated as the Nissl substance, is a form of chromidial apparatus, that is, extranuclear functioning nuclear material in the sense of Richard

Goldschmidt. The particular distinctions he would make are that such material denotes the capacity for function and that it is used up in the course of work and replenished during rest. It represents a mechanism by which metabolic nuclear substance is brought into direct relation with the cell body, a more highly advantageous means for the quick discharge of energy. It is no peculiar characteristic of nerve cells, for a similar material is possessed by widely varying types of cells, of muscle, gland, and even epithelium. Of all such substances, the chromatic substance of nerve cells, so labile in character and so sensitive in reaction, already long connected in the opinion of many directly with the functional expression of the cell and so readily subjected to experiment, is perhaps the best and most obvious type for investigation. The greater advance in functional knowledge regarding this as compared with those peculiar substances not thus identified in purpose, as for example, the mitochondria, make it the best criterion of the essential principles of Goldschmidt's doctrine.

Equally consistent in their application to the most highly specialized of all cells appear Hertwig's ideas regarding the relation between the cell and its nucleus. The side pertaining to the definite size relation which exists between them, so far as concerns the nerve cell, will be brought out sufficiently in the restatement which is necessary for the human case. Associated with this nucleus-plasma relation theory are the mutual interdependence between cell body and nucleus and the interchange of material between them whose visible result is the formation of chromatin. How these doctrines may be correlated to explain the mechanism of the process of functional activity has been attempted in detail ('10). However, without unnecessary repetition, one observation not then fully available may be mentioned. This comes from the study of the process of recuperation ('11). According to Hertwig, the substance from which chromatin is derived is formed in the cytoplasm and is taken up by the nucleus; by means of the nucleolar substance, the plastin, it is organized into chromatin, thereby becoming visualized;

from the nucleus, chromatin or its derivatives return to the
cytoplasm to be used in the vegetative functions of the cell.
Not only is the chromatin a shifting quantity, used up and
replenished according to need, but the behavior of the
nucleolar substance is entirely in accord with this theory.
In its continued activity the supply of nucleolar substance,
in the writer's opinion, follows the course of the chromatin,
and its gradual increase in the recovery from extreme
exhaustion is even more apparent. Not only is this so, but,
like the chromatin in virile animals, it tends to swing back to
a distinct excess.

The process has been divided into thirteen stages, comprising both main and subsidiary types, for objective description and the numerical statement of the size changes. These
are arbitrary in the measure natural to a continuous process.
Superficially, this division might present a misleading
appearance of complexity. This is far from the fact as a
brief analysis may help to show. In terms of the chromatin
distribution, they may be summarized as follows, the numbers being uniform in all publications:

1. The resting cell. It is lacking in intranuclear chromatin except within the karyosome (nucleolus) and the
amount of extranuclear chromatin varies with the individual.

2. The stages of progressive hyperchromatism, in which,
in the pure type, the initial enlargement of the whole cell
reaches its maximum.

3. The stage of maximum hyperchromatism, which is
associated with the beginning of shrinkage.

4 and 5. The stages of regressive hyperchromatism
together with the maximum of shrinkage. Coincident in
place but separated originally to denote difference in shape,
Stage 4 being more attenuated and spindle. Both Stages
4 and 5 are to be further divided into an early, the pure
Hodge type, and a late division, characterized by the sharp
beginning of nuclear edema.

6. The return of the cytoplasmic chromatin in its continued reduction to the average normal level. This stage is
principally distinguished morphologically by the maximum

disproportion in the size of the nucleus owing to its much greater edema.

7 and 8.   Two stages leading to the primary disappearance of cytoplasmic chromatin.

9 and 10.   The stages of secondary restoration of cytoplasmic chromatin.   The chromatin is first piled about the nuclear membrane and then passes out.

11.   The stage of the secondary disappearance of cytoplasmic chromatin.   With the complete using up of the previous supply, the karyosome is left containing the only vestige of basic chromatin in a much more exhausted looking cell.

12.   The disintegration and passing out of the ultimate content contained within the karyosome.

13.   The exhausted cell.

The main ideas in appreciating these stages are that this extranuclear functioning nuclear material, the so-called Nissl substance, is derived through the nucleus, that it is used up in the course of work and continually replaced by the mediation of the nucleus; that at first the supply is in excess of the demand, the hyperchromatism, but that with long continued drain, the supply falls short of the demand and there results a progressive diminution in chromatin, the hypochromatism, and finally no chromatin at all.   In terms of the theories mentioned, the idea of purpose stands forth preëminently in the nature of the changes.   There is manifest throughout a coördinate effort whose sole and simple purpose is to maintain so far as can be the proper level of cytoplasmic chromatin.   The discharge of chromatin is orderly throughout.

With this idea of purpose in mind, which is certainly not necessarily dependent upon the use of the theories in explanation, the whole process may be simply separated into three main divisions as regards the amount and distribution of chromatin: First, the hyperchromatism, increasing to a maximum as a result of the stimulus to work, then diminishing as the supply becomes unequal to the long continued demand.   Second, the attempt toward the secondary restoration of the cytoplasmic chromatic element. Morphologically, there is a sharp distinction here between these divisions, — on the deficient side a cell with little or no chromatin (Stage 8), on the restorative side a definitely larger cell with a considerable amount of chromatin, first massed about the nuclear membrane (Stage 9), and then with further increase in size more or less disseminated through the cell body (Stage 10). That the secondary supply is used up is obvious, for again chromatic material disappears from the cell body, which is also associated with a still larger cell (Stage 11). The stage is not only so distinguished, but also its karyosome, though at the maximum of size is perfectly intact. Then ensues the third and last of the three divisions, the passing out of the formed chromatin of the karyosome with the evident purpose of constituting a tertiary renewal of the cytoplasmic supply necessary for availability. After this is used up, then and then only is the cell exhausted. But it is by no means finally exhausted, for its return after rest has been traced step by step to complete restoration both in size and chromatin. On the basis of purpose, it seems plain that the process is not complicated, but simple and straightforward, though it is equally plain that there is not only wonderful organization but reserve after reserve.

It is most particularly in the second division of the process that the purposeful nature of the reaction manifests itself objectively. Along with an increasing inadequacy of chromatin, a progressive enlargement of both cell elements has led up to this. The constant association of the further increase of size with the renewal of chromatin which invariably ensues as a distinct stage cannot be a mere coincidence. While in terms of Hertwig's theories there is a most rational explanation for it, nevertheless, outside of these, there can be nothing short of purpose interpreted in any explanation of the enlargement of the cell as part of a natural process. It is a true functional hypertrophy, a physiological enlargement to conserve the function of the cell. The constant association in itself would seem sufficient to connect it with chromatin without any hypothesis at all. The three go together, the enlargement of the cell body, the enlargement of the nucleus, and the renewal of chromatin. Granting this, if the basic material of the cytoplasm is derived from that element, why should the nucleus enlarge, or, on the other hand, if it is a nuclear derivative in the limited sense, why should the cell body enlarge? That both the cytoplasm and the nucleus take part in the elaboration of chromatin seems the only adequate answer. The onset of the hypertrophy means that with an undiminished demand upon the cell the consumption of chromatin is exceeding the supply. To keep it from falling below a proper working level, the cell body makes an increased demand upon the nucleus for the chromatin there synthesized. But it must in turn supply the nucleus with the essential elements for that synthesis. To do this it makes a constantly increasing demand upon its outside source of supply. More preforming materials must enter the cytoplasm, hence its enlargement; more pro-chromatin materials must enter the nucleus, hence the nuclear enlargement. The visible result is a renewed outpouring of chromatin.

Conclusions bearing upon or essentially identical with the foregoing regarding the nature of the Nissl substance are reached by several investigators. Scott ('99) first and later Hatai ('04) and Collin ('06) have traced the embryonic development of this substance from the nucleus. Holmgren ('00) strongly insists that there is an interchange of formed material between nucleus and cell body. He determined for the adult cells of Lophius piscatorius a migration of chromatin granules not only from within the nucleus out but back again, the rays of the centrosome being regarded as the pathway. Hatai denies such return of chromatin and it would seem with right, considering the purpose of its discharge. Rohde ('03) maintains the migration of nucleoli and the diffusion of the chromatic substance from the nucleus.

The contribution of Hatai is especially noteworthy. Partly from the morphology of spinal ganglion cells from

intrauterine embryos of white rats, cats and pigs, and partly
from the chemical reaction, he concludes that in the embryo
the Nissl granules when first formed are derived either by
the diffusion of the nucleins or by a migration of the acces-
sory nucleoli.   In the adult waste is repaired by diffusion of
dissolved nucleins.   His pictures of the microchemical test
for this diffusion could not but be convincing even to one not
predisposed to agree.   It is worthy of note that to explain
the processes he depends upon the well-known scheme of
cell metabolism given by Verworn ('99), an abstract generalization to which R. Hertwig has given concrete and
specific expression.

In the embryonic state, Hatai lays great stress upon
nuclear pseudo-podia in connection with the centrosome as
a mechanism for the interchange of formed material.   While
the possibility in this state is not denied in the slightest, his
extension to the adult on the basis of the " stronger evidence" of pathological and experimental studies cannot
hold.   The changes he would so correlate are the irregularities of contour first described by Hodge and universally
found as will be brought out later.   Such changes only
occur in a single phase of the whole cycle of activity, and
consequently it would seem that at other times the cell gets
along without them, which excludes such an interpretation
for the state of activity proper.   Formed chromatin certainly passes out anywhere about the nucleus in the adult
Purkinje cell, though for the most part it passes out in solution.   In the early hyperchromatic stages, formed chromatin
appears within the nucleus, during the secondary restoration
just within, the preformed chromatin passes out at the end of
the process, and, finally, in an energetic recovery, it is frequently precipitated just within the nuclear membrane.   In
the last, one might most likely expect pseudo-podia if they
are necessary, and in fact irregularities do occur in a certain
number of cells, but not in the most virile.   They mean loss
of substance and senility.   The more irregular the cell, the
less the chromatin, and the alteration becomes permanent.

As regards the migration of nucleoli as such, for which Hatai, Collin, and Rohde offer strong evidence during the developmental period, it may be said for adult cells so far investigated that misplaced nucleoli are to be found with such exceeding rarity that they cannot be conceived as having connection with the usual process at least.

The comparison of the process in the two species. — The morphological comparison is one of similarity, for stage by stage the changes in the dog find their counterpart in those of man. In their essential nature there is obviously no difference, and, considering the fixation of the material, there is a remarkable lack of difference in the actual details. It is so little in fact that it does not seem worth while to discuss them.

There are differences, however, when the response is considered collectively. In this one human case there is a much greater degree of uniformity in the collective degree of reaction than in any dog ever studied. The great bulk of cells throughout the cerebellum fall within very closely connected stages. There is in short a striking solidarity in the reaction in man. In a moderately exercised dog, even in dogs that have not been exercised at all, the common finding has been of a much wider distribution among the stages of the process from one extreme to the other, so far as exhibited by the particular animal. While no differential counts were made in the present instance, probably a better general survey has resulted in the search for cells suitable for measurement. The quota of cells belonging to Stages 2, 3, and 5 was the last to be filled up and evidently they are relatively more scant than would be expected in the dog. Resting type cells, however, just as in the dog which is working capably, are in greater number — representing undoubtedly reserve cells which are not yet in action. But the ending of the process in man is much more convincing. Totally unlike the dog, in which at any time cells in complete exhaustion are to be expected, the reaction in the human case practically stops short at one stage and no exhausted cells whatever were to be found. Specifically, there are an abundance of cells belonging to Stage 9, but in looking over fully one hundred sections only twelve of Stage 10 were found to measure. Not a single stage beyond this could be found. Incidentally, while advancement of the process to this point could hardly leave any doubt as to the nature of the remaining course in man, resort was made to human material from two cases, one of tetanus, the other of post-operative hemorrhage, in both of which the further advanced stages toward exhaustion were readily to be identified.

But beside this difference in the solidarity of the collective cellular process, there is another, in some ways less tangible. The indications of it that can be put in words do not fully express it and its recognition could only come from long experience regarding special phases and minor eccentricities in the sequence of changes. To sum it up, the indication is that there is a better intracellular coördination between cell body and nucleus in man than in the dog. At any given stage, the ratio of the chromatin content on the one side and the apparent amount of nucleolar substance on the other does not seem to vary as much in man as in the dog. There is much more of irregularity, of jerkiness as it were, in the dog. The amount of nucleolar substance particularly would seem to be more uniform and better balanced in the human. On the other hand, neither is the supply of extra-nuclear chromatin likely to fall to such a low ebb as it does commonly in the dog before its resupplial. The supply is more continuous. Still further, there is more of evenness, of uniformity in the amount among the cells of any given stage.

Turning next to the shifting size relations in the course of the process, the complete identity between the two species receives a measured confirmation. The numerical data and the curves plotted therefrom are set forth in Table I. and Figure I respectively. There is no need to reproduce curves or tables for the dog. Since the technical procedures of measurement, of calculation and of plotting of curves which were presented in the numerical statement for the dog ('10) are essentially identical, they need not be discussed save only to make the context clear. The cells measured are only selected to the extent that choice was largely limited to those conforming well to a pear shape with solitary dendrite. The calculations of volume and nucleus-plasma coefficients are simplified by not computing them in actual figures. As was pointed out, the ratio of change is given equally as well by eliminating the constants in the formula, so that the calculation for such bodies ranging from the pear through the ellipse to the sphere resolves itself into the multiplication of the major axis (or radius) by the square of the minor axis (or radius). In the present case the diameter figures as obtained in millimeters from the projected cell images are used. While it is true that the longitudinal diameter of the cell body, involving as it does the separation of the cell body proper from its dendrite, is not always definitely determinate, though measured according to a constant rule, it may be pointed out that it is not only of less value in the computation, but that also the trend of the curves would not be altered if the figures for it were eliminated. These calculations permit as well the expression of the nucleus-plasma relation in a proper ratio. The total cell mass less the nuclear mass gives the plasma mass, and this in turn by division by the nuclear mass the size relation between the cell body and the nucleus. Twenty-five cells, which have been thoroughly proved to give a fair average, were measured for each stage save the last one as already indicated.

The course of the size changes of cell body and nucleus, both independent and relative to one another, may here be conveniently summarized together. The transition from the resting stage to activity is characterized by a marked increase of size which reaches its maximum in the pure type of Stage 2. In its size relations, the remarkable fact for this stage is that both the cell body and the nucleus increase in the same proportion. The nucleus-plasma relation remains unchanged, that is, the cell body is just as many times larger than the nucleus in this stage as it was in the resting cell and consequently the curve of this relation is represented by a straight line from Stage 1 to Stage 2. This is decidedly not the effect of trivial differences lost in the general averages. In the series here presented the percentage of increase of volume above the resting stage is practically thirty-seven for both elements. The nearly exact identity in the actual figures representing the nucleus plasma relation for these two stages which happened to result in this series is of course to some degree a coincidence. But because the measurements on the dog for these particular cells were somewhat fewer, a second series was obtained from the man, using an entirely different set of sections. In this pure types were selected with more care, and the volume increase for the nucleus was fifty-two per cent and for the cell body forty-eight per cent — that is, both were practically half again as large as in the resting type. The difference for such a mass increase is slight compared with that between cell body and nucleus in other stages and can have no significance other than to represent the variation to be expected in such technical procedures. It shows further the constant way in which the calculations run in that the nucleus-plasma quotient obtained from these more selected cells differed from the figures given in Table I. by only one and a half. In fact, the variation which brought about this negligible difference resulted largely from the more unsatisfactory longitudinal measurements.

From this point of maximal size attained by the early hyperchromatic cell, both the nucleus and the cell body diminish until a minimum is reached in the shrunken irregular cell of Hodge's type (Stage 5'). The decrease, however, is not proportionate and the nucleus progressively becomes smaller relative to the cell body. The figure representing this relation therefore becomes larger, that is, the nucleus-plasma relation is disturbed in favor of the cytoplasm. Consequently, the curve rises above the base line of the normal relation and the extent of the rise is equal to the difference from the normal.

It is in this part of the process that the only apparent

dissimilarity between the present nucleus-plasma curve and those of the dogs is to be found.   In this case, the height of the initial change in favor of the cytoplasm seems to be reached before Stage 5', while invariably in the dogs this stage with its analogue, Stage 4, marked its culmination. However, the actual difference here between the figures for Stages 3 and 5' is barely over one.   This is not only within the natural limits of technical variation, which would make them practically identical, but also the fact is to be remembered that these cells are extremely irregular so that their actual sizes are smaller than the computation given by the extreme diameters, which would raise the ratio.   Even if the dog work did not prove their place in the sequence of the nucleus-plasma changes, observation alone would leave little doubt of its correctness, since the nucleus is evidently more affected.

For the remainder of the process, the simple statement suffices that both elements progressively increase in their absolute size.   But this increase does not proceed at the same rate in both, from which there result other fluctuations in the nucleus-plasma relation.   The increase in size begins distinctly in the nucleus and becomes apparent in the latter part of Stage 5.

The effect of this larger nucleus in a cell considerably below the normal size is to lower their resultant quotient, so that in the present case the figure is diminished by four and the curve drops to just above the level of the normal relation.   In the next stage, the disproportion reaches its maximum, so that the cell body, instead of being nineteen and one-half times greater than the nucleus is only eleven and one-half times greater.   It is to be noted that this is notwithstanding the fact that the increase in size of the cell body is beginning to show appreciably, though it is yet notably smaller than that of the resting cell.   The curve, therefore, drops in Stage 6 well below the normal level, that is, the nucleus-plasma relation has gradually come to favor the nucleus.   The figures for this stage are by no means obtained from selected cells.   To test the results obtained from

averages further, the nucleus-plasma quotients of certain
individual cells differing most widely were calculated.   The
two largest cells, the one with the largest, the other with the
smallest nucleus, and the two smallest cells with the same
extremes of nuclei were taken.   The important fact is that there
was no overlapping, and for Stage 6 every nucleus-plasma
quotient was lower than the lowest of Stage 1.   It is obvious
in regard to size that it has nothing in common with the
resting cell.   It is only in its chromatin content that there is
more or less of a common appearance, while the beginning
edema and paleness of the cell body, the lack of chromatin
toward the dendrite pole and the loss of nucleolar substance
all mark the difference.   Yet if these size changes be dis-
regarded, these cells, which are among the predominant
types in this human case, would have to be considered as
nearly normal, showing only a slight chromatolysis.

Following this stage, the further increase in the nucleus is
relatively less in amount and proceeds at a slower rate as
compared with the cell body — it is approaching the limit of
its capacity.   On the contrary, the ratio of enlargement of the
cell body approaches, passes, and continues to exceed that of
the nucleus.   The result is that from the point last mentioned
the nucleus-plasma relation becomes progressively in favor
of the cell body and the curve rises to the end of the
process.   This means that the final balance of capacity
belongs to the cell body, which in its continued demand
upon a failing nucleus brings about its complete dechroma-
tinization.   Confirmation of these objective and measured
findings of the relatively greater exhaustion of the nucleus
has been obtained in the greatly slower rate of nuclear
recovery both as to size and chromatin ('11), which agrees
with Hodge's ('92) results for the spinal ganglion cell.

An analysis of the literature in its bearing upon the
sequence of changes. — Almost without exception, all who
have investigated the effects of functional activity upon nerve
cells have found very decided changes.   They differ widely
in their interpretation of the correspondence of these changes

with functional states of activity, fatigue, and exhaustion, but within the natural limitations of the part studied, their findings are practically identical. To sum up the relation to this work, all the changes which they describe fall naturally within the course of events as traced out in sequence for a cell which can readily exhibit the complete range of possibilities from activity to absolute exhaustion.

The explanation for the lack of appreciation of the extent of the response as anatomically considered, and, consequently, the explanation of the low and doubtful estimation in which the changes are commonly held, is to be found preëminently in the low order of cells which have been studied. To bring out this point, the authors will be cited in connection with the part studied: sympathetic ganglia only, Eve ('96), Lambert ('93), Lugaro ('95), Vas ('92); spinal ganglia only, Levi ('96); cord only, Holmgren ('00), Luxenburg ('99), Odier ('98), Pick ('98); retina only, Pergens ('96, '97), Chiarini ('06); spinal ganglia and cortex (latter not in detail), Hodge ('92, '94); cortex with or without attention to other parts, Demoor ('96), Guerrini ('99, '02), Mann ('95), Nissl ('94, '96), Pugnat ('01), Van Durme ('01). Of these last, four appear to have carried their experiments to a sufficient extent, namely, Guerrini, Mann, and Pugnat, using muscular work, and Van Durme, using electrical stimulation. In these it is very obvious that a much more advanced degree of changes is dealt with and it is in their results that a close parallel with the writer's is apparent. For instance, Van Durme portrays in his figures of the Purkinje cell (1–18, Plate VII.), all the essential types here set forth.

In further explanation of differences of interpretation, the following points are applicable in one case or another: (a) The neglect of the consideration of size changes, either relative or absolute. (b) The failure to use a counterstain to the basic dye. (c) Peculiar specifications for the state of repose: for example, by Lugaro, the delay until a number of hours after death (from his results, a negligible or unimportant factor); by Odier, as obtained by chloroform anesthesia; by Levi, as obtained by prior cutting of the peripheral nerve. (*d*) Disregard of effects of previous natural activities or of the concomitant operative procedure in exciting to activity. With a fixed type from which all activity starts, whether it means absolute rest in the case of the nerve cell or not, it is apparent that the difficulties of *c* and *d* are obviated. (*e*) The conception that hard and fast lines must separate the morphology of activity and fatigue and not that progressive fatigue, and exhaustion as well, are coincident with the progress of work. (*f*) Failure to recognize the inherent and absolute exhaustibility of the nerve cell. (*g*) The placing of definite stages, universally found whatever the fixation or stain and proved to correspond to functional states, in the category of artifacts.

Notwithstanding their differences, however, confirmation for every important phase of the process as offered by the writer is afforded on detailed analysis by at least two and usually more observers in their express or sufficiently implied statements. These parallels will be taken up first for the chromatin and then for the size changes.

Hyperchromatism is regarded as an indication of initial activity, in agreement with the writer, by Valenza (electric lobe of torpedo, '96), Holmgren, Lugaro, Odier, and Vas. Of the others, some explicitly, as Mann, Nissl ('96), and Van Durme refer this state to repose while the others appear tacitly to regard it as the starting point for activity if not itself active. Nissl ('94) earlier regarded the pyknomorphic smaller cell as indicating fatigue.

Hypochromatism, "chromatolysis," is universally considered to accompany in progressive degree activity and fatigue, in agreement with the writer.

Lugaro and Vas determine an initial augmentation of size, in association with the hyperchromatism, which is in agreement with the writer.

What corresponds to the secondary enlargement of the cell as determined by the writer is to be found in the statement of the enlargement of the cell as occurring with its chromatolysis — Guerrini (nucleus), Holmgren, Mann (nucleus only mentioned for cortical cells), Odier (nucleus), Pugnat, Van Durme. It is to be noted that this includes all the men who did the main work on the cortex. Here was one of the sources of greatest misunderstanding, which of course is cleared up by the knowledge that there are two distinct periods of enlargement, at the very beginning and toward exhaustion, the one hyperchromatic, the other hypochromatic.

Regarding the shrinkage and the shrinkage with irregularity of form, there is the greatest diversity of opinion, though again it is to be noted as a universal finding. The different places assigned are: aggregated with resting or starting point types, by Demoor, Van Durme; as a relatively early expression of fatigue coincident with work, by Hodge, Lugaro (strictly without irregularity), Odier, Pergens, Pick, Valenza, in agreement with the writer; as a late and separate manifestation of fatigue, by Holmgren, Mann, Pugnat; explicitly as artifact, by Lugaro (when distorted), Nissl; ascribed to extrinsic causes such as trauma and bleeding, by Luxenburg; noted, but not placed, by Guerrini.

Regarding the whole, it seems fair to say not only that there is lack of serious disagreement but that there is a very striking connection and consecutiveness when the individual opinions are thus correlated. The correlation has its fundamental basis in the unity of the anatomical response for all different types of cells which will be next discussed.

The universal unity of the anatomical process which underlies function in all nerve cells. — Whether it be ganglion cell or whether it be Purkinje cell, the anatomical response to stimulation as exhibited in the orderly discharge of chromatin is one and the same in nature and purpose. This discharge of chromatin is the mechanism through which the cell reaches its object, the performance of its specialized function. The associated changes in size and shape are contributory to this. As the Purkinje cell can be organically, which means functionally, exhausted, no nerve cell could go further, and in its lability it stands as a representative

of the highest type of specialization. This highly specialized and comparatively easily exhausted Purkinje cell
includes in its course of activity and first passes through the
changes to which the lower cells are generally limited. So
far as it goes, the mechanism in the lowest ganglion cell is
identical — it only stops short of going as far. It is more
primitive, less easily exhaustible. But that it is no less
capable of final exhaustion with adequate stimulation cannot
be doubted. A sufficient basis for this is to be found in the
results obtained from natural stimulation by Hodge and
Guerrini and from artificial stimulation by Lugaro, Mann,
Odier, Pick, and others. It need only be mentioned that
unpublished observations upon the relative states of Purkinje, cord and ganglion cells in surgical shock and in acute
anemia as well as in muscular exertion are in corroboration.
In this particular human case, spinal ganglion cells show in
general no more than advanced changes of the irregular
Hodge type with beginning deficiency of chromatin. There
is then a unity in the anatomical process from the lowest to
the highest cell throughout the ontogenetic scale in respect
to its inherent nature and as regards its ultimate possibilities.

Considering next the agents which are effective in exciting
to activity and the results which they call forth, there is the
same mechanism, the same result in all. From the identical
changes within the limitation of the part produced by the
electrical stimulation of ganglion cells by Hodge, Lugaro,
Mann, and Vas; of spinal cord cells by Holmgren, Pick,
Odier, and Luxenburg; of cortical cells by Van Durme;
from the identical changes in sequence produced by the
mechanical (traumatic) stimulation leading to surgical
shock and by the metabolic stimulation due to variously
induced acute anemia recorded in detail for the Purkinje cell
by the writer and observed upon the lower cells within the
limitations mentioned: from the identical changes produced
by intrinsic and normal activities, as recorded by Mann,
Guerrini and Pugnat and arranged in sequence by the writer,
but one conclusion can follow — there is the same unity in
the anatomical process after all possible stimuli.

Finally, the same unity holds between man and dog, which has its further support in the findings described for other animals more or less phylogenetically related. There is a unity in mechanism and in purpose universally displayed. It is upon this unity, so simple in principle, that function fundamentally rests.

The conception of Marinesco ('09), which is supported by others, that the chromatic substance is the immediate source of energy, in his term, a kinetoplasm, is the correct one. But the ideas as to the origin of such substances originated by Hertwig and as to their nature elaborated by Goldschmidt put the conception just mentioned upon a different plane. The chromatic substance is not merely an independent cytoplasmic derivative, however highly it be esteemed as a "superior" protoplasm. The elaboration and discharge of chromatin and its transmutation into the energy of work represent in themselves the integral activities of the nerve cell. To this end the whole cell integrates, this is its single response to any stimulus. Anatomically, specialization would seem to be a more elaborate and delicate mechanism to accomplish the same thing more rapidly and more efficiently. The liberation of the energy from this source manifests itself, whether directly or through a more indirect influence, as the work of the cell, its functional expression. Certainly, the discharge of chromatin is the basis of function and in its successive stages the index of functional capacity.

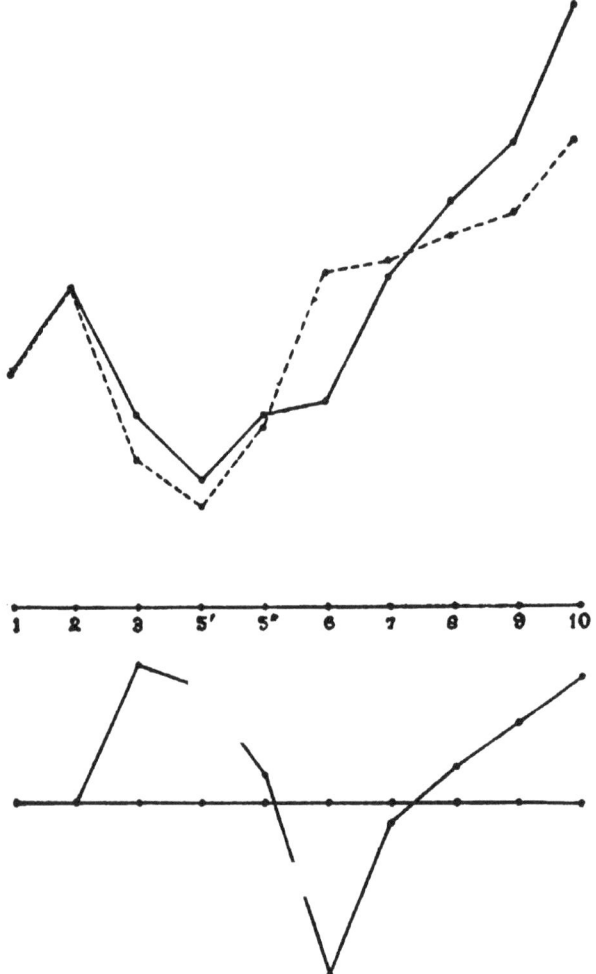

Figure 1. Curves of cell volumes (upper) and of nucleus-plasma coefficients (lower). In the upper curve, the solid line represents the cell body, the broken line the nucleus, which are constructed upon the upper base line. By making the reduction of the volume figures (Table I.) in the ratio of the resting cell to its nucleus, the figures being 40880 : 2000, the sizes relative to the resting cell as well as the absolute sizes are indicated. The resulting figures are taken as the ordinates in centimeters.

The nucleus-plasma curve is drawn about the lower base

line, which represents the level of the relation for the resting cell, by constructing the ordinates above or below according as the figures for the different stages as given in Table I. are greater or less than for the resting cell. These figures are taken as centimeters and reduced one-third.

TABLE I.

The cell measurements and the volumes and nucleus-plasma coefficients calculated therefrom for the human case.

Stages of Activity.	Average Diameters of Cell Body and Nucleus in Millimeters.	Relative Volumes of Cell Body and Nucleus.	Nucleus-plasma Coefficients.
1	80.92 × 42.54 22.66 × 17.78	146,437.04 7,163.46	19.44
2	87.12 × 47.94 24.88 × 19.88	200,223.02 9,832.92	19.36
3	78.92 × 39.12 20.66 × 14.76	120,777.12 4,500.95	25.83
5'	76.7 × 32.18 21.06 × 12.1	79,426.68 3,083.39	24.76
5''	84.1 × 37.78 23.12 × 15.46	120,038.29 5,526.42	20.72
6	73.98 × 41.8 24.24 × 20.64	129,260.81 10,326.48	11.52
7	81.34 × 50.64 24.48 × 20.86	208,589.11 10,652.23	18.58
8	85.68 × 54.54 24.86 × 21.44	254,864.76 11,427.50	21.30
9	91.1 × 56.64 25.34 × 21.86	292,257.00 12,108.97	23.14
10	99.75 × 61.63 25.6 × 22.00	378,666.69 14,417.92	25.26

BIBLIOGRAPHY.

Chiarini, P. 1906. Changements morphologiques qui se produisent dans la rétine des vertébrés par l' action de la lumière et de l' obscurité. Arch. Ital. de biol., xlv.

Collin, R. 1906. Recherches cytologiques sur le développement de la cellule nerveuse. Névraxe, viii.

Demoor, J. 1896. La plasticité morphologique des neurones cérébraux. Arch. de biol., xiv.

Dolley, D. H. 1909 (a). The pathological cytology of surgical shock. I. Jour. Med. Research, xx.

1909 (b). The morphological changes in nerve cells resulting from overwork in relation with experimental anemia and shock. Jour. Med. Research, xxi.

1909 (c). The neurocytological reaction in muscular exertion. I. The sequence of the immediate changes in the Purkinje cells. Amer. Jour. Phys., xxv.

1910. The pathological cytology of surgical shock. II. The numerical statement of the upset of the nucleus-plasma relation. Jour. Med. Research, xxii.

1911. Studies on the recuperation of nerve cells after functional activity from youth to senility. I. Jour. Med. Research, xxiv.

Eve, F. C. 1896. Sympathetic nerve cells and their basophile constituent in prolonged activity and repose. Jour. Phys., Camb., xx.

Goldschmidt, R. 1904. Der Chromidialapparat lebhaft functionierender Gewebezellen. Zool. Jahr., Abt. f. Anat. u. Ontog. der Thiere, xxi.

1904. Die Chromidien der Protozoen. Arch. f. Protistenk , v.

Goldschmidt, R., and Popoff, M. 1907. Die Karyokinese der Protozoen und der Chromidialapparat der Protozoen- und Metazoenzelle. Arch. f. Protistenk., viii.

Goldschmidt, R. 1909. Das Skelett der Muskelzelle von Ascaris nebst Bemerkungen über den Chromidialapparat der Metazoenzelle. Arch. f. Zellf., iv.

1908. Das Nervensystem von Ascaris lumbricoides und megalocephala. I Teil. Zeitschr. wiss. Zool., xc.

1909. Dasselbe. II. Teil. Zeitschr. wiss. Zool., xcii.

1910. Dasselbe. III. Teil. Festschrift zum sechzigsten Geburtstage Richard Hertwigs, II.

Guerrini, G. 1899. De l' action de la fatigue sur la structure des cellules nerveuses de l' écorce. Arch. Ital. de biol., xxxii.

1902. Action de la fatigue sur la fine structure des cellules nerveuses de la moelle épinière. Arch. Ital. de biol., xxxvii.

Hatai, S. 1904. A note on the significance of the form and contents of the nucleus in the spinal ganglion cells of the fetal rat. Jour. Comp. Neurol. and Psych., xiv.

Hertwig, R. 1902. Die Protozoen und die Zelltheorie. Arch. f. Protistenk., i.

1903. Ueber das Wechselverhältniss von Kern und Protoplasma. Sitzungber. d. Gesellsch. f. Morph. u. Physiol., xviii.

1903. Ueber Korrelation von Zell- und Kerngrösse und ihre Bedeutung für die geschlechtliche Differenzierung und die Teilung der Zelle. Biol. Cent., xxiii.

1908. Ueber neue Probleme der Zellenlehre. Arch. f. Zellf., i.

Hodge, C. F., 1892. A microscopical study of changes due to functional activity in nerve cells. Jour. Morph., vii.

1894.   A microscopical study of the nerve cell during electrical stimulation.   Jour. Morph., ix.

Holmgren, E.   1900.   Studien in der feineren Anatomie der Nervenzellen.   Anat. Hefte, xv.

Lambert, M.   1893.   Notes sur les modifications produites par l' excitation électrique dans les cellules nerveuses des ganglions sympathiques. C. R. Soc. de biol., xxxi.

Lee, A. B.   1906.   Vade Mecum, Sixth Edition, 155.

Levi, G.   1896.   Contributo alla fisiologia della cellula nervosa (cited by Van Durme) Riv. di patol. nerv. e ment., v.

Lugaro, E.   1895.   Sur les modifications des cellules nerveuses dans les divers états fonctionnels.   Arch. Ital. de biol., xxiv.

Luxenburg, J.   1899.   Ueber morphologische Veränderungen der Vorderhornzellen des Rückenmarks während der Thätigkeit.   Neur. Centralb., xviii.

Mann, G.   1895.   Histological changes induced in sympathetic, motor and sensory nerve cells by functional activity.   Jour. Anat. and Phys., Lond., xxix.

Marinesco, G.   1909.   La Cellule Nerveuse.   Chap. 11.   O. Doin et fils, Paris.

Nissl, F.   1894.   Mittheilungen zur Anatomie der Nervenzellen.   Allg. Zeitschr. f. psych., l.

1896.   Die Beziehungen der Nervenzellensubstanzen zu den thätigen, ruhenden und ermüdeten Zellzuständen.   Neur. Cent., xv.

Odier, R.   1898.   Recherches expérimentales sur les mouvements de la cellule nerveuse de la moelle épinière.   Rev. méd. de la Suisse romande, xviii.

Pergens, E.   1896 and 1897.   Action de la lumière sur la rétine.   Ann. de la Soc. royale des Sciences Med. et Natur. de Bruxelles, v and vi.

Pick, F.   1898.   Ueber morphologische Differenzen zwischen ruhenden und erregten Ganglienzellen.   Deutsch. med. Woch., xxii.

Pugnat, C. A.   1901.   Modifications histologiques des cellules nerveuses dans la fatigue.   Jour. de Phys. et de Path. générale, iii.

Rohde, E.   1903.   Untersuchungen über den Bau der Zelle.   Zeitsch. Wiss. Zoöl., lxxii.

Scott, F. H.   1899.   The structure, micro-chemistry and development of nerve cells with special reference to their nuclein compounds.   Trans. Canadian Institute, vi.   Also University of Toronto Studies, 1900.

Valenza, G. B.   1896.   I cambiamenti microscopici della cellula nervosa nell' attivita funzionale e sotto l'azione di agenti stimolanti e distruttori. (quoted by Luxenburg and Marinesco).   Atti. R. Acad. scienze fisiche e nat. di Napoli, vii.

Van Durme, P.   1901.   Etude des différent états fonctionnels de la cellule nerveuse corticale.   Le Nevraxe, ii.

Vas, F.   1892.   Studien über den Bau des Chromatin in der sympathischen Ganglienzelle.   Archiv. f. mikrosk. Anat., xl.

Verworn, M.   1899.   General physiology.   English translation.   The Macmillan Company.

THE RELATIVE IMPORTANCE OF THE BOVINE AND HUMAN TYPES OF TUBERCLE BACILLI IN THE DIFFERENT FORMS OF HUMAN TUBERCULOSIS.[*]

WM. H. PARK AND CHARLES KRUMWIEDE, JR.

Assisted by

BERTHA VANH. ANTHONY, MARIE GRUND, AND LOUISA P. BLACKBURN.

(*From the Research Laboratory of Department of Health, New York, N.Y.*)

The results of the examination of various cases as published in the Journal of Medical Research (Volume XXIII., page 205), showed conclusively that infection with bovine tubercle bacilli in New York City and its vicinity is essentially a disease of children, and a serious menace to life in infants. Because of this, it was decided to examine additional material, especially from fatal cases of tuberculosis in children. These additional cases have given results closely approximating those already published and, therefore, serve to verify the reliability of these figures.

No further details are given of technical procedures or methods of differentiation. These have been fully discussed and there is nothing to add. The only point that needs mention and cannot be too often reiterated is, that correct cultural differentiation depends largely on suitable culture media. Even with long experience one finds irregularities creeping in, viz., minimum growth of human viruses on some batches of media, or irregularities in individual tubes planted from the same material. We have prepared egg media in numerous ways and with widely varying reactions and have been unable to find anything that accounts for these variations. In the routine examination of a series of cases these irregularities are quickly noted and controlled. They would be especially dangerous in the cultural differentiation of isolated cases if rabbits were not used to control the results.

The above remarks must not be interpreted as a criticism of the egg medium, for there is no doubt that it is very

---

[*] Received for publication Oct. 23, 1911.

nearly perfect for isolation and cultural differentiation and shows a minimum of variability.

The study of the Smith Reaction has been continued and the complete results are given separately (see page 335 of this number).

Differentiation in the following additional cases has, in some instances, been on cultural characteristics alone. The bovine viruses have, however, been fully tested in rabbits.

TABULATION* OF ADDITIONAL CASES EXAMINED.

Diagnosis of Cases Examined.	Adults 16 Years and Over.		Children 5 to 15 Years.		Children Under 5 Years.		Notes.
	H.	B.	H.	B.	H.	B.	
Pulmonary tuberculosis	3	—	—	—	2	—	
Tuberculous adenitis, cervical	—	—	—	—	—	1	Case from Babies' Hospital.
Abdominal tuberculosis	—	—	—	—	1	—	
Generalized tuberculosis, alimentary origin	—	—	—	—	—	1	
Generalized tuberculosis	—	—	—	—	6	—	
Generalized tuberculosis, including meninges	1	—	—	—	7	—	
Tubercular meningitis	1	—	1	—	12	1	
Tuberculosis of bones and joints	—	—	—	—	1	—	
Genito-urinary tuberculosis	3	—	—	—	—	—	
Tuberculosis of skin.	1	—	—	—	—	—	
Totals	9	—	1	—	29	3	Total cases, 42.

* For method of tabulation, see previous article, Vol. XXIII., page 338.

Including these cases in the tabulations already published gives the following tabulation of cases examined:

FINAL TABULATION OF CASES EXAMINED.

Diagnosis of Cases Examined.	Adults 16 Years and Over.		Children 5 to 16 Years.		Children Under 5 Years.		Notes.
	H.	B.	H.	B.	H.	B.	
Pulmonary tuberculosis.	281	—	8	—	7	—	Clinical diagnosis only known, and therefore no positive details as to the extent of lesions elsewhere.
Tuberculous adenitis, inguinal and axillary.	1	—	4	—	—	—	(See next.)
Tuberculous adenitis, cervical.	9	—	19	8	6	13	In two cases cultures were from axillary nodes, but the primary focus was cervical. Another case died shortly afterward with pulmonary tuberculosis.
Abdominal tuberculosis.	1	.	1	1	1	3	Milk supply of one child subsequently examined. Tubercle bacilli isolated.
Generalized tuberculosis, alimentary origin.	1	.	—	.	1	2	Only three cases given under this heading. Many of the cases in the following subdivisions showed marked intestinal lesions and possibly some were of alimentary origin.
Generalized tuberculosis.	2	—	1	—	18	4	One bovine case had tuberculous osteomyelitis of the metatarsal bone.
Generalized tuberculosis including meninges.	1	—	—	—	25	1	

FINAL TABULATION OF CASES EXAMINED. — *Continued.*

Diagnosis of Cases Examined.	Adults 16 Years and Over.		Children 5 to 16 Years.		Children Under 5 Years.		Notes.
	H.	B.	H.	B.	H.	B.	
Tubercular meningitis.	1	—	2	—	26	2	No autopsy. Extent of lesions elsewhere unknown.
Tuberculosis of bones and joints.	1	—	10	—	7	—	
Genito-urinary tuberculosis.	6	1	1	—	—	—	The adult bovine case was tuberculosis of kidney. Removal of kidney. Complete recovery.
Tuberculosis of skin.	1	—	—	—	—	—	
Tuberculous abscess.	1	—	—	—	—	—	Possibly primary in bone.
Totals...........	305	1	46	9	91	25	

Double infection, one case. Both types isolated. Generalized tuberculosis including meninges, thirteen months. Mesenteric nodes gave human type. Meningeal fluid gave bovine type.

TOTAL CASES—478.

The additional cases necessitate some change in the mortality tabulations given in Volume XXIII., page 361. As it is noted, there is a slight lowering of the percentages in the Babies' Hospital series and also in the cases considered as a whole.

TOTAL FATAL CASES IN CHILDREN.

Diagnosis.	Children 5 to 16 Years of Age.		Children under 5 Years.		Notes.
	H.	B.	H.	B.	
Pulmonary tuberculosis.	—	—	7	—	One case included, probably fatal, data incomplete.
Tuberculous adenitis.	1	—	—	—	Other cases recovered as far as is known.
Abdominal tuberculosis.	—	—	—	3	Three other cases, one bovine and two human, were operative cases with recovery as far as known.
Generalized tuberculosis.	1	—	19	4	Two other bovine cases died directly of exanthemata with complications.
Generalized tuberculosis including meninges.	—	—	25	2	One case gave both type of bacilli, included under bovine, as this type caused the meningitis.
Tubercular meningitis.	2	—	26	2	
Totals . . .	4	—	77	11	or 12⅘ due to bovine type under 5 years.

FATAL CASES. BABIES' HOSPITAL.

Diagnosis.	Children 5 to 16 Years.		Children Under 5 Years.		Notes.
	H.	B.	H.	B.	
Pulmonary tuberculosis.	—	—	6	—	One case included, probably fatal, data incomplete.
Generalized tuberculosis.	1	—	14	2	
Generalized tuberculosis including meninges.	—	—	21	1	One case gave both types of bacilli, included under bovine, as this type caused the meningitis.
Tubercular meningitis.	—	—	18	1	
Totals . . .	1	—	59	4	or 6⅘ due to bovine type under 5 years.

No additional cases from the Foundling Hospital have been examined, but the table is repeated here for completeness. As has been noted this is really a special series of children, fed exclusively cow's milk, giving an excessively high incidence of bovine infection. They should be so considered and general deductions cannot be drawn from them and applied to infant mortality at large.

CASES. FOUNDLING HOSPITAL.

Diagnosis.	Children Under 6 Years.		Notes.
	H.	B.	
Abdominal tuberculosis.	—	1	
Generalized tuberculosis.	3	3	Two of the bovine cases died directly of exanthemata; the tuberculosis was not altogether negligible. These two cases not included in the total fatal cases.
Generalized tuberculosis including the meninges.	1	1	
Totals	4	5	Total cases, 9.

If we consider the Babies' Hospital cases alone the mortality due to bovine infection is six and one-third per cent. The total mortality cannot be considered normal while the Foundling Hospital cases are included. Exclusive of these the general mortality would be a small fraction under ten per cent. These percentages are believed to give a truthful picture of the percentage limits of mortality and are applicable to the general population of New York City and on the whole to most American cities. The children received at the Babies' Hospital represent the average New York City child, while those at the Foundling Hospital represent the portion fed on cow's milk.

AGE INCIDENCE OF TOTAL FATAL CASES UNDER FIVE YEARS.

Diagnosis.	Under 1 Year.		Between 1 and 2 Years.		Between 2 and 3 Years.		Between 3 and 4 Years.		Between 4 and 5 Years.	
	H.	B.	H.	B.	H.	B.	H.	B.	H.	B.
Pulmonary tuberculosis	4	—	2	—	—	—	1	—	—	—
Abdominal tuberculosis	—	—	—	2*	—	—	—	—	—	—
Generalized tuberculosis	10	1	5	2	3	—	—	1	1	—
Generalized tuberculosis including meninges	14	—	8	2†	2	—	1	—	—	—
Tubercular meningitis	9	—	10	1	5	1	2	—	—	—
Totals	37	1	25	7	10	1	4	1	1	—

* One case, bovine type, definite age not given, infant. Not included.
† Case of double infection included here.

If we further correct the application of the percentages deduced, it is evident from the above table that we should apply them mainly to the first three years of life. The cases examined between three and five years are few in number. Further, the pulmonary type of disease becomes more evident at this age and if more of this type of disease had happened to fall in this age period the percentage would have been reduced. The source of our material did not include many cases about this age and only four cases came to us between five and sixteen years, not including the pulmonary cases, the outcome of which we could not determine. We are inclined to believe that around the ages of four to five years the incidence of fatal bovine infection rapidly falls and that one factor in this fall is the increase in the pulmonary types of infection.

Since the completion of the article in Volume XXIII. a complete report has been published by Burckhardt of his investigation of surgical tuberculosis. The following tabulation gives the results of this important work. As will be noted the percentage of bovine infection is higher than one would expect.

Diagnosis of Cases Examined.	Adults 16 Years and Over.		Children 5 to 16 Years.		Children Under 5 Years.		Notes.
	H.	B.	H.	B.	H.	B.	
Tuberculous adenitis, cervical.	5	—	3	1	—	—	
Abdominal tuberculosis.	3	—	1	—	—	1	One case originally classified here, changed to next heading.
Generalized tuberculosis.	1	—	—	—	—	—	Clinically the abdominal symptoms only were marked.
Tuberculosis of bones and joints.	9	1	12	2	4	—	One case (child) age not stated, therefore not tabulated, gave human type.
Genito-urinary tuberculosis.	3	—	1	—	—	—	
Tuberculosis of skin.	1	—	—	—	—	—	
Totals............	22	1	17	3	4	1	(1) Total cases, 49.

The observations of Burckhardt on the comparison of the pathological and clinical details in his bovine and human infections in joints is interesting. Of twelve cases of tuberculosis of the knee, two were noteworthy by the absence of marked bone lesions. One of them had lasted one year, this was of the human type. The other, which showed a very small bone focus in the tibia only after careful search, had lasted for thirteen years. This case was of bovine origin. The age of the patient was nineteen years. Recovery in both instances was complete.*

Of the three cases of tuberculosis of the hip there was one bovine infection. Here again the history was of long duration, viz., eight years. This case (eleven years old) also showed only superficial granulations. The two other cases, which were human infections, showed deep bone involvement. The operative results in the bovine case were very good. Another bovine infection of the elbow joint showed only granulations on the capsule. This case, however, had lasted only one year.

Because of these peculiarities in the bovine cases Burckhardt thinks it would be well to investigate those joint cases showing practically no bone involvement and especially those cases giving a long history of infection. From the cases already cited one might expect many if not the majority of such cases to be bovine infections. This is only a suggestion, and only further cases could give the necessary evidence to prove this possibility.

It is of interest to note that the only distinct bone infection without joint complication due to the bovine bacillus is one reported by Oehleker. This case gave a long history of six and three-quarter years. The disease was limited to one metacarpal bone.

Another report is that of Rothe, who publishes a continuation and amplification of Gaffky's work. The bronchial and

---

* On the appearance of the preliminary report we had written to Dr. Burckhardt, who kindly sent us details of his cases prior to his final publication which would have given us the opportunity of including them in our preceding report had the letter reached us. Due to some clerical error the letter was filed away and not brought to our notice till too late.

mesenteric nodes of one hundred children up to five years of age were examined. The material was from successive autopsies on children dying from any cause. Guinea-pigs were inoculated and if the inoculations were positive, cultures were isolated and tested. The results of his work and Gaffky's are given in the following table copied with slight change from Rothe's article:

Author.	Number of Cases and Age.	Number Positive.	Both Nodes.	Tuberculosis Present in Mesenteric Nodes Only.	Tuberculosis Present in Bronchial Nodes Only.	Human Infections.	Bovine Infections.
Gaffky	300; under 14 years.	57 = 19%	29	11	17	55	(2?) See next paragraph.
Rothe	100; under 6 years.	21 = 21%	13	3	5	20	1
Totals	400.	78 = 19.5%	42	14	22	75	3 (?)

In the two cases in which Gaffky failed to isolate cultures, the great difficulty encountered is good presumptive evidence of their being bovine. They cannot be excluded and if the figures are used at all they should be considered as bovine, failing proof to the contrary. If these cases are not included Gaffky's figure should be completely excluded from any consideration of the incidence of bovine infections. Rothe says: " Lässt man die beiden Fälle der 1. Untersuchungsreihe, in denen die Gewinnung einer Reinkultur nicht gelungen war, als unentschieden ausser Rechnung, so verbleiben 76 tuberkulöse Fälle, darunter. 75–98.68 per cent mit humaner und 1 : 1.32 per cent mit bovine infection." What he should have said was, excluding Gaffky's cases altogether because they are incomplete and inclusive, there are left twenty-one cases of tuberculosis, of which twenty or 95.24 per cent are human and one or 4.76 per cent are bovine. If Gaffky's cases are considered and the two doubtful cases are classed as bovine the percentage is very much the same. The following tabulation gives the results of Rothe's work. It is

difficult to tabulate his cases, as the details of the post-
mortem examinations are not given.

| Diagnosis of Cases. | Children Under 5 Years. | | Notes. |
	Human.	Bovine.	
Pulmonary tuberculosis.	5	3	1	The bovine case had swollen mesenteric nodes which were considered non-tuberculous on macroscopic examination. One of four inoculated guinea-pigs became tuberculous.
Abdominal tuberculosis.	—	2	—	Slight latent tuberculosis of mesenteric nodes.
Generalized tuberculosis.	6	3	—	Three cases were latent or slight tuberculosis of lymph nodes.
Tubercular meningitis.	1	—	—	
Totals	20		1	

The above table brings out very strongly the difficulty
encountered in tabulating cases of different authors who
have had different aims in view.   Thus in one instance slight
tuberculosis of the lymph nodes in the thoracic and abdominal cavities is placed under generalized tuberculosis to bring
out the fact that the bacilli have disseminated.   On the other
hand, a more marked tuberculosis involving the lungs with
what has theoretically been assumed to be secondary involvement of the intestines and possibly the mesenteric nodes, is
classified under pulmonary tuberculosis.   Although this
seems contradictory we have attempted to keep certain
distinct types of disease separate, but where such types do
not exist to classify on the basis of the amount of dissemi-
nation.   In this case, as in our previous report, the degree
of involvement is noted by placing the slight infections to
the right of a verticle line subdividing the space for these
cases.

One fact in Rothe's article is very valuable for comparison
of the value of different media.   Gaffky failed to isolate two
viruses as has been noted.   Rothe states that isolation was

successful from the first guinea-pig passage seventeen times, from the second guinea-pig passage ten times, from the third, four times and in one case each, only after the fifth, eighth, and ninth passage. The last two were the bovine strains from the case of pulmonary tuberculosis. He used two per cent glycerine beef serum for isolation. This compares very badly with results obtained with egg media. Except for unavoidable variations in media and in individual technic, the positive results from the first pig, if the pigs are sufficiently tuberculous to give the necessary material, should be ninety-five per cent at the very lowest. This includes human and bovine viruses. As a matter of fact we have found it to be very exceptional to fail to isolate from the first pig, even with bovine viruses.

Finally, the results of Möllers are to be added. He reports the study of cultures isolated from the sputa of fifty-one cases of pulmonary tuberculosis. In ten, three isolations were made at different times, in nine, cultures were isolated twice, in the remaining thirty-two one culture only was isolated. Fifty of these cases were adults, one was an infant (the age groups were kindly sent us by Dr. Möllers).

Bullock also reports twenty-three cases of pulmonary tuberculosis yielding cultures of the human type. Adding the preceding reported cases and those of Fabyan (see foot note at end of previous report) gives the following tabulation:

TOTAL SUMMARY OF CASES REPORTED.

Diagnosis of Cases Examined.	Adults 16 Years and Over.		Children 5 Years to 16 Years.		Children Under 5 Years.	
	Human.	Bovine.	Human.	Bovine.	Human.	Bovine.
Pulmonary tuberculosis . .	363	(1?)*	3	—	13 | 3	1 | —
Tuberculous adenitis, axillary	1	—	—	—	2	—
Tuberculous adenitis, cervical	18	1	17	13	9	8
Abdominal tuberculosis . .	13	4	1 | 6	3 | 3	3 | 5	6 | 4
Generalized tuberculosis, alimentary origin	6	1	2 | —	3 | —	12 | —	10 | —
Generalized tuberculosis .	27	—	3 | —	1 | —	20 | 5	1 | —
Generalized tuberculosis including meninges, alimentary origin	—	—	1	—	3	8
Generalized tuberculosis including meninges . .	4	—	7	—	27	—
Tubercular meningitis . .	—	—	1	—	1	2
Tuberculosis of bones and joints	26	1	28	3	19	—
Genito-urinary tuberculosis	11	—	1	—	—	—
Tuberculosis of skin . . .	2	—	1	—	1	—
Miscellaneous cases:						
Tuberculosis of tonsils .	—	—	—	1	—	—
Tuberculosis of mouth and cervical nodes . .	—	1	—	—	—	—
Tuberculous sinus or abscess	1	—	—	—	—	—
Sepsis, latent bacilli . . .	—	—	—	—	1	—
Totals	472	9	71	27	124	40

Mixed or double infections: — 3 cases:

Generalized tuberculosis. Alimentary origin. 30 years. Human and bovine type in mesenteric nodes. Human type in bronchial node.

Generalized tuberculosis. Alimentary origin. 5½ years. Human type in spleen. Bovine type in mesenteric node.

Generalized tuberculosis including meninges. Alimentary origin. 4 years. Human type in meninges and bronchial nodes. Bovine type in mesenteric nodes.

TOTAL CASES — 746.

* See Addenda for additional cases of Royal Commission not included here, due to lack of complete details.

COMBINED TABULATION CASES REPORTED AND OWN SERIES OF CASES.

Diagnosis.	Adults 16 Years and Over.		Children 5 to 16 Years.		Children Under 5 Years.	
	H.	B.	H.	B.	H.	B.
Pulmonary tuberculosis . .	644	(1?)	11	—	23	1
Tuberculous adenitis, axillary or inguinal	2	—	4	—	2	—
Tuberculous adenitis, cervical	27	1	36	21	15	21
Abdominal tuberculosis . .	14	4	8	7	9	13
Generalized tuberculosis, alimentary origin	6	1	2	3	13	12
Generalized tuberculosis .	29	—	4	1	43	5
Generalized tuberculosis including meninges, alimentary origin	—	—	1	—	3	8
Generalized tuberculosis including meninges . . .	5	—	7	—	52	1
Tubercular meningitis . .	1	—	3	—	27	4
Tuberculosis of bones and joints	27	1	38	3	26	—
Genito-urinary tuberculosis	17	1	2	—	—	—
Tuberculosis of skin . . .	3	—	1	—	1	—
Miscellaneous cases :						
Tuberculosis of tonsils .	—	—	—	1	—	—
Tuberculosis of mouth and cervical nodes . .	—	1	—	—	—	—
Tuberculous sinus or abscess	2	—	—	—	—	—
Sepsis, latent bacilli . .	—	—	—	—	1	—
Totals	777	10	117	36	215	65

Mixed or double infections, 4 cases.

TOTAL CASES — 1,224.

Taking the cases given in the total tabulation and combining the important diagnoses under one heading gives us the following table, which shows clearly the percentage incidence of bovine infection.   Caution is necessary in applying these figures, they tell nothing but the incidence; the seriousness of the infection is indicated in the preceding tables.

PERCENTAGE INCIDENCE OF BOVINE INFECTION.*

Diagnosis.	Adults 16 Years and Over.	Children 5 to 16 Years.	Children Under 5 Years.
Pulmonary tuberculosis	0% (?)†	0%	4.1%
Tuberculous adenitis, cervical.......	3.6%	36%	58%
Abdominal tuberculosis	22%	46%	59%
Generalized tuberculosis...........	2.7%	40%	23%
Tubercular meningitis (with or without generalized lesions)..........	0%	0%	13.6%
Tuberculosis of bones and Joints	3.5%	7.3%	0%

* Exclusive of cases of double infections. In considering pulmonary cases it must, however, be remembered that bovine tubercle bacilli have been isolated from the lung in cases of generalized tuberculosis in children.

The number of cases under some of the headings is too small to deduce percentages. Reference to the preceding table makes this evident.

† If the two bovine cases of the Royal Commission (see Addenda) were included, the percentage would be 0.3%. We have not included these two cases as the additional human cases could not be included. If we combine the pulmonary cases regardless of age we can then add these cases, giving us a total of 710, exclusive of the one doubtful case. Of these, 3 or 0.42% were bovine infections.

This table gives only the incidence of infection and nothing as to the severity of the disease. This is seen by referring to the main tables in which we divide the cases according to severity. Under certain diagnoses a great many latent or slight infections are included, which may never have had any effect on the health of the child had not some intercurrent infection lead to death. Furthermore, due to selection of material, the number of cases of generalized tuberculosis of alimentary origin is markedly out of proportion and bears no relation to the incidence of these cases in proportion to other types of disease. If we rearrange the figures under these headings leaving out all but severe types of disease and consider the selected cases of alimentary tuberculosis separately, the following table gives the results. Only cases under sixteen years are considered, as we have only noted the severity of disease in these cases in the tables. The percentage of our cases are given for comparison.

PERCENTAGE OF BOVINE INFECTION.   (*Revised.*)

Diagnosis.	Children 5 to 16 Years.		Children Under 5 Years.	
	Combined Figures.	Own Figures.	Combined Figures.	Own Figures.
Abdominal tuberculosis........	66%	50%	69%	75%
Generalized tuberculosis, alimentary origin	60%	—	48%	66%
Generalized tuberculosis.......	20%	—	11%	18%
Tubercular meningitis, secondary to tuberculosis of alimentary type..................	—	—	72%	—
Tubercular meningitis (other than preceding)............	—	—	6%	5½%

Revising the percentages in this way gives close agreement. The percentages are highest in the relatively less common types of tuberculosis. In the two types of tuberculosis, which mainly constitute the fatalities in young children, the percentages range from five and one-half per cent to eighteen per cent. All things considered, we feel safe in saying that the percentage of deaths from bovine tuberculosis in young children, viz., six and one-third per cent to ten per cent as deduced from our unselected cases in New York City, are applicable to most cities throughout the world whose milk supply is similar to ours.

Addenda. — Since the completion of this final summary of our work the "Final Report of the Royal Commission appointed to inquire into the relations of Human and Animal Tuberculosis" has appeared. Only Part I., viz., the Report, has reached us. The Appendices which will contain the experimental data and details of the cases examined are not available. Failing these details it is impossible for us to include their added cases in our tabulations, nor is it possible to discuss their final conclusions. For this reason we add simply a short summary of the conclusions given in the report.

Concerning the existence of different types of tubercle bacilli they say : " For purposes of description it is advantageous to distinguish three types of tubercle bacilli, recognizable by their individual characters. These are the human, bovine and avian types. The human type, although so named, is not the only one found in cases of tuberculosis in man. It is the organism present in the majority of such cases, but in some cases of human disease the bacilli present are of the bovine type, and in others the bacilli have special characters distinguishing them from each of the three principal types. In natural cases of tuberculosis in cattle the only type of bacillus present is the bovine type." The conclusions on the differential characters of the human and bovine type are practically identical with our own. All their viruses from man are tabulated as bovine or human in all their characteristics with the exception of the viruses from cases of lupus. The following gives a short summary of their final results as they tabulate them :

CASES OF HUMAN TUBERCULOSIS OTHER THAN LUPUS (108 CASES).

Nature of Case.	Bovine.	Human.	Mixed Viruses, Human and Bovine.
Primary pulmonary tuberculosis,	o	14	^
Sputum from individual cases of pulmonary tuberculosis	2	26	
General tuberculosis	o	3	
Tuberculous meningitis	o	3	ʊ
Bronchial gland tuberculosis....	o	3	2 (H. 13 A.D.; H. 60 W.B).
Cervical gland tuberculosis	3	6	o
Primary abdominal tuberculosis,	14	13	2 (H. 49 T.C.; H. 90 I. P.
Joint and bone tuberculosis.....	o	13	1 (H. 16 J. H.).
Tuberculosis of testicle, kidney, or suprarenal. One each	o	3	

The most astonishing thing is the presence of bovine tubercle bacilli in the sputum of two cases. One case was twenty-one years, the other was thirty-one years of age. In one case the examination was repeated after seventy-six, one hundred and seventeen, and one hundred and eighteen days with the same results. In the other case, a second specimen collected after one hundred and eighteen days gave the same results. Both cases subsequently died. The cause of death in one was given as phthisis, the other apparently had general tuberculosis with intestinal ulceration. At the time of collection of the specimens they were clinically cases of primary pulmonary tuberculosis. No autopsy could be obtained.

The other noteworthy point is the number of mixed viruses. In the preceding report the Commission gave the results on certain passage experiments without drawing any final conclusions. The changes found, they now conclude, were due to the presence of a mixture of human and bovine viruses in the original viruses. In the case of some of the viruses, for instance, H. 13 A.D., such a conclusion is justified. The extraordinary fluctuations in Virus H. 49 could also be accounted for in this manner. That it is justified in all cases, however, seems almost beyond the possibility of experimental proof. As soon as material has passed through calves a possible experimental error, viz., "spontaneous" tuberculosis of the calf, is encountered. That this error can be surely avoided in each case seems to us almost impossible. Critical judgment, however, must be withheld till every detail of the experiments is published.

The cultures from cases of lupus form a group by themselves. Twenty cases were examined. Culturally the viruses fell into either the human or bovine type. In virulence, however, only one was typically of the bovine type and two of the human type. The cultures were tested by inoculation into calves, rabbits, monkeys, and guinea-pigs. In some or all of these animals the virulence, with the exception of the above three cultures, was less than one would expect with one or the other type as determined culturally. The decrease of virulence from the type with some of the viruses was very marked.   In a few of these viruses it was possible to raise the virulence to that of the type.   It is of interest to note that the only other aberrant cultures isolated were from two horses and these like some of the lupus cultures were culturally of the bovine type, but of degraded virulence. Passage experiments resulted in an increase of virulence to that of the bovine type.

The aberrant cultures were restricted to those isolated from the cases of lupus and equine tuberculosis already described.   From the description they are neither typically bovine nor human in all their characters.   The most striking thing is that they should have mostly come from one type of disease in man.

As to passage experiments and other means to cause modification of type they report complete inability to cause any change; the only exceptions being the lupus and equine cultures, where it was possible to enhance the virulence of some cultures by passage experiments.

They concluded: " Thus we are inclined to regard transmutation of the bacillary type as exceedingly difficult if not impracticable of accomplishment by laboratory procedure, though in view of certain instances in which we obtained from one and the same human body both types of bacillus, we are not prepared to deny that the transmutation of one type into another may occur in nature."   The lupus cultures and the cultures from the horse they believe must be considered to be naturally modified human or bovine tubercle bacilli, as the only other alternative is to consider them added fixed types.

For the preceding reason virulence cannot be considered a fixed characteristic, which makes it impossible to regard difference of virulence for the calf and rabbit as sufficient to establish the non-identity of the human and bovine types.

" There would therefore remain only slight cultural differences on which to found the conclusion that the human and bovine types represent two distinct organisms.   We prefer to regard these two types as varieties of the same bacillus.   .   .   ."

CONCLUSIONS.

Bovine tuberculosis is practically a negligible factor in adults. It very rarely causes pulmonary tuberculosis or phthisis, which causes the vast majority of deaths from tuberculosis in man and is the type of disease responsible for the spread of the virus from man to man.

In children, however, the bovine type of tubercle bacillus causes a marked percentage of the cases of cervical adenitis leading to operation, temporary disablement, discomfort, and disfigurement. It causes a large percentage of the rarer types of alimentary tuberculosis requiring operative interference or causing the death of the child directly or as a contributing cause in other diseases.

In young children it becomes a menace to life and causes from six and one-third to ten per cent of the total fatalities from this disease.

BIBLIOGRAPHY (ADDITIONAL).

Bullock, W. The problem of pulmonary tuberculosis considered from the standpoint of infection. Horace Dobell Lecture delivered before the Royal College of Physicians, London, Nov. 10, 1910.

Burckhardt, H. Bakteriologische Untersuchungen über chirurgische Tuberkulosen, ein Beitrag zur Frage der Verschiedenheit der Tuberkulose des Menschen und die Tiere. Deutschen Zeitschrift f. Chirurgie, cvi, 1, 1910.

Möllers, B. Uber den Typus der Tuberkelbazillen im Auswurf der Phthisiker. Veröffentlichungen der Robert Koch-Stiftung zur Bekämpflung der Tuberkulose. Heft 1, 1911.

Rothe. Untersuchungen uber tuberkulöse Infection im Kindesalter. Veröffentlichungen der Robert Koch-Stiftung zur Bekämpflung der Tuberkulose. Heft 2, 1911.

THE REACTION CURVE IN GLYCERIN BROTH AS AN AID IN DIFFERENTIATING THE BOVINE FROM THE HUMAN TYPE OF TUBERCLE BACILLUS.*

M. GRUND.

(From the Research Laboratory of the Department of Health, New York City.)

From the first, the division of tubercle bacilli into a human and a bovine type has been based upon cultural characteristics and virulence for laboratory animals, especially rabbits. While endeavoring to establish further biological differences between the two types, Theobald Smith in 1903 called attention to an apparently constant dissimilarity in reaction curves obtained when both types were cultivated on glycerin broth. There was, at the start, a reduction in the acidity of the medium, but the human type of tubercle bacilli soon produced enough acid to bring the reaction of the medium up to or beyond the original level, while bacilli of the bovine type continued to reduce the acidity until the broth became neutral or even alkaline. Although there was at times a slight acid production following this stage, the final reaction for a bovine virus never approached the initial degree of acidity. Further observations, which Smith published later, tend to illustrate the same difference between the two types.

Relatively few observers have undertaken the study of this reaction curve. Bang obtained results in the main identical with those of Smith. He extended his tests to twelve avian viruses, and observed that they followed the bovine type of curve, but produced more alkali than his bovine controls.

Lewis, at the suggestion of Theobald Smith, studied the reaction curves of six viruses of the human and nine of the bovine type, isolated from fifteen cases of cervical adenitis. For two of these viruses he obtained reversed reaction curves; Vir. 07.8, culturally and in virulence of the bovine type gave a typical human reaction curve. Vir. 07.54,

---

* These observations form a part of the studies on the Relative Importance of the Bovine and the Human Type of Tubercle Bacilli in the Different Forms of Human Tuberculosis by Dr. William H. Park and Dr. Charles Krumwiede, Jr., and have been carried out at their suggestion and under their direction at the Research Laboratory of the New York City Department of Health. Received for publication Oct. 23, 1911.

human in its virulence and cultural characteristics, gave in one test a human curve, in three other tests a curve of the bovine variety. In addition to these, there were four viruses of the bovine type, which in the early generations gave a human reaction curve, but later produced a typical bovine curve.

Duval obtained very atypical results with two of his viruses. These came from cases of a peculiar type of the disease and his results seem to require further proof.

Mohler and Washburn had among their viruses four which gave reversed curves; two of these were of the human and two of the bovine type.

Fibiger and Jensen reported thirty cases, and concluded that while the majority of human viruses produced one kind of reaction curve and most of the bovine the other, the number of both types of virus giving an intermediate curve was too great to make a division satisfactory. They found that the amount of growth and the acid production were not directly related. This was contrary to results obtained by Beitzke, whose tests, however, are few and rather incomplete.

The Royal Commission found that the final reaction of all the viruses could be so grouped as " to form an unbroken series in which there is nowhere a gap that would justify the conclusion that two essentially different kinds of organisms are dealt with." In the four viruses which produced atypical reaction curves, they noted variations in cattle virulence; a bovine curve resulting when the particular strain was highly virulent, a human reaction curve accompanying low virulence. To facilitate observations of the change in reaction, the Royal Commission have used, in addition to glycerin broth, glycerin litmus milk. On this medium, all the rapidly-growing viruses of the human type produced an initial increase in alkalinity, after which the milk turned acid and coagulated. When growth was less vigorous, the milk became acid but failed to coagulate. Poorly-growing viruses of the bovine type left the milk alkaline, while those which grew well turned it acid but never coagulated it.

It was thought desirable to test a large number of the viruses which had been isolated at the Research Laboratory from four hundred and seventy-eight unselected cases of tuberculosis in man, and also viruses from cattle, to ascertain how far this variation in acid production could be depended upon to aid in the differentiation of types. The tests were begun after both types had been cultivated on artificial media for some time and had become thoroughly accustomed to them. A few of the viruses, all of the human type, were transferred to glycerin broth after having been cultivated on artificial media for three or four generations only.

Unfermented one per cent peptone beef broth containing

five per cent of Merck's glycerin was used as a culture medium; the reaction varied between 1.9 per cent and 2.4 per cent acid to phenolphthalein. In the first three sets of tests Lehn and Fink's glycerin has been used; the results seem to be identical.

As soon as a virus had acquired the ability to form a good pellicle on this medium, it was planted out in three Erlenmeyer flasks of one hundred cubic centimeters capacity, each containing thirty-five cubic centimeters of the glycerin broth. This amount gives a layer of fluid about 1.5 centimeters deep and offers the largest possible surface for growth. Tin-foil caps proved generally efficient in checking undue evaporation. In looking over the report of other observers, one is struck by the fact that no two agree on the length of time during which they continued the tests, nor has any one titrated at regular intervals of time. Thus " end-reactions " are recorded after sixty-five, eighty-nine, or two hundred and twenty-five days. In no instance could a pellicle be induced to grow on either solid or fluid media after ninety days, and it was therefore decided to make this period the end of a test. Titrations of the three flasks were planned so as to use fluid from flask A at the end of fifteen and forty-five days, from flask B at the end of thirty and sixty days, and from A, B, and C at the end of ninety days. The final reaction therefore represents the average of the three flasks, except in a few instances where contamination, excessive evaporation, or very poor growth caused a flask to be ruled out. Although two of the flasks for each virus were opened twice during the experiment, the number of contaminated flasks was not large. For about two-thirds of all the viruses tested, the final growth in each flask was collected on dried and weighed filter paper and was weighed, after drying in the incubator for forty-eight hours. As the relative growth of the several viruses was wanted, rather than the absolute weight of any one pellicle, this method was deemed sufficiently accurate. But there was often considerable difference between the actual weight of a pellicle and its estimated growth. The latter is expressed in grades, of which Grades 1-3 indicate poor growth, Grades 4-6 moderate development, Grades 7-9 luxuriant growth.

Of the one hundred and seventy-three viruses which have been cultivated on glycerin broth, fifty-three were tested twice, seven three times and one four times on different lots of broth. Generally speaking, the human viruses have reduced the reaction of the broth to nearly neutral, and later have produced enough acid to bring the reaction back somewhere near the initial. The bovine viruses have rendered the medium alkaline and the final reaction, though often above the lowest reaction recorded, has been considerably lower than the initial acidity of the medium. According to their final reactions both human and bovine viruses may be divided into three groups each, which are not sharply separated but merge gradually into one another.

Group I. of the human type includes by far the largest number of all the human viruses; its reaction curve corresponds with the classical " human curve " of Smith. Judged by the amount of alkali produced during the earlier growth, the viruses of Group I., human type, may again be divided into three sub-groups, of which HIa is the largest. The viruses included in HIa reduce the acidity of the medium to a point between 0 and 1.0, later raising it above the original. The viruses belonging to HIb produce more alkali, and the reaction of the broth becomes slightly alkaline; later its acidity again increases and finally exceeds the initial per cent. HIc includes those viruses which produce relatively little alkali; the reaction of the broth never falls below 1.0 + and soon rises from this, above the degree of initial acidity.

Group II. of the human type is also recorded by Smith. It comprises those viruses which under the same conditions produce twice as much acid as the viruses of Group I. But while Smith found this peculiarity a constant phenomenon of a virus, this has not been our experience throughout, as Table III. of duplicate tests shows. According to the greater or less amount of alkali produced in the initial stage, Group II., human type, may be divided into two sub-groups, of which HIIa includes those viruses which do not produce any considerable amount of alkali, while to HIIb belong the viruses which markedly reduce the acidity of the broth but later manufacture enough acid to double the initial reaction. Two viruses of this sub-group, pulmonary tuberculosis, adults, 485, and general tuberculosis, children, 161, have rendered the broth slightly alkaline.

Group III., human type, is the smallest of the three, but it is of especial interest since its viruses approach most nearly to the bovine type of curve. They first reduce the acidity of the medium, and later produce only enough acid to bring the end reaction back to or slightly below the initial. HIIIa includes those viruses which produce relatively little alkali, while those which during the early stage reduce the reaction of the broth below o constitute HIIIb.

The same division into three groups holds for the viruses of the bovine type. Again the largest number belong to Group I., for which the end reaction is somewhat higher than the lowest reaction recorded, but much below the initial acidity; this reaction curve agrees with Smith's classical " bovine curve."

Group II., bovine type, is the next largest, and because of the relatively high end reactions the viruses included in it approach very closely — and in some individual cases even overlap — the reaction curve of the third group of the human type. Group III., bovine type, is very small and comprises those bovine viruses for which the end reaction is the lowest reaction recorded.

In the following table are given the reactions of one virus representative of each group or sub-group:

TABLE I.

HIa.	Virus 574. Pulmonary Tuberculosis, Adults.					
Days of growth	o	15	30	45	60	90
Grade of growth	o	2	4	5	6	7
Reaction of broth ...	2.2+	1.1+	.2+	1.3+	1.8+	3.1+

TABLE I. — *Continued.*

HIb.	Virus 665. Tuberculosis of Bones and Joints.					
Days of growth	o	15	30	45	60	90
Grade of growth	o	2	4	6	7	8
Reaction of broth ...	2.2+	1.6+	.3+	.7+	.1—	2.8+

HIc.	Virus 692. General Tuberculosis, Children.					
Days of growth	o	15	30	45	60	90
Grade of growth	o	4	6	7	7	8
Reaction of broth ...	1.9+	1.2+	2.4+	2.7+	2.9+	3.0+

HIIa.	Virus 582. Pulmonary Tuberculosis, Children.					
Days of growth	o	15	30	45	60	90
Grade of growth	o	2	5	5	5	5
Reaction of broth ...	2.1+	1.7+	3.1+	4.2+	3.6+	5.1+

HIIb.	Virus 713. General Tuberculosis, Children.					
Days of growth	o	15	30	45	60	90
Grade of growth	o	2	3	5	6	7
Reaction of broth ...	1.9+	1.2+	.2+	2.0+	3.4+	4.0+

HIIIa.	Virus 215. Genito-urinary Tuberculosis.					
Days of growth	o	15	30	45	60	90
Grade of growth	o	1	1	2	3	4
Reaction of broth ...	1.9+	1.8+	1.8+	1.4+	1.0+	1.8+

TABLE I. — *Concluded.*

HIIIb.	Virus 650. General Tuberculosis, Children.					
Days of growth	0	15	30	45	60	90
Grade of growth	0	4	4	5	6	7
Reaction of broth ...	2.1+	.6+	0	.5—	.8—	1.6+

BI.	Virus 12. Milk Virus.					
Days of growth	0	15	30	45	60	90
Grade of growth	0	2	3	4	4	5
Reaction of broth ...	1.9+	.3—	.6—	.7—	.5—	.2—

BII.	Virus 216. Tuberculous Adenitis, Children.					
Days of growth	0	15	30	45	60	90
Grade of growth	0	2	3	5	7	7
Reaction of broth ...	2.1+	.6+	.3+	.2+	.3+	1.2+

BIII.	Virus 491. From Cattle.					
Days of growth	0	15	30	45	60	90
Grade of growth	0	3	3	5	5	6
Reaction of broth ...	2.2+	1.0+	.3+	.4+	.2—	.4—

Table II. shows the number of reaction curves included in each group or sub-group, classified according to source of viruses. The viruses numbered 173, the tests 243.

TABLE II.

Source.	HIa.	HIb.	HIc.	HIIa.	HIIb.	HIIIa.	HIIIb.	Total H.	BI.	BII.	BIII.	Total B.	Total.
Pulmonary tuberculosis, adults,	47	6	29	4	4	3	0	93	2	0	0	2	95
Pulmonary tuberculosis, children	1	2	3	1	2	0	1	10	0	0	0	0	10
Tuberculous adenitis, adults . .	2	0	2	0	0	0	0	4	0	0	0	0	4
Tuberculous adenitis, children .	10	2	6	1	6	1	0	26	9	5	0	14	40
General tuberculosis, children,	6	1	9	2	5	4	1	28	4	3	1	8	36
Tubercular meningitis, children,	8	2	1	1	0	2	0	14	0	0	0	0	14
Abdominal tuberculosis, children	0	0	1	0	0	0	0	1	0	0	0	0	1
Genito-urinary tuberculosis . .	0	0	0	0	0	1	0	1	1	0	0	1	2
Tuberculosis of bones and joints,	9	2	1	0	1	0	0	13	3	0	0	3	16
Miscellaneous	2	0	0	0	1	0	0	3	0	0	0	0	3
Milk viruses	0	0	1*	0	0	0	1*	2*	1	2	0	3	5
Bovine virus from cattle	4	0	2	0	0	0	0	6	6	4	1	11	17
	89	15	55	9	19	11	3	201	26	14	2	42	243

*Virus No. 111, human type.

Five human viruses, constituting one set of tests, gave such curious returns that they may well be considered apart. As far as can be ascertained the broth on which they were grown was in all respects like the other lots used throughout these experiments, but on it all the viruses grew poorly and the reaction of the medium at different times was either just above or just below the initial reaction. This is unlike anything noted in our other tests. Smith reports a similar experience and confesses that he is unable to account for it, but warns against using only one lot of broth in determining the type of a virus by the glycerin reaction curve. The fact that in our experience the reaction curve was thus irregular for all the viruses grown on that lot of broth, while duplicate tests of several of the viruses on different lots of broth failed to show anything out of the ordinary might perhaps mean that there had been an error in the preparation of the medium.

Virus 609. Tuberculous adenitis, adults, may be taken to represent these five viruses.

Days of growth	o	15	30	45	60	90
Grade of growth	o	2	2	3	3	4
Reaction of broth ...	2.4+	2.1+	2.3+	1.8+	2.1+	2.6+

The majority of viruses which were repeatedly tested remained constant to one group of curve; in twenty-two of them, however, nineteen of the human and three of the bovine type, there were variations, the curve falling now into one group, and now into another. In most of these cases the greater acid production went hand in hand with more luxuriant growth, but this was not constant. Thus Virus 583 in one test reached Grade 8 of development, without producing as much acid as in another test, when it failed to develop beyond Grade 4. Virus 536 in one test caused the broth to become .4 acid, while in another test the final reaction reached 1.7+, yet in both tests the growth was Grade 6.

Details of these duplicate tests will be found in Table III., showing reaction curve of nineteen human and three bovine viruses, which on repeated tests gave curves not belonging to the same group.

TABLE III.

Virus (Source and Number).	Days of Growth.	0	15	30	45	60	90	Group.
Pulmonary tuberculosis, adults:								
509 (1)	Grade of growth.	0	3	3	5	5	6	
	Reaction of broth.	2.1+	1.4+	.4+	1.5+	3.7+	5.7	H2b.
509 (2)	Grade of growth.	0	1	2	2	3	4	
	Reaction of broth.	1.9+	1.7+	.6+	.5+	1.9+	3.1+	H1a.
485 (1)	Grade of growth.	0	5	7	7	7	7	
	Reaction of broth.	2.1+	.3—	4.3+	5.2+	6.2+	7.0+	H2b.
485 (2)	Grade of growth.	0	2	5	5	6	6	
	Reaction of broth.	2.1+	1.5+	3.2+	3.3+	—	3.8+	H1c.
437 (1)	Grade of growth.	0	3	7	7	7	7	
	Reaction of broth.	2.1+	1.2+	1.6+	4.0+	4.5+	5.2+	H2a.
437 (2)	Grade of growth.	0	1	2	3	5	6	
	Reaction of broth.	2.1+	1.5+	1.0+	1.4+	3.0+	2.8+	H1b.
545 (1)	Grade of growth.	0	2	3	5	6	6	
	Reaction of broth.	2.1+	1.8+	1.2+	1.1+	3.1+	5.4+	H2a.
545 (2)	Grade of growth.	0	1	4	5	6	6	
	Reaction of broth.	2.1+	2.0+	2.6+	4.2+	3.8+	4.2+	H1c.
595 (1)	Grade of growth.	0	2	6	6	7	8	
	Reaction of broth.	2.2+	1.5+	.2+	—	3.4+	4.5+	H2b.
595 (2)	Grade of growth.	0	2	4	5	5	5	
	Reaction of broth.	1.9+	1.8+	.7+	1.0+	1.0+	3.2+	H1a.
603 (1)	Grade of growth.	0	1	1	2	5	8	
	Reaction of broth.	2.2+	2.1+	2.1+	.6+	3.0+	3.7+	H1a.
603 (2)	Grade of growth.	0	0	2	2	2	3	
	Reaction of broth.	2.4+	2.2+	.9+	1.5+	1.0+	1.6+	H3a.

TABLE III. — *Continued.*

Virus (Source and Number).	Days of Growth.	0	15	30	45	60	90	Group.
Pulmonary tuberculosis, children:								
583 (1)	Grade of growth.	0	2	4	5	5	6	
	Reaction of broth.	2.1+	1.4+	.5+	1.5+	1.4+	3.5+	H1a.
583 (2)	Grade of growth.	0	1	2	3	3	4	
	Reaction of broth.	1.9+	1.8+	1.5+	.2—	1.2—	3.0+	H2a.
583 (3)	Grade of growth.	0	3	6	7	7	8	
	Reaction of broth.	2.4+	.2+	.6—	.5—	.7—	2.1+	H3b.
Tuberculous adenitis, children:								
161 (1)	Grade of growth.	0	3	4	6	7	7	
	Reaction of broth.	2.2+	1.4+	.3—	.3—	3.8+	4.2+	H1b.
161 (2)	Grade of growth.	0	4	6	7	7	7 .	
	Reaction of broth.	2.4+	.2—	4.0+	3.4+	4.8+	4.7+	H2b.
668 (1)	Grade of growth.	0	1	1	2	2	4	
	Reaction of broth.	2.2+	2.2+	2.3+	2.2+	2.1+	1.4+	H3a.
668 (2)	Grade of growth.	0	2	7	7	8	8	
	Reaction of broth.	1.9+	.8+	2.3+	2.5+	3.3+	4.1+	H2a.
668 (3)	Grade of growth.	0	5	5	6	6	7	
	Reaction of broth.	2.1+	.3+	2.0+	3.1+	3.5+	4.7+	H2a.
682 (1)	Grade of growth.	0	4	7	7	8	8	
	Reaction of broth.	2.0+	1.3+	2.2+	2.6+	2.5+	4.0	H1c.
682 (2)	Grade of growth.	0	0	6	6	7	7	
	Reaction of broth.	1.9+	1.8+	.3+	1.0+	3.5+	4.4+	H2b.
706 (1)	Grade of growth.	0	1	2	2	3	4	
	Reaction of broth.	2.4+	2.4+	3.0+	1.8+	2.6+	2.3+	H1c.
706 (2)	Grade of growth.	0	2	4	5	6	6	
	Reaction of broth.	2.1+	1.8+	.5+	3.5+	3.9+	1.7+	H2a.
536 (1)	Grade of growth.	0	1	3	4	5	6	
	Reaction of broth.	2.2+	1.9+	.4+	.1—	.5+	.4+	B1
536 (2)	Grade of growth.	0	2	6	7	7	8	
	Reaction of broth.	1.9+	1.0+	0	0	.4—	1.0+	B2
536 (3)	Grade of growth.	0	2	3	4	5	6	
	Reaction of broth.	2.1+	1.2+	.4+	.4—	.3—	1.7+	B2

TABLE III. — *Continued.*

Virus (Source and Number).	Days of Growth.	0	15	30	45	60	90	Group.
General tuberculosis, children:								
165 (1)	Grade of growth.	0	4	4	6	6	7	
	Reaction of broth.	2.4+	2.0+	2.5+	3.2+	2.7+	4.0+	H1c.
165 (2)	Grade of growth.	0	3	6	6	7	7	
	Reaction of broth.	2.4+	.9+	3.1+	4.3+	4.1+	4.9+	H2a.
276 (1)	Grade of growth.	0	1	2	3	4	5	
	Reaction of broth.	1.9+	1.9+	1.4+	0	1.0+	2.7+	H1b.
276 (2)	Grade of growth.	0	2	4	5	5	5	
	Reaction of broth.	2.1+	1.5+	1.7+	4.4+	5.2+	4.3+	H2a.
388 (1)	Grade of growth.	0	0	1	2	3	4	
.	Reaction of broth.	2.4+	2.4+	2.6+	2.1+	1.8+	2.3+	H3a.
388 (2)	Grade of growth.	0	2	4	5	5	6	
	Reaction of broth.	2.4+	1.7+	2.7+	4.4+	3.7+	3.8+	H1c.
66 (1)	Grade of growth.	0	2	2	3	3	3	
	Reaction of broth.	2.1+	1.1+	.3+	.2+	.3—	1.1+	B2
66 (2)	Grade of growth.	0	1	2	2	3	4	
	Reaction of broth.	1.9+	1.6+	0.2+	1.1+	.4+	.3+	B1
Tubercular meningitis, children:								
61 (1)	Grade of growth.	0	3	4	4	6	6	
	Reaction of broth.	2.0+	1.2+	2.5	2.8	3.2	4.0	H2a.
61 (2)	Grade of growth.	0	2	3	4	4	5	
	Reaction of broth.	1.9+	.7+	2.2+	2.1+	2.4+	3.2+	H1a.
684 (1)	Grade of growth.	0	2	2	3	3	4	
	Reaction of broth.	1.9+	1.7+	1.0+	.4+	.3+	.9+	H3a.
684 (2)	Grade of growth.	0	2	3	4	5	5	
	Reaction of broth.	2.1+	1.9+	2.1+	.3+	.4+	2.2+	H1a.
Tuberculosis of bones and joints:								
664 (1)	Grade of growth.	0	5	5	6	6	6	
	Reaction of broth.	2.1+	.5+	2.3+	3.8+	4.1+	6.1	H2a.
664 (2)	Grade of growth.	0	2	4	4	5	6	
	Reaction of broth.	1.9+	—	.5+	2.1+	3.5+	3.8+	H1a.

TABLE III. — *Concluded.*

Virus (Source and Number).	Days of Growth.	0	15	30	45	60	90	Group.
Miscellaneous :								
661 (1)	Grade of growth.	0	2	3	5	6	7	
	Reaction of broth.	2.2+	1.5+	.8+	1.8+	2.8+	3.4	H1a.
661 (2)	Grade of growth.	0	1	3	5	6	6	
	Reaction of broth.	2.4+	.8+	.8+	3.1+	3.7+	4 9	H2a.
Milk virus :								
III (1)	Grade of growth.	0	2	3	3	4	5	
	Reaction of broth.	2.2+	1.7+	1.0+	.1—	0	1.4	H3b.
III (2)	Grade of growth.	0	1	2	3	4	5	
	Reaction of broth.	2.4+	2.7+	1.6+	1.1+	2.4+	3.9+	H1c.
Bovine virus from cattle :								
491 (1)	Grade of growth.	0	3	3	5	5	6	
	Reaction of broth.	2.2+	1.0+	.3+	.4—	.2—	.3—	BIII.
491 (2)	Grade of growth.	0	3	4	5	6	6	
	Reaction of broth.	2.0+	0	.2—	.3—	0	0	BI.

There remain nine viruses, four of the human and five of the bovine type. Of these, Virus 122, Generalized Tuberculosis, Children ; Virus 247, Tuberculous Adenitis, Children ; Virus 323 and Virus 490 from cattle belonged, culturally, to Group B1, that is, they grow poorly on glycerin egg in the first three generations, and growth remained sparse in the first five generations. Growth on glycerin potato in the early generations was also poor. About a year after the viruses had been isolated they grew well, or moderately well, on glycerin broth, reaching Grades 5, 6, and 7. Virus 314, from cattle, belonged culturally to Group B3, that is, in the earlier generations the growth on glycerin egg and potato was slight, but increased rapidly, reaching Grade 7 in the fifth generation, resembling the human viruses in the amount of growth. All five viruses were virulent for cattle, Virus 122 showing a special virulence ; but recent experiments with rabbits point to a loss of virulence for Virus 490. Each of these viruses on glycerin broth produced a reaction curve of the human type, as will be seen from the table of reactions following. Virus 122 and Virus 247, however, produced bovine curves in one test each.

Of the four human viruses included in this number, Virus 551, Pulmonary Tuberculosis, Adults, belongs culturally to Group H1; that is, it grew actively in the first three generations on glycerin egg and also on potato, and early showed pigmentation.

Virus 517, Pulmonary Tuberculosis, Adults, culturally belongs to Group H2; its growth on glycerin egg in the first three generations was also abundant, but it lacked the power to grow well on glycerin potato.

Virus 558, Tuberculosis of Bones and Joints, culturally belongs to Group H3. The viruses included in this group grew moderately well on glycerin egg in the first three generations, reaching Grades 4–6, and remained without pigmentation. Growth on glycerin potato = Grades 3–4 in the first four generations.

Virus 377, General Tuberculosis, Children, belongs, culturally, to Group H4; on glycerin egg it grew to Grades 4–6 in the early generations, but failed altogether to grow on glycerin potato in the first four generations.

With the exception of Virus 517, which reached Grade 7 on glycerin broth, the viruses grew only fairly well on that medium, reaching Grades 4–6 only. Virus 377 and Virus 517 were tested only once, and gave the bovine type of reaction curve; Virus 551 gave on one test a bovine, on another a human type of curve, while Virus 558 produced once a human and three times a bovine type of curve.

TABLE IV.

Showing reactions produced by viruses which gave a reversed type of curve.

Virus Number and Cultural Type.	Days of Growth.	0	15	30	45	60	90	Type of Curve.
377 H4	Grade.	0	1	2	2	3	3	
	Reaction.	2.1+	1.5+	0	.2—	.6+	.1+	BI.
517 H1	Grade.	0	4	5	6	7	7	
	Reaction.	2.1+	0	.1+	.1—	.2—	0	BI.
551 H1	(1) Grade.	0	3	4	4	5	5	
	Reaction.	2.2+	.2—	.5—	.4—	.2—	.3—	BI.
	(2) Grade.	0	1	1	3	6	7	
	Reaction.	2.1+	2.0+	1.9+	1.1+	2.8+	3.1+	HIc.
588 H3	(1) Grade.	0	3	3	3	3	4	
	Reaction.	2.2+	.3—	.6—	.2—	.6—	.3—	BI.
	(2) Grade.	0	1	3	3	4	5	
	Reaction.	1.9+	1.5+	.3—	.8—	.2—	3.2+	HIb.
	(3) Grade.	0	0	1	3	4	4	
	Reaction.	2.4+	2.4+	.6—	.6—	.3—	.2—	BI.
	(4) Grade.	0	3	4	5	5	5	
	Reaction.	2.1+	.9—	1.0—	.8—	.6—	.3—	BI.
122 B1	(1) Grade.	0	5	5	6	6	7	
	Reaction.	2.0+	.6+	.4+	1.0+	.6+	1.8+	HIIIb.
	(2) Grade.	0	4	6	6	7	7	
	Reaction.	1.9+	0	.4—	.1—	0	1.4	BII.
323 B1	Grade.	0	0	1	4	5	6	
	Reaction.	2.1+	2.1+	2.6+	.5+	3.5+	3.8+	HIa.
490 B1	(1) Grade.	0	3	4	4	5	5	
	Reaction.	2.0+	1.3+	2.8+	.6+	3.0+	2.4+	HIa.
	(2) Grade.	0	3	4	4	5	5	
	Reaction.	1.9+	.6+	2.2+	1.9+	3.0+	3.2+	HIa.
	(3) Grade.	0	4	5	5	6	6	
	Reaction.	2.1+	2.8+	3.2+	4.0+	4.1+	3.9+	HIc.

TABLE IV.— *Continued.*

Virus Number and Cultural Type.	Days of Growth.	0	15	30	45	60	90	Type of Curve.
314	(1) Grade.	0	6	7	7	7	7	
	Reaction.	2.4+	.1+	2.3+	1.0+	3.2+	4.2+	HIa.
	(2) Grade.	0	3	4	4	5	6	
	Reaction.	2.4+	2.0+	2.5+	3.4+	3.4+	3.0+	HIc.
247	(1) Grade.	0	0	1	1	2	5	
	Reaction.	2.2+	1.9+	1.9+	2.0+	1.3+	.3—	BI.
	(2) Grade.	0	2	4	5	6	7	
	Reaction.	2.1+	1.6+	.5—	.1—	.4—	2.6	HIIIb.

An examination of these nine viruses might lead one to conclude that the reaction curve was influenced, to a certain extent, by the abundance of growth, the poorly growing human viruses tending toward a bovine type of curve and the vigorously growing bovine viruses producing a human type of reaction curve. Such has not been our experience throughout these tests. A number of our bovine viruses have reached Grade 7 of growth, and the weight of the dried pellicle has been equal to, and even higher than the average weight of the human viruses, yet their reaction curves have fallen into one of the three bovine groups. On the other hand, some of our human viruses which have produced a high acid end reaction, have, when dried, barely exceeded a one hundred milligram weight. In other words, we have not found that the amount of growth and the degree of acid production stand in definite and constant relation to each other. Compare the following table, which shows a combination of high acid production and relatively poor growth in a human virus, and the maximum of growth, without corresponding acid production in a bovine virus:

TABLE V.

Virus.	Days.	0	15	30	45	60	90
545 (human) Pulmonary Tuberculosis, Adults	Grade.	0	2	3	5	6	6
	Reaction.	2.1+	1.8+	1.2+	1.1+	3.1+	5.4+
334 (bovine) from cattle..	Grade.	0	3	7	7	8	8
	Reaction.	2.1+	.1—	.8—	.4—	.3—	.4—

Why some of these viruses have been constant in producing the opposite type of reaction curve, while others varied in repeated tests, we do not know; every effort was made throughout these tests to keep all conditions absolutely uniform. Most of the other observers have had a similar experience, but few offer any explanation. Smith has thought that it might be due to variations of the glycerin and suggests careful analysis to determine the glycerin consumption and decomposition. Lewis believes that one will always encounter irregular and atypical curves when dealing with recently isolated cultures. This affords an explanation for four of his bovine viruses, which in earlier tests produced human reaction curves and later typical bovine curves. It would not explain the results of some of the other workers who were dealing with older cultures. All nine of our atypical viruses had been cultivated on glycerin media for a year or more. Only one of our bovine viruses, 490, had lost much of its rabbit virulence when it produced a human reaction curve in broth.

The following table shows the viruses reported by different observers, for which the glycerin reaction curve does not correspond with the type established by cultural or virulence characteristics:

TABLE VI.

Observer.	Virus.	Culture.	Virulence.	Reaction Curve.	Remarks.
Bang.	H I.	No details given.		Bovine.	In one of three tests.
Mohler and Washburn.	Girl I.	Human.	Human.	Bovine, 3 tests. Human, 3 tests.	
	Sputum C.	Bovine.	Bovine.	Human.	
	Human 33.	No details given.		Bovine.	
	Bovine 3.	" "	"	Human, 1 test. Bovine, 3 tests.	
Fibiger and Jensen.	5	Human.	Bovine.	Human.	
	12	"	"	Bovine.	
	16	"	Human.	"	
	17	"	"	"	
	20	Bovine.	"	"	
	27	Human.	"	"	
Royal Commission.*	B IX.	Bovine.	"	Human.	
Lewis.	07.8	"	Bovine.	"	
	07.54	Human.	Human.	Human, 1 test. Bovine, 3 tests.	
	07.9	Bovine.	Bovine.	Variable.	Human, 2 tests early. Bovine, 1 test late.
	07.53	"	"	"	Human, 2 tests early. Bovine, 1 test late.
	07.73	"	"	"	Human, 2 tests early. Bovine, 3 tests late.
	07.94	"	"	"	Human, 2 tests early. Bovine, 3 tests late.
Grund.	377	Human.	Human.	Bovine.	
	517	"	"	"	
	551	"	"	Human, 1 test. Bovine, 1 test.	
	558	"	"	Human, 1 test. Bovine, 3 tests.	
	122	Bovine.	Bovine.	Human, 1 test. Bovine, 1 test.	
	247	"	"	Human, 1 test. Bovine, 1 test.	
	314	"	"	Human.	
	323	"	"	"	
	490	"	"	"	Lately has lost virulence.

*The other three viruses giving irregular curves as noted in their report on the glycerin reaction curve have been excluded because of possible mixture of types. See pages 14 and 15, Final Report, 1911.

Of the thirty-six viruses tested by us which have shown variant characters either culturally or in virulence, only nine were at all irregular in their glycerin reaction curve. Taking away from these eleven the five which have been discussed before, as giving the opposite type of reaction curve, namely, Viruses 377, 517, and 558 of the human, and Viruses 122 and 314 of the bovine type, there remain four, all of the human type, Viruses 437, 485, and 545, Pulmonary Tuberculosis, Adults; Viruses 434 and 689, General Tuberculosis, Children, and Virus 661, Miscellaneous Lesions. Four of these, Viruses 434, 545, 434, 661, grew rather poorly in early generations, while Viruses 485 and 689 showed a very slightly increased tendency to produce more generalized tuberculosis in rabbits.   These peculiarities indicate, if anything, a leaning toward the bovine type, yet all four of these viruses belong, as far as their reaction curve in glycerin broth is concerned, to Group HI, which is the typical human curve, or to Group II, human type, the viruses of which are distinguished by their high acid end reactions, their only irregularity being the fact that in repeated tests they give reaction curves belonging now to one, and now to another group.   None of the thirteen viruses included in Group HIII, with very low end reactions approaching the bovine type of curve, showed irregularities either culturally or in regard to virulence.

TABLE VII.

Showing glycerin reaction curves of the four viruses which showed variations in cultural or virulence characteristics.

Virus.	Days.	0	15	30	45	60	90	Culture.	Virulence.
437 ..	(1) Grade.	0	3	7	7	7	7	Growth slow on potato. Otherwise typical.	Typical.
	Reaction.	2.1+	1.2+	1.6+	4.0+	4.5+	5.2+		
437 ..	(2) Grade.	0	1	2	3	5	6		
	Reaction.	2.1+	1.5+	1.0+	1.4+	3.0+	2.8+		
485 ..	(1) Grade.	0	5	7	7	7	7	Typical.	Slightly increased for rabbits.
	Reaction.	2.1+	.3—	4.3+	5.2+	6.2+	7.0		
485 ..	(2) Grade.	0	2	5	5	6	6		
	Reaction.	2.1+	1.5+	3.2+	3.3+	—	3.8+		
545 ..	(1) Grade.	0	2	3	5	6	6	Poor on potato.	Typical.
	Reaction.	2.1+	1.8+	1.2+	1.1+	3.1+	5.1+		
545 ..	(2) Grade.	0	1	4	5	6	6		
	Reaction.	2.1+	2.0+	2.6+	4.2+	3.8+	4.2+		
661 ..	(1) Grade.	0	2	3	5	6	7	Moderate growth on all media.	Typical.
	Reaction.	2.2+	1.5+	.8+	1.8+	2.8+	3.4+		
661 ..	(2) Grade.	0	1	3	5	6	6		
	Reaction.	2.4+	.8+	.8+	3.1+	3.7+	4.9+		

Concerning our one bacillus of the avian type, it was with difficulty induced to grow on glycerin broth and has therefore been tested once only. Its curve shows the same general direction as that of the bovine viruses, but it does not produce as much alkali. Bang's twelve avian viruses produced more alkali than his bovine controls, but that is apparently due to the fact that his observations extend over a longer period than ours. At the end of ninety days his avian cultures are still more acid than the bovine controls.

A few other acid-fast organisms were included in our tests, and so was a sub-culture of the original T. B. Koch. Both the timothy and the smegma bacillus reacted like the bovine type of virus, the latter producing more alkali than any other bacillus tested. The curve of T. B. Koch was rather irregular, following at first the human type, but later it again produced some alkali and left the medium nearly neutral at the end of ninety days.

On glycerinated litmus milk ninety tests have been carried out, with seventy-eight viruses of the human and twelve of the bovine type. The results have corresponded fairly well with those obtained in broth. But pellicle formation on the milk was in many cases delayed, and not once as vigorous as on broth. Thirteen of our human viruses, which grew poorly on the milk, did not produce acid; this includes the atypical human Viruses 551 and 558; Viruses 377 and 517 have not been grown on milk. The two atypical bovine Viruses 314 and 490 turned the milk acid, but did not coagulate it.

CONCLUSIONS.

Broadly speaking, the reaction curve in glycerin broth divides tubercle bacilli into two types. The bacilli which possess a low degree of virulence for rabbits and the power to grow well on glycerin media in the early generations, produce one type of reaction curve, while those which are virulent for rabbits and which, in the early generations grow slowly and with difficulty on glycerin media, form the other type of curve in glycerin broth. These two types of glycerin reaction curve are again divisible into groups according to their final reactions. The curves of adjacent groups show much the same general direction and there is a gradation from one group to the next; but the reaction curves of the groups at both extremes are widely divergent. When any large number of viruses is examined there will be found a small percentage of cases, which, by cultural characteristics and virulence, belong to one type of tubercle bacilli while they would be classed with the opposite type of bacilli if judged by their glycerin reaction curve alone. On repeated tests this reversed glycerin reaction curve may, or may not, be a constant feature of these particular viruses, although the conditions under which they have been cultivated are apparently the same in the several tests. Undetected variations of the culture medium must be taken into consideration; it is not advisable to depend on the reaction curve obtained from one lot of broth only, but several examinations of a virus are desirable. In from thirty to forty per cent of the viruses retested, the reaction curves belong to different groups, that is, the end reaction may be high in one test, and low or medium in the next. In only three instances was the variation so great as to justify the classification of the reaction curves into different types. In about half the cases the degree of acidity produced has been in direct ratio to the amount of growth.

There is also no constant relation between irregularities of culture and virulence on the one hand, and irregularity of the glycerin reaction curve on the other. Some viruses which culturally and in virulence showed nothing unusual have given very atypical curves, while perfectly normal reaction curves were produced by viruses which from cultural and virulence tests could not be called quite typical.

The glycerin reaction curve is undoubtedly a valuable corroborative evidence of a division of tubercle bacilli into two types. Its value is lessened, however, by the number of irregular and atypical reactions encountered, while as a practical aid in determining the type of an individual virus, it is also much handicapped by the length of time required to carry it out.

BIBLIOGRAPHY.

Bang, O. Einige vergleichende Untersuchungen über die Einwirkung der Säugetier-und Geflügeltuberkelbazillen auf die Reaktion des Substrats in Bouillon-kulturen. Centralbl. Bakt. I. Abt., 1907, xliii, 34.

Beitzke, H. Uber die Infektion des Menschen mit Rindertuberkulose. Tuberkulosestudien. Virchow's Archiv. f. path. Anat. Beiheft. z. 190, Bd. 58.

Duval, C. W. Studies in atypical forms of tubercle bacilli isolated directly from the human tissues in cases of primary cervical adenitis. Jour. Exp. Med., 1909, xi, 403.

Fibiger, J., and Jensen, C. O. Untersuchungen über die Beziehungen zwischen der Tuberkulose und den Tuberkelbazillen des Menschen und der Tuberkulose und den Tuberkelbazillen des Rindes. Berl. Klin. Wchnschr., 1908 (1876-1926-1977-2026).

Lewis, P. A. Tuberculous Cervical Adenitis. A study of the tubercle bacillus cultivated from fifteen consecutive cases.   Jour. Exp. Med., 1910, xii, 82.

Mohler, J. R., and Washburn, H. J.   A comparative study of tubercle bacilli from varied sources.   Bull. No. 96, Bureau of Animal Industry. Dept. of Agriculture, U.S.A., 1907.

Royal Commission on Human and Animal Tuberculosis.   Second Interine Report, Part II., Appendices.   Vol. 3, Additional Investigations of Bovine and Human Viruses.   Wyman and Sons, London, 1904-1907.

Smith, T.   Studies in mammalian tubercle bacilli, III.   Description of a bovine bacillus from the human body.   A culture test for distinguishing the bovine from the human type of bacillus.   Jour. Med. Research, 1904-1905, xiii, 253.

Smith, T.   The reaction curve of tubercle bacilli from different sources in bouillon containing different amounts of glycerin.   Jour. Med. Research, 1904-1905, xiii, 405.

Smith, T.   The reaction curve of the human and the bovine type of tubercle bacillus in glycerin bouillon.   Jour. Med. Research, xxiii, 1910, 185.

GRUND.

FATTY COMPOUNDS AS A FACTOR IN THE ETIOLOGY OF APPENDICITIS.[*]

BERTHA VAN HOUTEN ANTHONY.

(*From the Research Laboratory of the Health Department of New York and the Pathological Laboratory of New York Infirmary for Women and Children.*)

The following work was suggested after bacterial examination, with both aërobic and anaërobic cultures,[†] of the contents of ten appendices had resulted in very irregular findings. This irregularity corresponds with the results of all other investigators who have attempted to place the cause of appendicitis on a bacteriological foundation alone. The following review of the work of Owen T. Williams suggests a most interesting problem. The attempts to verify his theory are given further on.

In the Bio-Chemical Journal of Oct. 31, 1908, Owen T. Williams discussed the micro-chemical changes occurring in appendicitis. Clinical and chemical observations tend to show that there are conditions in the intestines in which certain soaps or other insoluble compounds of fatty acids are not absorbed by the intestinal mucous membrane and that, with the presence of these, certain diseases of the alimentary tract are related — as true intestinal sand, appendix concretions, etc. He had shown previously in an article on abnormal fat assimilation (British Medical Journal, July 27, 1907) that intestinal sand and appendix concretions are allied in chemical composition and are made up of compounds of saturated fatty acids (palmitic and stearic). Calcium and fatty acids are normal constituents of excretion of intestinal mucous membrane and it would seem that excess in secretion or defect in absorption might be the origin of true intestinal sand and appendix concretions. Thus the chemical constitution of these bodies may throw much light on the etiology of appendicitis, mucous colitis and similar processes.

In a large number of appendix concretions that he examined all showed concentric lamination and could be cut without difficulty. They are to be distinguished from masses of inspissated feces, inorganic salts, gall-stones, false intestinal sands, and remains of undigested food which have no relation to the pathological processes accompanying fat digestion. He found true feces in only three out of one hundred adult appendices at postmortem. In the remainder occurred a dark slimy mucoid material similar

---

[*] Received for publication Oct. 27, 1911.

[†] The cultural work was done under the direction of Dr. Anna W. Williams in 1908, with material obtained from the New York Infirmary for Women and Children.

to feces in consistency but not in pigment nor in chemical make-up. In one appendix were small granules which showed a likeness to true intestinal sand both in composition and in formation.

Although concretions occur by no means in all cases of appendicitis, yet Owen T. Williams shows that the process by which they are formed may go on in the wall of the appendix and give rise to morbid conditions which allow of easy invasion by microörganisms, and thus lead to acute and chronic forms of appendicitis. These changes are the formation of calcium soaps in the sub-mucosa and mucosa of the appendix. Normally these soaps are expelled into the lumen and then into the cecum, but in appendicitis they block the lumen of the gland or form a ring in the sub-mucosa. The micro-chemical changes in the walls of the appendix are demonstrated by certain staining reactions of:

Osmic acid which blackens neutral fats and not palmitic and stearic acids; Soudan III. or Scharlach R. which stain neutral fats bright pink, soaps a pinkish yellow; nile blue sulfate, neutral fats pink, fatty acids violet; silver nitrate impregnates calcium compounds; petroleum ether dissolves the neutral fats but not the soaps from a specimen.

Owen T. Williams found that appendices so stained show a large increase in the production of calcium soaps in the mucous membrane and sub-mucosa, observable also in sections stained in bulk. In some, almost a complete ring is visible in the sub-mucosa. "Marked thickening," a change frequently recorded of the sub-mucosa, is due to the laying down of these soaps. Small amounts of fatty acids are also demonstrated by nile blue sulfate, and calcium is found in definite amounts in the sub-mucosa of diseased appendices. A few show the excretion of these soaps into the lumen. Chemically, the excretions and the concretions show the same reaction.

Zuckerland (Nothnaegel — Diseases of Intestines and Peritoneum, 1905, p. 382) states that, as a result of age, with the atrophy of the mucous membrane and gland structures, the sub-mucosa undergoes thickening and the fat accumulates at that point. Yet in appendicitis the change in fatty compounds far exceeds (as Owen T. Williams has observed) any result of further involution changes. The formation of calcium soaps is a very slow process and must begin long before the acute attack of the disease but, by producing altered conditions of the mucous membrane or by acting as a barrier in the sub-mucosa and thus cutting off the nutritive supply, it renders easier the invasion by microörganisms.

To test the above reactions a set of forty-eight appendices was stained. They were obtained chiefly from the New York Infirmary for Women and Children (four cases came from Bellevue Hospital morgue). The material was partly selected and falls in four groups:

(a) Acute appendicitis   .   .   .   .   . 10 cases
(b) Chronic appendicitis   .   .   .   .   . 7 "

(c) Chronic appendicitis with involvement of
     other organs . . . . . . 18 cases
(d) Autopsy appendices (seven from infants) . 13 "

In class (c) the appendix was removed usually as a routine measure during operation for other causes. In only two of these cases was a diagnosis of chronic appendicitis recorded before operation.

Four of the autopsy cases, class (d), serve as controls, as they showed no sign of intestinal lesions of any kind on examination at the morgue.

Method. — Since the alcohols dissolve out fats, the material after fixation in ten per cent formalin for one or more days was cut on the freezing microtome in the Pathological Laboratory of the New York Infirmary for Women and Children. The sections were rather thick, usually fifteen to twenty microns, and could be handled with ease. They were left in the stain about two or three minutes, washed in water and mounted in glycerine. Each section was treated with but one stain, yet as the pieces came from one small area of an appendix, the different pictures were practically comparable.

As an additional control human fat from the mesentery of a case with no intestinal lesions (Bellevue Hospital) was stained in smears made both from the fresh material and from a portion kept in formalin twenty-four hours. There was no difference in the staining of the two lots. A part of the fresh fat was broken up into fatty acids; calcium soap was also made from it. This work was very kindly done for me by Mr. J. P. Atkinson.

CONTROLS FROM HUMAN FAT.

Stain.	Fatty Acids.	Calcium Soaps.	Human Fat (Untreated).
Scharlach R.	Bright pink.	Dirty pink or yellowish pink.	Bright pink.
Soudan III.	Yellowish pink.	Pinkish yellow.	Orange.
Nile blue sulfate.	Blue.	Blue.	Rose pink (occasional purple tinge).
Osmic acid.	Black or gray.	Black or gray.	Black.
Silver nitrate.	Brown.	Grayish.	

From the beginning there has been a good deal of difficulty in interpreting the colors obtained with the above
stains when applied to appendix material. Fatty acids are
undoubtedly present to a greater or less degree in all cases
except those of the infants. The presence of calcium soaps
has not been so easy to determine, for to distinguish between
a pink and a pinkish yellow (or a yellowish pink) is not
always possible with the Scharlach R. or Soudan III. stains.
Besides in the controls the nile blue sulfate gives both fatty
acids and calcium soaps a blue color. The violet shade
described by Owen T. Williams is due doubtless to an
admixture of the pink of the neutral fats present in varying
amounts in many cases. Osmic acid was of very little
value as it was found to stain fatty acids, calcium soaps
and unsaturated neutral fats indiscriminately. Petroleum
ether, in my hands, did not dissolve out the neutral fats even
though used with thin sections for an hour or more. The
silver nitrate, on which so much depends for detection of
calcium soaps, was also most unsatisfactory. In some cases
calcium compounds were doubtless blackened by it but
usually not to the extent one would be led to suppose after
observation of the Scharlach R. or Soudan III. slides. It
may be that the presence of other soaps gave an exaggerated
picture with these stains.

The normal presence of fat in the sub-mucosa of the
appendix is given in a number of histologies and anatomies.
Adami, Delafield and Prudden, Bailey, Stoehr (Lewis' addition) show it in the pictures but make no mention of it.
Ferguson, in the description of a cut, gives " many fat cells."
Woodhead mentions fat cells and their stains, while Gray's
Anatomy gives " at times adipose tissue." Neither Osler
nor Cunningham makes any mention of fat occurring in the
appendix. Zuckerland, as noted before by Owen T. Williams,
says the laying down of fat in the appendix occurs with old
age, coincident with atrophy of the mucous membrane. It
is interesting to note that in nine cases of this test where
very large amounts of fatty material were found, seven were
from patients between thirty-two and fifty-one years of age,

of the remaining, one was from a girl of fifteen, the other from an adult control case (exact age unknown). On the other hand, five of the seven appendices of infants, two days to two and a half years old, showed no signs of fat. One infant, four months old, had a slight trace, the other, two and a half years, showed a small amount of fatty compounds present. This last case had been diagnosed " acute suppurative appendicitis," microscopically. Of the nine cases in which a large amount of fatty material was found, only one, that of the young girl, was acute appendicitis, the others were two chronic cases come for operation, five appendices removed during operation for other causes, and one from a control case (autopsy case at the morgue) where no intestinal lesions were noted.

While the results obtained by this method of staining correspond in some measure to those of Owen T. Williams, the difficulty in interpreting the colors is a drawback to a satisfactory determination of the presence or absence of calcium compounds. It may be said, however, that in this set of forty-eight cases the presence of fatty acids and soaps increases with the age of the patient, especially between twenty-five and fifty years, particularly in chronic cases. One marked exception to the general theory is that of Case 633, acute purulent appendicitis in a girl nineteen years old, where the neutral fats predominate instead of fatty acids or soaps. Nevertheless further work on micro-chemical lines may give even more promise of discovering the real cause of appendicitis and allied diseases. The field is one of the widest involving not only histological, pathological, and bacteriological changes but the most variable question of diet. To review Owen T. Williams a little further :

The nature of fats in food determines to some extent the nature of fats in the tissues. It would seem, therefore, that as these abnormal processes are set up by saturated fatty compounds of palmitic and stearic acid (not by the unsaturated compounds of oleic acid) the nature of fat in the diet might have some influence in their production, since the absorbability of the two is so different. A diet of olive oil, especially as used in salads, has been found beneficial in cases of intestinal lithiasis and mucous colitis, which seem so closely related to appendicitis, having in common abnormality of fat absorption. A further study of these diseases as occurring in those countries which use more oil in the diet is suggested.

SUMMARY.

One of the factors in the cause of appendicitis, as suggested by Owen T. Williams, may be the laying down of soaps and other insoluble fatty compounds in the mucous membrane and sub-mucosa of the appendix.

Conclusions. — That irregular cultural findings in various investigations (including our own) of the cause of appendicitis further this view.

That, as judged by the staining reactions of Scharlach R., nile blue sulfate, silver nitrate and osmic acid applied to a set of forty-eight appendices, fatty acids and calcium soaps predominate in the acute and chronic forms of appendicitis, while only neutral fats and slight amounts of fatty acids occur in the controls. In general, these findings agree with those of Owen T. Williams.

That in this series the presence of calcium soaps and fatty acids seems to increase with the age of the patient, especially between twenty-five and fifty years.

That experience in the above work does not fully bear out the assumption of Owen T. Williams as to the ease with which calcium compounds are distinguished by means of these stains since they bring in the error of correct interpretation of color.

That diet, as suggested by Owen T. Williams, and the occurrence of appendicitis, especially with regard to the use of fats and oils, should be closely studied in the various countries.

[Acknowledgment is due Dr. Katharine R. Collins, who called my attention to the article on which this work is based.]

ACUTE APPENDICITIS. TEN CASES.

(The use of the symbols plus and minus is an attempt to indicate the amount of material taking the selective stains for fats and their compounds.)

KEY.

| indicates trace.
± indicates present.
+ indicates moderate amount.
+ + indicates marked amount.
+ + + indicates more marked.
+ + + + indicates very marked.

Case.	Age (Years).	Scharlach R.	Nile Blue Sulfate.	Osmic Acid.	Silver Nitrate.	Interpretation of Stain.
476	11	Yellow pink ±	Blue +	Black		?	Calcium soaps. Fatty acids.
395	15	Yellow pink + + +	Bluish violet + + +	Black + +	o	Calcium soaps. Fatty acids.
468	15	Bright pink + +	Blue + +	Black ±	?	Fatty acids.
591	15	Pink + / Dirty pink		Bluish violet +	Black + +	±	Calcium soaps(sl.). Fatty acids.
382	18	Pinkish yellow ±	Blue +	Black ±	?	Calcium soaps. Fatty acids.
633	19	Bright pink ±	Pink + / Violet		Black ±	?	Calcium soaps and Fatty acids (sl.) Neutral fats.
613	22	Bright pink ± / Dirty pink		Pink	/ Violet ±	Black ±	±	Calcium soaps(sl.). Fatty acids. Neutral fats (tr.).
620	29	Bright pink ± / Dirty pink		Pink ± / Violet (sl.)		Black +	i	Calcium soaps(sl.). Fatty acids (sl.). Neutral fats.
592	33	Pink ± / Yellow pink		Purple ± / Blue +	Black + +			Calcium soaps(sl.). Fatty acids.
454	42	Yellow pink ±	Violet ± / Blue ± / ;Pink		Black ±			Calcium soaps. Fatty acids. Neutral fats (tr.).

CHRONIC APPENDICITIS. SEVEN CASES.

Case.	Age (Years).	Scharlach R.	Nile Blue Sulfate.	Osmic Acid.	Silver Nitrate.	Interpretation of Stain.
488	25	Yellow pink +	Violet + / Blue	/ Pink		Black +	?	Calcium soaps(sl.). Fatty acids. Neutral fats (sl.).
478	27	Yellow pink ± / Bright pink ±	Blue ± / Violet ±	Black ±	?	Calcium soaps. Fatty acids.
405	29	Yellow pink ±	Violet +	Black ±	+	Calcium soaps. Fatty acids.
363	36	Yellow pink + + +	Violet + + +	Black +	±	Calcium soaps. Fatty acids.
390	37	Yellow pink +	Bluish violet +	Black +	?	Calcium soaps. Fatty acids.
470	41	Yellow pink ±	Violet ± / Pink		Black ±	o	Calcium soaps. Fatty acids. Neutral fats (tr.).
439	44	Bright pink + + +	Violet + + +	Black + + +	?	Fatty acids.

CHRONIC APPENDICITIS WITH INVOLVEMENT OF OTHER ABDOMINAL ORGANS.
EIGHTEEN CASES.

Case.	Age (Years).	Scharlach R.	Nile Blue Sulfate.	Osmic Acid.	Silver Nitrate.	Interpretation of Stain.
397	24	Yellow pink |	Violet |	Black |	|	Calcium soaps (sl.). Fatty acids (sl.).
475	24	Yellow pink + +	Blue ± Purple violet ±	Black ±	|	Calcium soaps. Fatty acids. Neutral fats (sl.).
403	26	Bright pink ±	Blue ±	Black +	+	Calcium soaps (sl.). Fatty acids.
404	26	Bright pink |	Blue |	Black |	?	Calcium soaps? Fatty acids (sl.).
453	28	Bright pink ±	Pink ± Violet ±	Black ±	o	Fatty acids. Neutral fats.
398	30	Bright pink |	Violet | Blue |	Black o	?	Fatty acids (sl.).
389	About 30.	Bright pink +	Bluish violet ±	Black |	?	Fatty acids.
385	32	Bright pink + + +	Blue + + +	Black + + +	o	Fatty acids.
513	35	Yellow pink + +	Purple blue + + Pink ±	Black + +	?	Calcium soaps. Fatty acids. Neutral fats (sl.).
420	37	Bright pink ±	Blue +	Black +	o	Fatty acids.
388	40	Bright pink ±	Blue +	Black |	o	Fatty acids.
371	40	Yellow pink + + +	Violet + + +	Black + +	+	Calcium soaps. Fatty acids.
377	40	Yellow pink +	Blue ±	Black +	?	Calcium soaps. Fatty acids.
457	41	Yellow pink + + + +	Bluish violet + + + +	Black + + +	o	Calcium soaps. Fatty acids.
399	43	Bright pink + + +	Violet + + +	Black + + +	±	Calcium soaps (sl.). Fatty acids.
409	43	Yellow pink ±	Violet + Pink |	Black |	±	Calcium soaps. Fatty acids. Neutral fats (sl.).
391	43	Yellow pink ±	Bluish violet ±	Black ±	?	Calcium soaps. Fatty acids.
406	51	Bright pink + + +	Blue + + +	Black +	?	Calcium soaps? Fatty acids.

AUTOPSY MATERIAL. THIRTEEN CASES.

Case.	Age.	Scharlach R.	Nile Blue Sulfate.	Osmic Acid.	Silver Nitrate.	Interpretation of Stains.
220A	2 days.	o	o	o	o	No fat present.
219A	4 months.		o	o	o	No fat present.
244A	4 months.		o	Black |	?	Calcium soaps? Neutral fats (sl.).
221A	6 months.	Yellow pink on m. m. tips.	o	o	o	Calcium soaps (tr.).
228A	7 months.		o	o	?	Calcium soaps?
245A	10 months.	o	o	o	o	No fat present.
227A	2½ years.	Yellow pink |	Pink |	Gray on m. m. tips.	o	Calcium soaps (sl.). Fatty acids (sl.). Neutral fats (sl.).
248A	37 years.	Pink ±	Pink | Purple +	Black +	?	Calcium soaps? Fatty acids. Neutral fats (sl.).
217A	70 years.	Bright pink +	Violet + Purple ±	Black +	o	Fatty acids.
*CA1	Adult.	Bright pink +	Pink + Purple ± Violet ±	Black +	o	Neutral fats. Fatty acids (sl.).
*CA2	Adult.	Bright pink + + +	Pink + + + Violet |	Black + + +	o	Neutral fats. Fatty acids?
*CA3	Adult.	Bright pink ±	Pink + + Violet ±	Black + +	o	Neutral fats. Fatty acids (sl.).
*CA4	Adult.	Bright pink |	Pink | Blue |	Black |	o	Neutral fats (sl.). Fatty acids (sl.).

* Controls.

CONDENSED TABLE.

Diagnosis.	Number of Cases.	Fatty Acids.			Calcium Soaps.			Neutral Fats.		
		+	Sl.	None.	+	Sl.	None.	+	Sl.	None.
(a) Acute appendicitis,	10	8	2	o	4	5	1	2	2	6
(b) Chronic appendicitis,	7	7	o	o	5	1	1	o	2	5
(c) Chronic appendicitis, (with other organs involved)	18	15	3	o	6	6	6	1	3	14
(d) Autopsy	13	2	5	6	o	5	8	3	4	6
(Including controls),	(4)	(o)	(3)	(1(?))	(o)	(o)	(o)	(3)	(1)	(o)

BIBLIOGRAPHY.

Smith, j. L. On simultaneous staining of neutral fat and fatty acid by oxanine dyes. Jour. of Path. and Bacteriol., Cambridge, 1907–8, xii, 1–4.

Staining of fat with basic aniline dyes. Jour. of Path. and Bacteriol., Edinburg and London, 1906, ix, 415.

Preliminary note on further staining of fat with aniline dyes. Med. Chron., Manchester, 1906, 4th series, xii, 283.

Klotz. Micro-chemical tests for soaps. Jour. Exper. Med., 1905 (November 25), ii, No. 6.

Williams, Owen T. Abnormal fat assimilation. Bio-chemical Journal, 1907, ii, 395; Brit. Med. Jour., July 27, 1907.

A NOTE ON THE REGENERATION OF RENAL EPITHELIUM IN THE INTACT CAT KIDNEY.*

WM. deB. MacNider, M.D.

(*From the Laboratory of Pharmacology of the University of North Carolina.*)

In a previous number of this journal [1] in an article entitled "The Pathological Changes which Develop in the Kidney as a Result of Occlusion by Ligature of One Branch of the Renal Artery," the regenerative capacity of the kidney formed part of the subject under investigation. A continued study of the material employed in this investigation and of subsequent material similarly prepared, furnishes such additional evidence, not only of the regenerative capacity of the kidney but also of the method of regeneration, that the following note on this subject seems advisable.

The present communication is limited to the observations made on the regenerative capacity of the stroma and of the epithelium of the cat kidney.

Following the ligation of the posterior branch of the renal artery, there develops in the posterior one-third of the kidney a zone of necrosis variable in its extent. This necrotic zone is surrounded by renal tissue the integrity of which is preserved by an adequate blood supply. It is in this area of necrosis that the connective tissue, vascular, and epithelial ingrowths have been studied.

Connective tissue changes in the zone of necrosis. — The first evidence of cell life in the necrotic zone is seen in the ingrowth from the surrounding kidney tissue of connective tissue nuclei. These nuclei are usually elongated, stain lightly, and contain as compared with the nuclei of the renal epithelium a small amount of chromatic substance.

This invasion takes place before the existence of a capillary ingrowth can be demonstrated. The preservation of the integrity of these nuclei likely depends upon the seepage in

---

* Received for publication Nov. 13, 1911.

of lymph into the necrotic zone from the surrounding kidney
tissue.

The connective tissue ingrowth takes place in two forms.
Either it appears as a tongue-like ingrowth of connective
tissue nuclei, or the nuclei use the dead tubules as a supporting structure and are found arranged around the tubules;
the necrotic tubules are thus given more or less definite outline.

With the ingrowth of capillaries the invasion of stroma
cells is more rapid and more diffuse and is invariably in
excess of the normal.

The utilization of a supporting structure by regenerating
cells has recently been shown by Carrell and Burrows [2] in
their experiments on the cultivation of fetal chicken skin.

Epithelial regeneration in the zone of necrosis. — The
earliest indication of the regeneration of tubular epithelium
is seen on the medullary side of the necrotic area and consists in an increase in the number of nuclei in the preserved
tubular epithelium. This increase is not only numerically
striking but the staining capacity of these nuclei is much
greater than that of the same nuclei in surrounding cells; so
that, in the tubules which are undergoing proliferative
changes, the deeply staining hyperchromatic nuclei stand
out in rather striking contrast to the nuclei in those tubules
which are, comparatively speaking, quiescent.

In tubules such as have just been described, where there
is undoubted evidence of nuclear division, the demonstration
of mitotic figures is exceptionally rare.

Following these changes, which take place in the tubular
epithelium which has not entered into the necrotic process,
the epithelial invasion of the zone of necrosis occurs in two
forms.

The first and more unusual form of invasion is that of a
syncytial-like tongue or bud of poorly staining cytoplasm,
containing hyperchromatic, spherical, epithelial nuclei which,
as it grows, pushes its way into the necrotic cytoplasm of
the old tubule (Plate V., Fig. 1). As this ingrowth pushes

its way deeper into the necrotic zone, its syncytial structure becomes lost, and the nuclei, either surrounded by a very narrow zone of feebly staining cytoplasm, or having no demonstrable cytoplasmic structure, are found arranged with a fair degree of regularity in parallel rows in the old tubules.

Early in this nuclear invasion the necrotic cytoplasm of the original tubules surrounds the invading nuclei. As the nuclei become arranged in the periphery of the necrotic material, within the old tubules, the dead material assumes a position in the center of the tubule, surrounded by the parallel rows of nuclei.

The outline of the tubular wall is formed by connective tissue nuclei which have previously invaded the zone of necrosis.

The second and by far the most usual type of epithelial ingrowth is in the form of epithelial nuclei which have arisen from a proliferation of the nuclei in the tubular epithelium surrounding the area of necrosis. These nuclei stain with the same intensity, are hyperchromatic and regularly spherical as were the nuclei in the syncytial ingrowth previously described.

So far as I have been able to determine, when the invasion of the necrotic tubules by these nuclei commences, the nuclei are not surrounded by cytoplasm. But the cytoplasm is acquired later after the nuclei have a fairly definite arrangement within the necrotic tubules (Plate V., Fig. 2). The time at which the nucleus acquires its cytoplasm varies considerably. In the kidney of one animal it was well defined by the twenty-sixth day, while in another animal the acquisition was only partly completed by the forty-fourth day (Plate V., Figs. 3, 4).

The distribution of the necrotic cytoplasm of the old renal epithelium within the regenerated tubule in this type of regeneration is similar to the distribution described when the regeneration begins as a syncytial ingrowth. The necrotic material remains in the lumen of the tubule forming a cast.

This newly formed epithelium has two characteristics. It

is invariably flattened, and its avidity for cytoplasmic stains is much less than that of ordinary renal epithelium. It stains poorly and appears pale even when every precaution is taken in the technic of staining.

The regenerative changes which develop in the intact cat kidney as enumerated above are strikingly confirmed by the experiments of Carrell and Burrows[3] and of Fleisher and Loeb[4] on the cultivation of adult kidney tissue outside of the body.

BIBLIOGRAPHY.

1. MacNider. Journal of Medical Research, xxiv, No. 2, 425–445.
2. Carrell and Burrows. The Journal of Exp. Med., xiv, No. 3.
3. Carrell and Burrows. Journal Amer. Med. Ass., lv, 1379–1381.
4. Fleisher, M. S., and Loeb, L. Proceedings Soc. Exp. Biol. and Med., viii, No. 5.

DESCRIPTION OF PLATE.

PLATE V.

FIG. 1. — Camera Lucida drawing. Leitz, obj. $\frac{1}{12}$, oc. 2. The drawing represents the ingrowth into the necrotic tubule (a) of a syncytial tongue of cytoplasm (b) containing hyperchromatic nuclei (c). Fibroblasts (d), embryonic fibers (e), and red blood cells (f) are seen surrounding the necrotic tubule.

FIG. 2. — Camera Lucida drawing. Leitz, obj. $\frac{1}{12}$, oc. 2. The drawing shows three necrotic tubules (a, a¹ and b) into which spherical, epithelial nuclei (c) are growing in more or less parallel rows. The nuclei are hyperchromatic as compared with the connective tissue nuclei (d) between the tubules. Two capillaries (e and e¹) are seen in this area of beginning regeneration.

FIG. 3. — Camera Lucida drawing. Leitz, obj. $\frac{1}{12}$, oc. 2. A longitudinal sketch of a regenerated tubule (a). In the center is seen the necrotic cast-like formation (b).

FIG. 4. — Camera Lucida drawing. Leitz, obj. $\frac{1}{12}$, oc. 2. A transverse sketch of a regenerated tubule. The sketch shows especially well the flattened epithelium (a), and the cast of necrotic material (b).

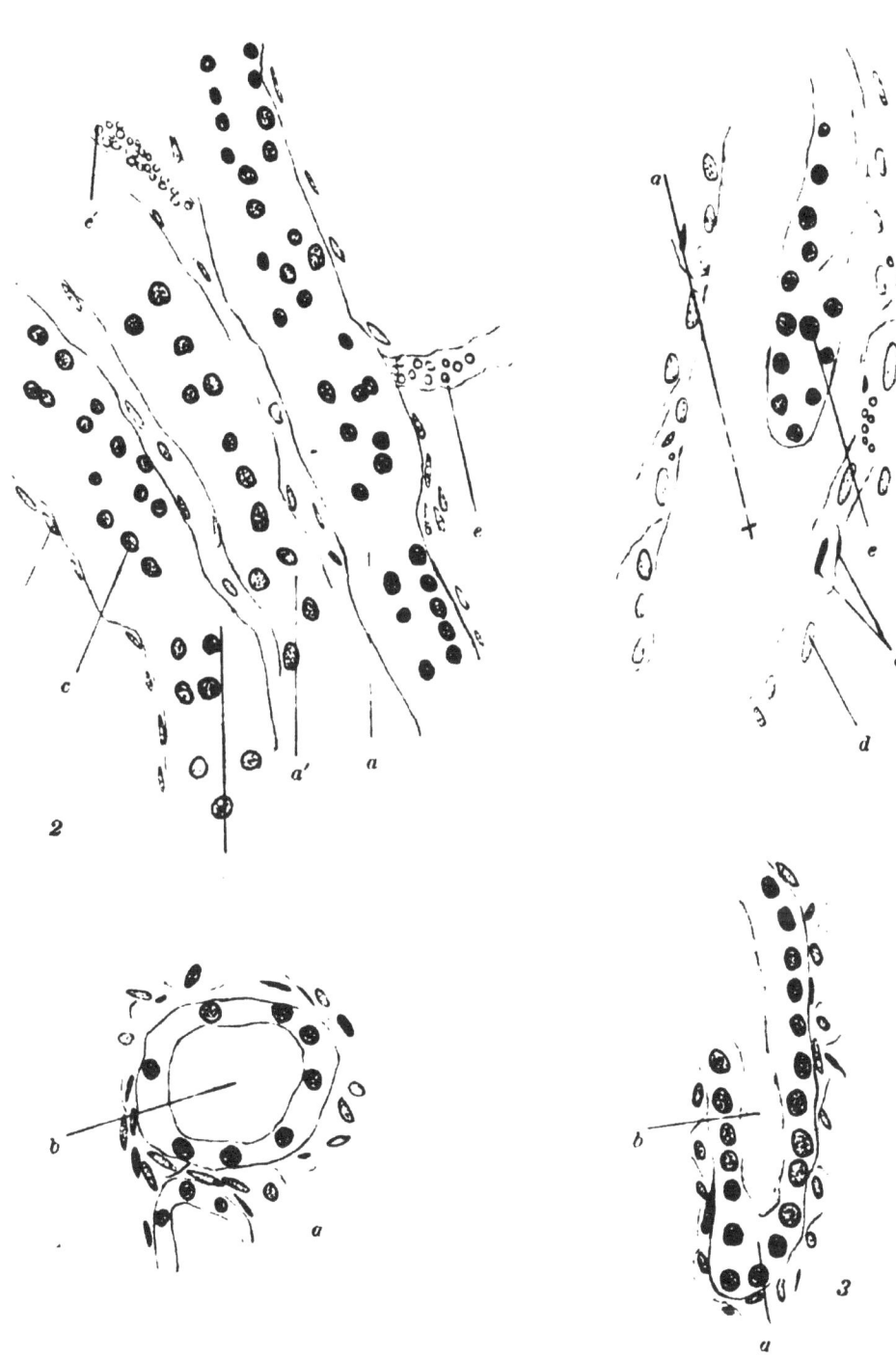

STUDIES ON CALCIFICATION AND OSSIFICATION. IV.[*]

H. GIDEON WELLS, HARRIET F. HOLMES, AND GLADYS R. HENRY.

(*From the Pathological Laboratory of the University of Chicago.*)

A relatively large proportion of the research upon experimental calcification has concerned itself with calcification of the kidney, and hence this work must receive consideration more extensive than its real significance in the general problems of calcification probably justifies, for, as will be explained later, the applicability of the results obtained in these experiments to calcification in other tissues is more than doubtful. Historically the earliest experimentally produced calcification of the kidney of which we know was reported in 1859 by Blessig;[1] in rabbits which were killed four and six days after ligation of the left renal artery. In 1866 Saikovsky observed in the kidneys of rabbits poisoned with $H_g^1Cl_2$, deposits of calcium which he stated consisted chiefly of phosphate with a little carbonate. This observation was confirmed by several observers,[2] but Litten[3] in 1880 conducted a series of experiments which were the real foundation of the more recent work. He found that while necrosis of the renal epithelium could be produced by many means, in only certain cases was this necrosis followed by calcification, and this was when the necrosis took the form of coagulation necrosis with the transformation of the cells into a coagulated non-nucleated protein mass; therefore he assumed that the state of aggregation of the degenerated protoplasm is the determining factor in calcification, an idea quite in harmony with recent developments. He also found that necrotic epithelium and cylinders, when of a variety that would subsequently calcify, stained with indigo-carmine, whereas if not capable of calcification they would not stain, suggesting that the affinity of this material for calcium and indigo-carmine depends upon the same or related factors. A most effective method of inducing calcification in the renal

[*] Received for publication Nov. 18, 1911.

epithelium was found to be temporary ligation of the renal artery for one or two hours, followed by restitution of circulation. After the second day following such circulatory disturbances calcification could always be found in the injured kidney, increasing in amount up to the tenth day. The deposits involved only the epithelium and casts formed from it, chiefly in the convoluted tubules, and according to his microchemical tests consisted of a mixture of phosphate and carbonate of calcium ; the stroma, vessels, and glomerules were never calcified. Blood casts and the hyaline casts found in collecting tubules did not calcify. When the ligation of the vessels was permanent Litten failed to observe calcification, from which he concluded that without restitution of circulation to bring in the blood with its supply of calcium salts, no calcification could take place.

V. Werra[4] repeated Litten's ligation experiments and confirmed the results of the latter in most respects. He found, however, that seven to eight days after a one-hour ligation the deposited calcium begins to be absorbed and the kidney is eventually restored to normal ; from this he assumes that the calcified cells could not have been necrotic, despite the loss of nuclear stain, but he does not seem to have excluded desquamation of these calcified cells and their replacement by new formed epithelium. He also disagreed with Litten in that he observed marked calcification of kidneys after permanent ligation of their vessels. The deposition took place in two zones, one beneath the surface and the other at the base of the pyramids, these sites corresponding to the zones · of collateral circulation.

Neuberger[5] in 1890 found that calcification could be induced by various poisons (phosphorus, aloin, bismuth sub-nitrate), and that the deposits thus produced are identical in location and character with those obtained after temporary ligation of the vessels, and consist of a " mixture of carbonate and phosphate of lime." He agrees with v. Werra that permanent occlusion of the renal vessels may cause calcification of the renal epithelium, which had also been observed by Blessig in 1859.

Leutert[6] in an extensive investigation of H_gCl_2 poisoning devotes considerable attention to renal calcification and discredits Litten's view that coagulation necrosis is the necessary basis of this form of calcification; on the contrary he maintains that completely necrotic epithelium does not calcify. The epithelial cells become calcified, he states, while still living and attached to the basement membrane, later desquamating and fusing into casts; cylinders already present in the tubules may also take up calcium salts from the secreted urine. The process is the same whether caused by poisons or by temporary ligation of the renal artery. Gebhardt[7] agrees with Leutert, stating that calcification may be observed in cells which are not necrotic.

The most elaborate study of experimental calcification of the kidney since Litten's article has been contributed by v. Kossa.[8] He found that in calcification of the renal epithelium, from whatever cause, the deposit consisted of calcium phosphate alone, never observing evolution of gas after addition of acids to the deposits, thus contradicting Litten, Saikovsky and Neuberger. Analysis of one specimen of calcified kidney did not yield sufficient P_2O_5 to combine with the Ca found, hence v. Kossa concluded that part of the calcium must be combined with some other radical, perhaps as an albuminate, but we may properly question the value of a single analysis of such small quantities of material. He failed to find magnesium, but the validity of this observation may also be doubted, since he also failed to find magnesium in the calcified aorta wall, where Baldauf[9] and Barillé[10] show it to exist in considerable amounts. As poisons which cause renal calcification do not cause demonstrable changes in the bones, and do not decrease the elimination of calcium in the urine, v. Kossa concludes that the calcification is the result of defective excretion. He found that as high as .1138 gram of CaO might be present in the two kidneys of a rabbit after poisoning with aloin. In agreement with v. Werra he found that permanent ligation of the vessels (and ureter) was followed by extensive calcification, in such a kidney after twenty-two days .0288 gram

CaO being found. To the list of poisons previously known to cause calcification of the kidney he added $CuSO_4$, iodin, and iodoform, the latter of which also caused calcification in the liver of the rabbit. (We have observed calcification in the rabbit's liver involving areas of focal necrosis produced by hemagglutinating serum; the calcium occurred as large crystals staining amethyst blue with hematoxylin. It is a very common occurrence in areas of focal necrosis in the livers of mice.)

Calcification may be produced in dogs' kidneys by one-hour ligation (v. Werra) and human kidneys may show calcium deposits after poisoning with H_gCl_2 or phosphorus (Neuberger) and not infrequently independent of such poisoning.[11, 12]

It was found by Tartarini-Galleani[13] that injection of H_gCl_2 directly into the kidneys causes coagulation necrosis, followed after several days by calcification which is not preceded by any fatty change. The deposits dissolve with effervescence of gas.

Brodersen[14] studied the calcification following temporary ligation of two hours' duration, and attempted to show that the localization of the calcium depends upon the vascular distribution. This same point is particularly emphasized by the experiments of Sacerdotti and Frattin[15] and of Liek,[16] who described ossification as a conspicuous result of permanent ligation of the renal vessels. This process takes place chiefly in the connective tissue of the pelvis, and its localization here depends upon the collateral circulation from the ureteral vessels, as shown by direct experiments. Liek observed such ossification constantly in sixteen to twenty-one days if sufficient collateral circulation was introduced by wrapping the ligated kidney in the omentum.

In an investigation of the occurrence of calcium soap formation as a preliminary step in calcification, Klotz[17] studied the calcium deposits produced in the kidney of the rabbit by various poisons. He found that the same areas which stained with silver nitrate (v. Kossa's method of demonstrating calcium salts) also stained with Sudan III, showing that fatty changes accompany calcification in the renal epithelium. As silver nitrate detects not calcium but certain acid radicals, especially P_2O_5, the blackening by this reagent of the areas stained with Sudan III is evidence that such radicals, presumably combined with calcium, are present here together with the fat, but is not proof of the presence of calcium soaps. Klotz states that "it has been possible to extract from the cortex of the kidney potassium, sodium and calcium soaps," but an extensive experience with these substances compels doubt of the possibility of proving the presence of minute quantities of calcium soaps in the simultaneous presence of fats, other soaps and other salts of calcium. It may also be added that Roehl[18] was unable to demonstrate the presence of calcium soaps by Fischler's method (whatever that may be worth[19]) in the calcium deposits resulting from temporary ligation of the renal artery, nor could he find fatty acids. He also found that the calcium was combined with P_2O_5, but not as the carbonate, agreeing with v. Kossa. He agrees with v. Werra and Leutert that the calcium is deposited in cells that are still living but injured, and ascribes the fact that certain poisons destroy the cells but cause no calcification to a too rapid action of these poisons, believing that the calcium is deposited during the interval between the beginning of protoplasmic injury and total death of the cell.

More recently Eisendrath and Strauss[20] have studied the changes taking place in the kidneys of rabbits after temporary occlusion of the vessels. They found that fifteen to thirty minutes' occlusion was not followed by calcification (four weeks later), but that after forty-five minutes' obstruction there was a little calcium deposit, and after one hour a great deal. In all respects their results agree with Litten's. The nature of the calcium deposits was not investigated.

Whatever value the foregoing observations may have concerning the specific problem of calcification in the kidney, we doubt that they can be applied without qualification to the problems of calcification as it occurs in other tissues.

This is because the calcification of renal epithelium and tube casts, through or over which is continuously flowing an aqueous solution, containing excreted inorganic calcium salts, chiefly the phosphate, is in all respects a different matter from the calcification of necrosed or hyaline tissue elements permeated by blood plasma, which is, in contrast to urine, a neutral fluid rich in colloids, in which are present most minute quantities of calcium salts, probably held partly in colloidal suspension by the proteins and partly as a double salt of calcium carbonate and phosphate in true solution.[21] There would seem to be no essential differences between the calcification of necrotic renal epithelium or tube casts and the calcification which takes place in any pervious dead organic material present in the urinary passages, $e.g.$, a blood clot or a gauze sponge left in the bladder; in other words, the process is comparable to the formation of urinary concretions, but not to pathological calcification of the tissues. That this is true is supported by the fact that v. Kossa and Roehl found that these calcium deposits contained P_2O_5 but no CO_2, for in the urine the calcium is chiefly in the form of the acid phosphate. Of similar significance is the fact that oxalic acid poisoning causes the formation of deposits of calcium oxalate upon the dead epithelium and casts present in the kidney. When the necrosis of the kidney has become extensive, so that glomerules, stroma, and vessels are involved, as after permanent occlusion of the vessels, we may then expect to find typical tissue calcification present, proceeding inward from the cortex and involving all the necrotic structures; such a process is a typical pathological calcification, and is entirely different from those calcium deposits which are limited to tube contents and necrobiotic epithelium, and which are properly incrustations.

In spite of the amount of work that has already been done, however, it seemed desirable to take up the subject again, with reference to the bearing of calcification of renal epithelium and casts upon calcification in other tissues, and especially because there exists much diversity of statement concerning certain essential points. The first of these is the question of the presence of carbonates in the deposits. Saikovsky, Litten and Neuberger all report that gas bubbles escape when the calcium deposits are dissolved in acid. Tartarini-Galleani also observed effervescence when the calcium deposits in his experiments were dissolved with acids; but as he produced calcification by injecting sublimate directly into the kidneys, causing injury to stroma as well as to epithelium, his material cannot properly be considered in this connection. V. Kossa and Roehl both expressly state that there is no gas evolved when acid acts upon experimental deposits. Although as a general rule a positive result is of more value than a negative one, yet reading the reports of the above mentioned observers one gets the impression that the negative results have the better support. Thus, Neuberger, on the affirmative, merely says that "bisweilen Gasblasen sich entwickeln," while v. Kossa states that he has repeatedly tested the deposits for CO_2 and has never observed any evolution of gas. Roehl seems also to have considered this matter carefully, without being able to confirm the positive results of Litten. As the presence or absence of CO_2 is a most essential point in interpreting this renal calcification, it has seemed worth while to again study experimental calcification of the kidneys with particular reference to this discrepancy of evidence.

Another item of disagreement is the calcium content of rabbit's blood, also a matter of fundamental importance. It is generally assumed that lesions in herbivora are particularly prone to calcification because their vegetable food introduces into the blood a much larger amount of calcium than the blood of carnivora contains; thus, bovine tuberculosis shows calcification in early and active stages, whereas in man calcification usually takes place only when the tubercle is encapsulated and quiescent. V. Kossa's article served to put this idea on a firmer footing, for he made the statement that the dried blood of five rabbits contained respectively 2.01, 1.56, 1.69, 1.80, and 1.06 per cent of Ca, an average of 1.62 per cent, whereas the dried blood of two chickens contained .0993 and .0876 per cent of Ca, and dog blood (according to Gerlach and Voit) contains .077 per cent of Ca. That is, rabbit's blood is said by v. Kossa to contain about twenty times as much calcium as dog's blood. On the other hand, Abderhalden's extensive series of analysis of blood showed that the amount of calcium in the blood of different animals is almost constant, namely, from .09 to .06 parts of lime for one thousand parts of blood (fresh), there being no calcium in the corpuscles; in Abderhalden's tables the rabbit serum is set down as containing practically the same as pig, cat, dog, ox, horse, bull, sheep, and goat blood. Examination of v. Kossa's article shows, however (P. 200), that his conclusions are based on an error in calculation, whereby the decimal point is put two places to the right of where it should be. According to the figures given in his tables the average amount of calcium in dried rabbit blood is .0162 per cent, and not 1.62 per cent as stated. The actual amount of calcium in rabbit's blood, according to the analytic figures given by v. Kossa (.0025 to .0044 gram Ca. in 15.9 to 27.66 grams dried blood) is therefore much lower than that found in the blood of other animals.

It also was attempted to determine the source of the calcium in the deposits by suppressing the excretion of urine by ligation of the ureter, with the idea that if the calcium came from the urine which was being secreted, such obstruction to the ureter should reduce the amount of calcium deposited in the injured epithelium.

Lastly, it was thought that possibly systematic quantitative analysis of the calcified kidneys might give information in addition to what can be obtained by microscopic examination alone.

The method of investigation adopted was as follows: Rabbits of nearly equal weight were used in pairs. In one of each pair calcification of the kidney was induced either by ligation of the vessels of the left kidney, or by poisoning with H_gCl_2, while the other rabbit was subjected to the same treatment and in addition the left ureter was ligated close to the kidney. After a suitable length of time the animals were killed by breaking the neck, the kidneys removed and weighed; portions taken for microscopic study were weighed and hardened in alcohol of increasing strengths.

The calcium was determined after combustion of the kidneys, and, on account of the minute quantities present in most of the analyses, was weighed as oxalate in Gooch crucibles; allowance was made for the portion removed for microscopic examination, and the figures given are corrected for the entire kidney.

The material taken for microscopical study was fixed in graded alcohols and imbedded in celloidin. After trying out a large number of calcium methods on kidney sections containing small and large amounts of calcium respectively, all sections were treated with the following solutions:

1.   For histological detail, Delafield's hematoxylin and eosin.

2.   For a nuclear as well as a calcium stain, a dilute solution of alum hematoxylin.

3.   To demonstrate calcium an aqueous solution of hematin with a trace of NH_3 in the wash water, an alkaline solution of pupurin and an alkaline solution of pyrogallic acid.

4.   To demonstrate calcium as phosphate, a dilute solution of $AgNO_3$.

5.   To demonstrate calcium as carbonate, a dilute solution of HCl.

The protocols of the experiments follow, in abbreviated form:

I.   Ligated left renal artery permanently, and killed after six days. Left kidney white, swollen, necrotic; weight, 10.5 grams. Right kidney weighed 5.8 grams. Left kidney contained .0058 gram CaO. Right kidney, no appreciable amount of Ca.

Histological examination showed left kidney enlarged, granulation tissue outside the capsule. Just beneath the capsule a narrow zone in which the epithelium of convoluted tubules shows marked calcification. Below this a dense leucocytic zone, sharply defined and five millimeters in width, bounds an area in which there is extensive necrosis of the epithelium of the convoluted tubules, the nuclei of the glomerules, and straight tubules, the stroma still staining well. There are occasional patches of calcification in the epithelium of the convoluted tubules. The blood vessels at the pelvis of the kidney contain large thrombi, some of which show considerable calcification.

Iᴀ. Ligated left renal artery for two hours and removed ligature, but the artery was not permeable to blood and was found permanently thrombosed when the animal was killed six days later. Left kidney was almost entirely necrotic, weighed ten grams, and contained .011 gram CaO. Right kidney weighed eight grams.

Histological examination showed granulation tissue outside the capsule. Narrow zone of normal kidney tissue beneath the capsule showing a trace of calcification in the epithelium. Dense leucocytic zone bounding a large infarcted area showing only slight calcification. At the junction of the cortex and medulla is an area where the cells are well preserved. Here there are many casts and extensive calcification.

II. Ligated left ureter permanently; killed after six days. Left kidney greatly swollen, pale, hydronephrotic; weight, 13.5 grams. Right kidney 8.5 grams. Left kidney contained .002 gram CaO. Right kidney no appreciable amount of Ca.

Histological examination showed left kidney greatly enlarged. The tubules are dilated, particularly the collecting tubules. Granulation tissue outside the capsule. Some albuminous material in the tubules. Trace of Ca in the epithelium of the convoluted tubules. Round cell infiltration.

IIᴀ. Ligated left ureter permanently, and killed after seven days. Left kidney showed marked hydronephrosis, weighed 12 grams, and contained .0035 gram CaO; there was 2.1 grams urine in the pelvis of the kidney which contained only a trace of Ca. Right kidney weighed 6.5 grams.

Histological examination showed left kidney enlarged, tubules dilated, particularly the collecting tubules. Albuminous material in the tubules. Some casts. Slight amount of Ca in the casts in the tubules just beneath the capsule.

III. Left ureter, artery, and vein ligated permanently; killed after six days. Left kidney weighed 20.5 grams, large, hard, and pale external surface; cut surface dark red and appeared necrotic. Right kidney weighed seven grams. Left kidney contained .013 gram CaO. Microscopic examination: Tubules dilated, particularly the collecting tubules. Aside from this the histological description of Rabbit I. answers in all respects. The leucocytic zone is rather broader here and the calcification in the tubules beneath the capsule is more pronounced.

IIIᴀ. Left ureter, artery, and vein ligated permanently; killed after six days. Left kidney weighed sixteen grams, cut surface dark in color, very bloody. Right kidney weighed 5.3 grams. Left kidney contained .022 gram CaO. Microscopically the same picture as above, with the addition of a heavily calcified thrombus in a vessel of the pelvis.

IV. Ligated left renal artery for óne hour. Died thirty-six hours later, no cause for death found. Left kidney looked normal, weighed 5.8 grams, and contained .0023 gram CaO. Microscopic: Kidney in good condition; some albuminous material in the tubules and a few casts. Only a trace of Ca.

IVᴀ. Ligated left renal artery for one hour, then removed ligature and let circulation return; killed seven days later. Left kidney the same size as right, but a little paler, artery patent. Weighed seven grams, contained .009 gram CaO.

Histological examination showed left kidney in good condition. No casts. Only a trace of Ca in the epithelium of the convoluted tubules. Marked round cell infiltration and interstitial increase.

IVB. Left renal artery occluded one hour and left ureter ligated permanently; killed the animal after seven days. Right kidney weighed 7.2 grams. Pelvis of left kidney distended with urine; kidney weighed fourteen grams, pale but apparently not necrotic, contained .0035 gram CaO. Three cubic centimeters of urine in pelvis contained .0016 gram CaO, or .0005 gram per cubic centimeter, while the urine in the bladder contained .0042 gram CaO per cubic centimeter.

Histological examination showed left kidney enlarged. The tubules are dilated, particularly the collecting tubules. Some albuminous material in the tubules. Casts more numerous than in IVA. Trace of calcification. Marked round cell infiltration.

V. Pregnant, weight two thousand grams. Left renal artery ligated two hours and circulation then restored. Killed at the end of seven days. Left kidney paler than right, but of the same size; contained .018 gram CaO. Microscopic: Numerous casts. Considerable calcification in casts and epithelium of the tubules, irregularly distributed throughout the cortex. Considerable interstitial increase. The marked degree of calcification in this experiment is of interest in view to the antagonism of pregnancy to arterial calcification in rabbits observed by Loeper and Boveri.[32]

VA. Left renal artery ligated two hours, and left ureter ligated permanently. Killed after seven days. Left kidney greatly enlarged and hydronephrotic, weighs 15.7 grams and contains .005 gram CaO. Right kidney weighs 6.9 grams. The urine in the left pelvis weighed 2.5 grams, but contained too little calcium to be determined, whereas the urine in the bladder contained .0022 gram CaO per cubic centimeter. Microscopic: Left kidney enlarged. The tubules are dilated, particularly the collecting tubules. Albuminous material in the tubule, and some casts. Quite marked calcification in the casts and in the epithelium of the tubules near the junction of the cortex and the medulla. Marked interstitial increase.

VB. Left renal artery ligated two hours, and circulation then restored. Killed after seven days. Left kidney slightly smaller and paler than the right, weighs six grams, contains .0013 gram CaO. Right kidney weighs 6.3 grams.

Histological examination shows slight amount of albuminous material in the tubules and a few casts. No Ca. Marked round cell infiltration.

VC. Ligated left renal artery two hours, and left ureter ligated permanently. Killed after six days. Left kidney showed several infarcts and but slight hydronephrosis; weighed 10.2 grams, and contained .011 CaO. Right kidney weighed 6.3 grams.

Histological examination showed left kidney enlarged. The tubules are dilated, particularly the collecting tubules. Many casts. Extensive

calcification in the casts and to a less extent in the epithelium of the convoluted tubules. Round cell infiltration. Infarcted area near the pelvis
with a leucocytic zone about it.

VI. Ligated left renal artery three hours, and killed after seven days.
Kidney showed little apparent change; weight, 9.2 grams, and contained
.028 gram CaO.

Histological examination showed left kidney increased in size. The
tubules are dilated, particularly the collecting tubules. Some casts.
Considerable calcification in the convoluted tubules near the capsule.
Round cell infiltration.

VIA. Ligated left renal artery three hours. Ligated left ureter permanently. Left kidney found enlarged, pale, with pin-point yellowish
white spots beneath the capsule, pelvis distended with urine; weighed
eleven grams and contained .002 gram CaO. The pelvis contained
three cubic centimeters of urine, which in all contained about one mg.
CaO. Right kidney weighed 7.4 grams and contained about .001 gram
CaO. Microscopic: Left kidney in good condition; slight amount of
albuminous material in the tubules; few casts. No calcification. Marked
round cell infiltration.

VII. Ligated left renal artery and vein, and killed after seven weeks.
The left kidney was shrunken, rough, stony hardness, brownish color on
cut surface which grated under the knife and contained visible spicules of
inorganic material; weighed 5.4 grams and contained .162 gram CaO.
Right kidney weighed ten grams (compensatory hypertrophy).

Histological examination showed right kidney increased in size. No Ca.
Left kidney smaller than normal. Just beneath the capsule a narrow zone
in which the tubules stain and show some calcification. Beneath this is
a wholly necrotic area, corresponding in position to the leucocytic zone of
the six-day infarcts. Beneath this again the greater part of the section
shows total necrosis, aside from occasional stroma cells, and very extensive calcification. Here the calcification involves all the tissues, stroma,
and glomerules as well as the epithelium. (In the six-day kidneys there
is no indication of calcification of the stroma.) There is also considerable calcification in the thrombi in the vessels at the pelvis. Near the
pelvis there is less complete necrosis of the epithelium. Ossification is
taking place in several areas at the pelvis.

VIIA. Ligated left renal vein, artery and ureter permanently. Killed
after seven weeks. Left kidney resembled that of No. VII., but was not
so hard and contained no visible inorganic material; weighed 7.9 grams
and contained .100 gram CaO. Right kidney weighed fifteen grams
(compensatory hypertrophy).

Histological examination showed right kidney enlarged. Some albuminous material in the tubules. No Ca. Interstitial increase. Left kidney much smaller than normal. Histological picture similar to No. VII.,
except that the calcification is less in amount and less regular in distribution. There is also rather more stroma retaining a nuclear stain. (Some
of this seems to be proliferating connective tissue.)

VIII. Ligated left ureter permanently, and gave .01 gram H_4Cl_2 on first, third, and fourth days. Died after five and one-half days. Left kidney greatly swollen, red, inflamed, but no evidence of infection ; pelvis dilated ; weighed 18.5 grams and contained .0025 gram CaO. Right kidney showed no gross changes, weighed 8.4 grams and contained .0089 gram CaO. Microscopic: Left kidney increased in size The tubules are dilated, particularly the collecting tubules. Albuminous material in the convoluted tubules. Epithelium not swollen. Only a trace of Ca. Round cell infiltration. Right kidney increased in size. Epithelium of convoluted tubules swollen. Many casts. Extensive calcification in the casts in straight tubules. Round cell infiltration.

VIIIA. Gave .01 gram H_4Cl_2 to two thousand grams rabbit, on first, third, and fifth days, and killed on the sixth day. The kidneys showed a few atrophic patches, weight together 17.5 grams, and contained in all .008 gram CaO.

Histological examination showed epithelium of convoluted tubules some-what swollen. Some casts. Only slight amount of calcification in the epithelium of the convoluted tubules. Round cell infiltration. Both kidneys alike.

VIIIB. Gave three injections of .01 gram H_4Cl_2 during four days, and killed on the fifth day. Kidneys weighed together 17.9 grams. Amount of calcium not determined because of error in analysis.

Histological examination showed epithelium of the convoluted tubules swollen. Many casts. Very extensive calcification in the casts and the epithelium, particularly at the junction of the cortex and the medulla. Considerable round cell infiltration. Both kidneys alike.

VIIIc. Duplicate of VIIIB. Kidneys weighed together 14.8 grams and contained .0026 gram CaO.

Histological examination showed slight amount of albuminous material in the tubules. Very few casts. Only a trace of Ca.

Two old rabbits were given one per cent emulsion of iodoform with gum acacia, by means of a stomach tube, as follows :

IX. Weight, two thousand three hundred and fifty grams, given twenty cubic centimeters on the first and third days, fifteen cubic centimeters on the fourth, fifth, and seventh days. Immediately after the injection on the seventh day this rabbit died from a rupture of the vena cava at the point of its passage through the diaphragm, probably during struggling while the rubber tube was in the esophagus. The liver was found to be very fatty, somewhat cirrhotic, with lighter yellow patches and also well defined pinker areas with pale margins like infarcts. No visible calcification. Weighed seventy-five grams ; ten grams taken for chemical analysis, and found to contain .013 gram CaO. The kidneys together weighed 13.7 grams, and were paler in the medulla than in the cortex. Contained together .013 gram CaO.

Histological examination showed kidney in fair condition. Slight amount of albuminous material in the tubules. No casts. Slight amount of calcification in the epithelium of the convoluted tubules. Liver showed marked fatty infiltration irregularly distributed, and increase of the periportal connective tissue. Trace of calcium in some of the liver cells. The light patches noted in the gross examination represent areas where there is a very extreme degree of fatty change.

IXA. Weight, two thousand seven hundred grams, given twenty cubic centimeters iodoform emulsion on the first and third, fifteen cubic centimeters on the fourth, sixth, eighth, and thirteenth days, and killed on the twenty-first day, apparently in good health throughout. No gross changes in the liver. Kidneys weighed eleven grams and showed no gross changes; 14.5 grams of liver tissue and ten grams of kidney tissue analyzed, but a bare trace of CaO was present in each sample, less than one milligram.

Histological examination showed the kidney and liver in good condition. No evidence of calcification. Whether this deficiency in No. IXA. is due to the eight days' interval after the iodoform was stopped has not been investigated. V. Werra found that calcium deposits in the kidney may be reabsorbed.

TABULATION OF RESULTS.

		Gram CaO in Kidney.	Amount of Ca According to Microscope.
I.	Artery ligated 6 days.	.006	+ +
IA.	Artery ligated 6 days.	.011	+ + +
II.	Ureter alone ligated 6 days.	.002	±
IIA.	Ureter alone ligated 7 days.	.0035	+
III.	Artery and ureter both ligated for 6 days.	.013	+ +
IIIA. . . .	Artery and ureter both ligated for 6 days.	.022	+ + + +
IV.	Artery occluded 1 hour; died 36 hours later.	.0023	+
IVA.	Artery occluded 1 hour; killed after 7 days.	.007	+
IVB. . . .	Artery occluded 1 hour, ureter permanently ligated; killed after 7 days.	.0035	+
V.	Artery occluded 2 hours; killed after 7 days.	.018	+ + +
VA.	Artery occluded 2 hours, ureter permanently ligated; killed after 7 days.	.005	+ +
VB.	Artery occluded 2 hours; killed after 7 days.	.0012	—
VC.	Artery occluded 2 hours, ureter permanently; killed after 6 days.	.011	+ + +
VI.	Artery occluded 3 hours; killed after 7 days.	.028	+ +

TABULATION OF RESULTS. — *Continued.*

		Gram CaO in Kidney.	Amount of Ca According to Microscope.
VIA.....	Artery occluded 3 hours, ureter permanently; killed after 7 days.	.002	—
VII.....	Artery ligated 7 weeks; animal then killed.	.162	(+) [10]
VIIA....	Artery and ureter both ligated 7 weeks; animal then killed.	.100	(+) [7]
VIII. ...	H$_2$Cl$_2$ poisoning 6 days; left ureter ligated; left kidney. Right kidney untouched; right kidney.	.0025 .009	± +++
VIIIA...	H$_2$Cl$_2$ poisoning 6 days; kidneys together .008 gram CaO; each.	.004	+
VIIIB...	H$_2$Cl$_2$ poisoning 5 days; analysis lost.	?	++++
VIIIC...	H$_2$Cl$_2$ poisoning 5 days; kidneys together .0026 gram CaO; each.	.0013	±
IX......	Iodoform poisoning 7 days; liver, weight 75 grams, contained in all. Kidneys together contained .013 gram CaO; each.	.0975 .0065	Trace. +
IXA.....	Iodoform poisoning 13 days, then 8 days' rest. Trace only of Ca in either liver or kidneys.	Trace.	—

Discussion of the results. — One of the most striking results of this series of experiments is the almost unfailing agreement of the microscopic and the chemical determination of calcium. The arbitrary designation of the amount of calcium present by " + " signs lacks the objectivity of the quantitative results of a chemical analysis, but the results expressed in this way were found to correspond, with few exceptions, to the analytical figures; they were all set down without knowledge of the results of analysis, to avoid subjective influences. It was found that a rabbit kidney containing less than two milligrams of calcium oxide did not show positive microscopic evidence of calcium (VB., VI., VIIc.); if two milligrams of calcium were present it might or might not be visible as fine, dust-like particles (II., IV.,

VIA., VIII.), but amounts larger than this were always detected. If we consider what fraction of an entire rabbit kidney a single section ten to twelve microns thick consti- tutes, the delicacy of the microscopic detection of calcium becomes apparent. There were few instances in which there was any serious discrepancy between analytical and chemical results.

It is evident that simple ligation of the renal artery leads, within six days, to a considerable degree of calcification (I. and IA.), which is contrary to Litten and agreeing with v. Werra. When the artery was left ligated for seven weeks a truly remarkable degree of calcification resulted (VII.) so that three per cent of the moist weight of the kidney was CaO. There was in this kidney considerable ossification. Occlusion of the renal artery for one to three hours, with subsequent restoration of circulation, leads to considerable calcification, as has been shown by many other experi- menters; even thirty-six hours after the temporary occlusion there is enough deposition of calcium to be visible micro- scopically and to be determined chemically (IV.). In one experiment (VB.), however, two hours' ligation was not followed by calcification; whether this failure was caused by an unusually abundant collateral circulation is not known. It is interesting to note that thrombus formation with marked calcification can take place in six days (IA., IIIA.).

Occlusion of the ureter alone leads to marked hydro- nephrosis, severe inflammatory changes in the kidney, and a barely visible deposition of calcium (.002 to .0035 gram CaO) in six to seven days (II., IIA.). The attempt to influence the effect of arterial ligation on calcification, by stopping urinary secretion by ligating the ureter, did not give conclusive results. In most of the experiments the urinary obstruction seemed to reduce the amount of calcifi- cation (IVA. and IVB.; V., VA., and VC.; VI. and VIA.; VII. and VIIA.). There were, however, certain contrary results (III., IIIA., and VB.). Furthermore, the positive results obtained may be ascribed as well to interference with collateral circulation by the ligation of the ureter and by the increased pressure of the hydronephrosis, as to the stoppage of the secretion of urine.   Therefore, although most of the results are in harmony with the hypothesis that decreased secretion of urine decreases the amount of calcification by providing less urine from which calcium can be deposited in the injured renal epithelium, yet on account of the simultaneous effect on circulation these experiments do not prove this hypothesis.

Poisoning with H_gCl_2 was found to produce less uniform degrees of calcification than temporary arterial occlusion. Calcification by this means was also found to be less in the kidney with ligated ureter than in the normal kidney (VIII.), but here, too, it is not possible to decide how much of the difference is due to the obstruction to circulation and how much to decreased secretion of urine.

The important question of the presence of carbonate as well as phosphate of calcium in the deposits may, according to these experiments, be settled as follows: When the calcification is limited to the renal epithelium and tube casts, as in H_gCl_2 poisoning or temporary arterial ligation, it is present in the form of phosphate, as shown by the positive reactions with silver nitrate and lead acetate, and by a total absence of effervescence when acted upon by acid.   When the injury and calcification involve the stroma, as long after ligation of the artery, as in experiments VII. and VIIA., then carbonate is present and demonstrable by the development of CO_2 on acidification.   These facts are in support of the hypothesis that in the lesser degrees of calcification the deposition of calcium is from the urine in the tubules into the injured epithelium and casts, and of different origin and composition from calcification taking place in other situations, where, deposited directly from the blood stream, part of the calcium is in the form of carbonate.   Possibly the divergent results obtained by previous experimenters concerning the presence of carbonate in experimental calcification of the kidney may be explained by these observations.

The calcium of rabbit blood and urine. — Analyses of

specimens of rabbit blood and of dog blood gave the following results:

Twenty-one grams dried normal dog blood contained thirteen milligrams CaO = .06 per cent CaO.

Fourteen grams dried normal dog blood contained 7.7 milligrams CaO = .055 per cent CaO.

These figures are a trifle lower than those obtained in most other analyses recorded in the literature; *e.g.*, .077, .084, .061 per cent Ca (Voit), .05 to .06 per cent Ca (Foster). In dog serum Abderhalden found .0113 per cent and Gerlach .014 per cent of the fresh weight to be Ca, which corresponds to .14 to .17 per cent of the dry weight, the serum containing all the calcium of the blood.

When compared with the amount of calcium in dog blood, the proportion of calcium in dried rabbit blood determined by the same method was always a little higher; thus in five analyses we found: .12, .10, .11, .12, .09 per cent CaO, an average of .11 per cent CaO, or .078 per cent Ca. This is somewhat higher than Abderhalden's figure for fresh rabbit serum, .0116 per cent, for, in view of the absence of Ca in the corpuscles which constitute half the blood, the proportion in dried entire blood on this basis would be about .05 per cent.* Our figures would indicate that rabbit blood does contain somewhat more calcium than does dog blood, a fact which is entirely in harmony with the known predisposition of herbivora to calcium deposition.

In the calcium deposition in renal lesions, it is probable that the amount of calcium in the urine will have particular influence. Analysis of samples of urine from the bladders of rabbits have shown a remarkably high calcium content, as follows: .0022 gram CaO per cubic centimeter, .042 gram per cubic centimeter, .008 gram per cubic centimeter. This, contrasted with normal, human urine, which ordinarily contains about .0001 to .0002 gram of CaO per cubic

---

* Klemperer in 1889 (Virchow's Archiv., 1889, cxviii, 445) found in samples of fresh rabbit blood .10, .15, .07, .13, .14 per cent of CaO, an average of .12 per cent, equivalent to about .7 per cent in the dry blood, while the blood of a dog contained about half as much calcium. Obviously these figures are much too high.

centimeter, shows that the concentration of Ca in rabbit urine is relatively very high, which must decidedly favor deposition of calcium in renal lesions of rabbits. The calcium content of the urine present in the hydronephrotic kidneys with ligated ureters was in each of three cases much smaller in amount than in the bladders of the same rabbits — thus, in three pairs of analyses the bladder urine contained eight, four and 2.2 milligrams per cubic centimeter, while the urine from the dilated pelvis was in each case less than one-half milligram per cubic centimeter. This may be ascribed either to an absorption of calcium from the dilated pelvis, or it may be that the fluid in the pelvis was more in the nature of an exudate, and not urine.

SUMMARY.

The calcification of renal epithelium and tube casts which is experimentally produced by temporary anemia and by certain poisons, is probably analogous to calcification of organic material in urine, rather than to calcification of tissues with lime salts direct from the blood. These epithelium and tube-cast deposits contain no demonstrable carbonate and seem to contain calcium only as phosphate. When the necrosis and calcification involve the interstitial tissues, carbonate can be demonstrated; here the calcium probably comes from the blood rather than from the urine.

Permanent ligation of the ureter alone leads to slight deposition of calcium. Ligation of the ureter usually decreases the amount of calcification which is produced by temporary anemia or poisoning, but whether by decreasing urinary secretion or by circulatory disturbances, is not determined.

Chemical and microscopic determination of the quantity of calcium in the kidney agree well with one another. Amounts of calcium under two milligrams CaO in an entire rabbit kidney cannot be detected microscopically, but any amount over two milligrams usually can be detected; ten to twenty milligrams CaO in a rabbit kidney constitute an extensive calcification.

Demonstrable deposition of calcium may be present as early as thirty-six hours after a temporary anemia of one hour. Permanent ligation of the renal artery causes extensive calcification of the kidney within a few days; ligation for seven weeks led to a deposition of .162 gram CaO in a kidney, so that about twenty-five per cent of the dry weight of the kidney was calcium phosphate. Thrombus formation with marked calcification of the thrombus can take place in six days.

The blood of rabbits contains a little more calcium than dog's blood, about .078 per cent of the dry weight.

REFERENCES.

1. Blessig. Virchow's Arch., 1859, xvi, 120.
2. See Weichselbaum (Cent. allg. Pathol., 1891, ii, 9), who gives a review of literature on sublimate poisoning to that date.
3. Litten. Zeit. kl. Med., 1880, i, 131; Virchow's Archiv., 1881, lxxxiii, 544.
4. Von Werra. Virchow's Archiv., 1882, lxxxviii, 197.
5. Neuberger. Arch. exp. Path. u. Pharm., 1890, xxvii, 39.
6. Leutert. Fortschr. d. Med., 1895, xiii, Nos. 3–11.
7. Gebhardt. Inaug. Dissert., Freiburg, 1897.
8. Von Kossa. Ziegler's Beitr., 1901, xxix, 163.
9. Baldauf. Jour. Med. Research, 1906, xv, 355.
10. Barillé. Jour. pharm. et de chem., 1910, cii, 342.
11. Beer. Jour. Path. and Bact., 1903, ix, 225.
12. Baum. Virch. Arch., 1900, clxii, 85.
13. Tartarini-Galleani. Lo Sperimentale, 1904, lviii, 371.
14. Brodersen. Inaug. Diss. Rostock, 1904.
15. Sacerdotti and Frattin. Virchow's Arch., 1902, clxviii, 431.
16. Liek. Arch. kl. Chir., 1908, lxxxv, 118.
17. Klotz. Jour. Exper. Med., 1905, vii, 633.
18. Roehl. Ziegler's Beitr., 1905, 7th Suppl., 456.
19. See Dietrich. Verh. Deut. Path. Gesellsch., 1910, xiv, 263.
20. Eisendrath, D. N., and Strauss, D. C. Jour. Amer. Med. Assoc., 1910, iv, 2286.
21. See Wells. Archives of Int. Med., 1911, vii, 721.
22. Loeper and Boveri. Presse Méd , 1907, xv, 401.

A CASE OF CEREBELLAR ABSCESS WITH ISOLATION OF MICROCOCCUS CEREUS ALBUS.*

HOWARD T. KARSNER, M.D.

(*Assistant Professor of Pathology, Harvard Medical School.*)

(*From the McManes Laboratory of Pathology, University of Pennsylvania.*)

The micrococcus cereus was first described by Passet[1] who considered it definitely pathogenic for man. Since that time Passet's statements have been concurred in by some writers and contradicted by others, the latter holding it to be an air or water organism of no pathological significance. This variance of opinion led to the detailed bacteriological study of the following case, the study being carried out in a series of experiments the record of which follows the clinical and autopsy notes.

Case. — F.H., white, male, sixty years of age, railroad brakeman. His family history is negative. His personal history includes three railroad accidents, in all of which he received severe head injuries. Shortly after an operation for gastric ulcer (January, 1909) he developed a right-sided basal pneumonia. From this time, although recovering sufficiently to return to work, he continued to have cough until his death, Nov. 16, 1910. Patient frequently expressed a feeling of suffocation, sometimes accompanied by pain in the right thorax, and followed by profuse expectoration, sometimes amounting to five hundred cubic centimeters of blood-stained, gray, purulent sputum of extremely putrid odor. Physical signs in the lungs were indefinite and frequent examination of the sputum failed to show the presence of tubercle bacilli.

On Nov. 1, 1910, patient noticed that in walking his body bent to the right and he had a tendency to turn to the right. After ten days he began to suffer with pain in occiput, back of neck and shoulders and went to bed because of profound general weakness. Examination showed that the pupils were normal, the tongue protruded in a straight line, coördination in the movement of the hands was poor, station was good, gait was as noted, knee jerks were exaggerated, ankle clonus was present, there was no Babinski, the grip was weaker in the right than in the left hand. Examination of thorax and abdomen was negative. The condition was afebrile and there was slight slowing of the pulse. Patient died November 16 and the autopsy was performed about twenty-four hours after death.

---

* Received for publication Nov. 13, 1911.

At the autopsy there was found chronic hypertrophic gastritis, recent round ulcer of duodenum, cloudy swelling of liver, chronic interstitial pancreatitis, acute splenic tumor, chronic interstitial nephritis with cloudy swelling, diminution of chromaffin tissue in adrenals, slight hypertrophy of heart, sclerotic mitral and aortic valvulitis, chronic adhesive pleurisy, chronic adhesive peritonitis about pylorus and liver. These diagnoses represent a composite of gross and histological findings. There was no evidence of old or recent gastric ulcer.

The base of the right lung was found densely adherent to the diaphragm and a moderately extensive, partly encapsulated, suppurating gangrenous area found, involving the base of the lung, penetrating the diaphragm and appearing between the diaphragm and capsule of the liver as a sub-phrenic abscess not invading the liver substance. Throughout both lungs numerous irregularly outlined areas of consolidation about one centimeter in diameter were found which histologically were seen to be foci of fibrino-purulent broncho-pneumonia. No cultures were made from the lungs because the organs had been freely punctured by the embalmer's needle.

The brain was found to be under considerable tension and showed much superficial congestion. There was no inflammation of the meninges and neither thrombosis nor suppuration in the large venous sinuses. The middle lobe of the cerebellum was almost completely replaced by a non-encapsulated deeply seated abscess which extended slightly and equally into both lateral lobes and contained a thick, creamy, viscid, slightly brownish pus. Bacteriological examination of the pus showed a pure culture of Staphylococcus cereus albus. Sections from the wall of the abscess showed only one type of organism, a micrococcus, arranged in staphyloid clusters from whose margins short chains of ten or twelve members stretched out. The cocci showed no flattening of adjacent surfaces. Although the chain-like arrangement was not found in bouillon or on solid media a semi-fluid gelatine bouillon mixture showed staphyloid groups with chain-like outgrowths quite similar to those seen in the sections of abscess wall. Sections from the abscesses in the lungs showed frequent groups of cocci similar to those found in the brain.

Passet described two forms of Staphylococcus cereus, the white form (albus), which he found twice, and the yellow form (flavus), which he found once among thirty-three human abscesses. In these three cases the organisms were present in pure culture. Passet states that he was able to infect animals with both forms, but that the animals were refractory to successful inoculation. Tils[2] found Staphylococcus cereus flavus in the water supply of Freiburg, and concludes that it is an air coccus free from pathogenicity. More recently several workers have isolated it in cases of various human infections.   Singer,[3] for example, found it in pure culture in the urine of four rheumatic patients and mixed with other organisms in the urine of three additional rheumatic cases.   Arkovy[4] found it mixed with other organisms in an alveolar abscess.   Ohlmacher[5] found it associated with Proteus vulgaris and Staphylococcus pyogenes albus in a cerebellar abscess of otitic origin.   Florentini[6] found it in pure culture in the blood of three cases of purpura hemorrhagica and declared it to be pathogenic for rabbits.   Pes[7] obtained a pure culture from one case of acute conjunctivitis. Lehmann[8] found it in the feces of young children but was of the opinion that it produced no pathologic alteration. Bajardi[9] found it to have distinct hemolytic activity.   Fermi[10] found that mineral and organic acids, alkalis, alkaloids, and certain salts influenced its biological characters in the same way as the pyogenic staphylococci.   Sulima[11] in his studies on the effect of increased body temperature on the growth of microörganisms used the micrococcus cereus as a pathogenic organism apparently without difficulty.

As might be expected from the diversity of findings reported in current literature the text-books vary considerably in their discussions of this organism.   Flügge[12] does not mention it nor does Park.[13]   Muir and Ritchie[14] mention it briefly but do not comment on its pathogenicity.   Kolle and Wassermann,[15] and Migula[16] speak of it briefly; they state that it is not pathogenic and evidently base their conclusions on the work of Tils.

The strain isolated from the case reported in this paper was studied for several weeks culturally, and then experiments were tried in rabbits, guinea-pigs, and one macacus rhesus monkey.

In culture the organism showed on nutrient agar typical white waxy colonies with elevated irregular edges which later became yellowish in color.   In gelatine stabs a few yellowish granules appeared after several days.   There is absolutely no liquefaction on the surface or in the depths.   On potato there was a grayish yellowish growth.

Rabbits were inoculated intracranially, intraperitoneally

and subcutaneously with one cubic centimeter twenty-four-
hour bouillon culture without observable effect. Another
rabbit was inoculated with one cubic centimeter twenty-four-
hour slant agar growth suspended in five cubic centimeters
salt solution and no effect noted.

Guinea-pigs were inoculated intraperitoneally and subcutaneously with one cubic centimeter twenty-four-hour slant
agar growth suspended in five cubic centimeters salt solution
and no effect appeared. Hoping to aid growth by producing a locus resistentiæ minoris a guinea-pig was deeply
etherized, its belly opened, the liver seared with a hot platinum needle, the abdomen closed and an entire twenty-four-
hour slant agar growth suspended in five cubic centimeters
salt solution injected into the peritoneum. In the same
animal also under deep ether anesthesia the skin was bruised
with a hemostatic forceps and an entire twenty-four-hour
agar slant growth suspended in three cubic centimeters salt
solution injected subcutaneously under the bruise. After a
week the animal was killed and the liver appeared perfectly
normal. An abscess appeared in the skin near the site of
subcutaneous inoculation, but Staphylococcus cereus was
not found in the pus.

Hoping to infect an animal of higher type one cubic centimeter twenty-four-hour slant agar growth suspended in five
cubic centimeters salt solution was injected into the brain of
a deeply anesthetized fairly large macacus rhesus monkey.
No ill effects followed. Thinking that this first dose might
act to sensitize the animal to re-inoculation, three cubic
centimeters twenty-four-hour slant agar growth suspended in
five cubic centimeters salt solution were injected into the
opposite side of the brain five weeks later. Again there
was complete failure to produce a demonstrable abscess.

It would appear then that the strain of Staphylococcus
cereus albus isolated from this case of brain abscess is not
pathogenic for the rabbit, guinea-pig, or monkey. On the
other hand, however, the organism seems to be pathogenic
in man, for both cultural and histological examination of
the cerebellar abscess in this case were in agreement and

showed the presence only of the micrococcus cereus albus. The likelihood of the presence of an anaërobe as the causative agent must of course be considered, but the fact that the histological sections on most minute study failed to show any other organism than the Gram positive coccus identical in arrangement with the micrococcus cereus growing in a semi-solid medium, would make it seem probable that no other organism played any part in the production of this abscess. In the stained section of the lung abscesses, in addition to various other organisms, micrococci identical in morphology with those of the brain abscess were found in considerable numbers. This circumstance would lead to the supposition that the lung abscesses possibly were caused by the same organism as the brain abscess and that the latter was probably metastatic in origin.

[The patient was attended by Dr. Holmes Walker, of Philadelphia, and was seen in consultation by Dr. Henry D. Jump. These gentlemen invited me to perform the autopsy, have permitted me to make an abstract of the clinical notes, and consented to my publication of this report. The bacteriological studies were made with the kind assistance of Dr. Herbert Fox, Dr. Rucker, and Mr. McCaskey of the Pennsylvania State Department of Health. They have this acknowledgment and my sincere thanks.]

REFERENCES.

1. Passet. Fortschr. d. Med., 1885, iii, 68.
2. Tils. Ztschr. f. Hyg. u. Infektionskrankh., 1890, ix, 282.
3. Singer. Centralb. f. Bakteriologie, 1899, xxv, 829.
4. Arkovy. Centralb. f. Bacteriologie, 1898, xxiii, 962.
5. Ohlmacher. Lancet-Clinic, 1897, N.S , xxxix, 221.
6. Florentini. La clin. med. Ital., 1902, No. 10, ref. Centralb. f. Bakteriolgie, 1903, xxxiii, 706.
7. Pes. Arch. f. Augenh., 1902, xlv, 205.
8. Lehmann. Innaug. Diss., Munich, 1903.
9. Bajardi. Ann. d'ig. sper., 1901, N.S. xl, Fasc. 3, ref. Centralb. f. Bakteriol., 1902, xxxi, 447.
10. Fermi. Centralb. f. Bakteriol., 1898, xxiii, 208.
11. Sulima. Centralb. f. Bakteriol., 1909, xlviii, 318.
12. Flügge. Handbuch der Hygiene, Leipsig, 1882.
13. Park. Pathogenic Bacteria and Protozoa, Philadelphia, 1908.
14. Muir and Ritchie. Manual of Bacteriology, 176, New York, 1907.
15. Kolle and Wassermann. Handbuch der pathogenen Mikroörganismen, iii, 141, Jena, 1903.
16. Migula. System der Bakterien, ii, 126, Jena, 1900.

THE ABSORPTION OF ALBUMIN WITHOUT DIGESTION.[*]

ELEANOR V. N. VAN ALSTYNE AND P. A. GRANT.

(*From the Department of Experimental Therapeutics, Cornell Medical School, New York.*)

The question of the extent to which proteid must be digested in the stomach and intestine before it is normally absorbed is a live problem in physiology. With the increased information in regard to the activity of proteid splitting enzymes in the intestine has come the belief held by the majority of physiologists that most if not all of the ingested proteid is reduced to the simpler fragments, the "bausteine," before leaving the digestive tube. These simpler compounds, the amino bodies of varying degrees of structure and complexity, may then in part be synthesized into the characteristic albumin of the particular species by the absorbing epithelium; or a large portion may be absorbed as cleavage products and be still further broken down, appearing within a very few hours as an increased nitrogen output in the urine. Most physiologists are agreed that the intestinal epithelium ordinarily prevents the entrance of foreign, undigested proteid into the circulation, in much the same way that the epithelium lining Bowman's capsule prevents the serum albumin from leaving the blood stream under normal conditions.

The chief obstacle to the determination of this question has been the difficulty of recognizing small amounts of foreign proteid in the blood serum. The question may not be solved by noting the disappearance of foreign proteid from an intestinal loop. There must be an actual recognition of the foreign proteid in the blood and urine by a method which is beyond question. The biological methods involved in various serum reactions have been employed recently in such lines of investigation, and some evidence has been obtained by these means which indicates that foreign albumin may

---

[*] Received for publication Sept. 11, 1911.

be absorbed as such.   The evidence is conflicting, however, and it seemed to us worth while to reinvestigate this question using the anaphylactic reaction as a method for detecting the proteid.

The questions which we set for ourselves in this connection are the following:

(1.)   May a foreign proteid in the circulation be recognized when present in relatively small quantities by means of the anaphylactic reaction?

(2.)   For how long a time does a foreign albumin remain in the circulation in such form that it may be recognized?

(3.)   May a foreign albumin be absorbed through the intact intestinal epithelium and be recognized as such in the circulation?

(4.)   Does a foreign albumin so introduced appear in the urine?

The anaphylactic reaction has been the subject of much investigation.   In this country Rosenau and Anderson and Gay and Southard have established the phenomenon.   In the work of Rosenau and Anderson we have evidence that the reaction is highly specific; guinea-pigs may be sensitized simultaneously with serum, milk and egg albumin and successively intoxicated by the same proteids.   The specificity of the reaction has not been questioned by the more recent work, and it is not our purpose to reinvestigate this phase of the matter further than may be involved in our control experiments.   It may be said at this point that we have had no reason to question previous conclusions, and all our results strongly confirm the statements made by Rosenau and Anderson regarding the specificity of this reaction.

From the work of Wells we know that so small a quantity of proteid as one twenty-millionth part of a gram of crystalized egg albumin may suffice to sensitize a guinea-pig. No other means of detecting a proteid in any way approaches this in delicacy.

In regard to the first question: " May a foreign proteid,

when introduced directly into the circulating blood in relatively small quantities be recognized by means of anaphylaxis?" the following experiments were made:

A healthy dog was placed in a cage a few days for observation and to obtain samples of urine. The animal was found to be well in all respects and the urine free from albumin. Under morphine and cocaine anesthesia the jugular vein was aspirated and one hundred cubic centimeters of blood drawn to obtain normal serum. Following this procedure twenty cubic centimeters of an egg albumin solution prepared as follows was injected into the vein: The whites of fresh eggs were thoroughly broken up and diluted with an equal volume of salt solution. The solution filtered through paper was ready for use. In all of these experiments egg albumin was the foreign proteid used because the contrast between it and animal serum permits very sharp distinctions and it gives especially good results in anaphylactic experiments. An egg albumin solution prepared by this method was used throughout the work.

Fifteen minutes after the albumin was injected twenty cubic centimeters of blood was withdrawn; and again at four hours, forty-eight hours, and seventy-two hours after the injection a further quantity was withdrawn. The serum from this blood was allowed to separate and kept for sensitizing purposes. The urine was collected daily and examined for albumin, and a portion of it kept for sensitizing purposes. The injection of the egg albumin into the vein caused a very marked depression of the animal (in some instances it was given without morphine) and was followed a few hours later by a diarrhea.

Guinea-pigs were sensitized with the serum and urine obtained as outlined above, the sensitizing dose varying from .1 cubic centimeter for the first twenty-four hours to .5 cubic centimeter for the seventy-two-hour period. One cubic centimeter of the urine was used for the same purpose. The sensitizing dose of serum was given subcutaneously and, after an appropriate period, in no case less than ten days, an intoxicating dose of the egg albumin solution, two and one-half, three cubic centimeters, was given intra-peritoneally.

In order to determine that our procedure was correct certain control experiments are included. In the following table is shown the results of this work.

TABLE I.

No. of Pig.	Preparation.	Results of Toxic Dose.
279......	Normal pig.	No symptoms.
282......	" "	" "
264......	Sensitized with serum of dog drawn previous to albumin injection.	" "
171......	Sensitized with serum of dog drawn previous to albumin injection.	" "
260......	Sensitized with egg albumin .1 cc.	Dead in 1 hour.
297......	Sensitized with normal serum and the (see note) calculated amount of egg albumin.	Severe symptoms.
246......	Sensitized with serum drawn 15 minutes after intravenous albumin injection.	" "
258......	Sensitized with serum drawn 4 hours after intravenous albumin injection.	" "
244......	Sensitized with serum drawn 48 hours after intravenous albumin injection.	Moderately severe symptoms.
945......	Serum drawn 72 hours after intravenous albumin injection.	Slight symptoms.
634......	Sensitized with urine collected from dog during 24 hours following intravenous infection.	Severe symptoms.

Note. — Pig Number 297 was sensitized with a mixture of egg albumin and normal dog serum of the same relative proportions as we supposed would be found in the blood of the dog to whom the egg albumin injection had been given. This amount was determined as follows: The experimental dog weighed 10 kilos; considering that 10 per cent of his weight was blood the 20 cc. of egg albumin solution was diluted with 1,000 cc. of blood. Accordingly, to 10 cc. normal serum, .2 cc. egg albumin solution was added, the mixture thoroughly shaken, and .1 cc. used for sensitizing Pig Number 297.

The results of this line of experimentation show that the serum of the dog to which the albumin was given intravenously was capable of sensitizing guinea-pigs to the subsequent dose of egg albumin. The egg albumin, therefore, was not immediately excreted or fixed in the tissues, but remained in the serum for at least three days in such concentration that .5 cubic centimeter serum would sensitize a pig to a slight degree.

The egg albumin must have been excreted in the urine, for qualitative tests made of the first twenty-four-hour urine showed a small amount of albumin, and a pig sensitized with one cubic centimeter of it showed a severe anaphylactic reaction.

The same method in regard to drawing the blood and sensitizing the guinea-pigs as described above has been followed throughout the work. Certain objections that may be raised to the conclusions drawn to this experiment will be discussed later.

In the next series of experiments the primary question at issue was the third point of our previous outline, viz.: "May a foreign albumin be absorbed through the intact intestinal epithelium and be recognized as such in the circulation?" To decide this point we made Thiry-Vella fistulas at three different portions of the intestine, first just below the duodeno-jejunal junction, second, about the middle of the ileum and, third, low down in the ileum. The wounds caused by the operative procedure were allowed to heal before any further experimental work was attempted on the animals; except that in two instances the fistula was made and tested to see whether it would leak and then without waiting for the wound to heal the albumin was injected into the fistula.

Fistulæ were made in a total of forty-eight dogs, but comparatively few of these could be used for this experiment because of the high mortality. In this connection it is interesting to note that the complete removal of a section of the gut was a simple operation with a low mortality while the making of a Thiry-Vella fistula by closing one end of the resected piece and bringing the other end open to the surface was accompanied by a very high mortality. Other investigators have encountered the same difficulty and we are now studying the cause of this high death rate.

In table No. II. given below is shown the results of the control experiments made to determine whether the serum from normal dogs would sensitize pigs to a subsequent injection of egg albumin and to determine that our technic was correct.

TABLE II. — *Fistula dogs.*

No. of Dog.	No. of Guinea-pig.	Substance used for Sensitizing Dose.	Substance used for Toxic Dose.	Results of Toxic Dose.
	284	Nothing.	Normal dog serum.	No symptoms.
	290	"	" " "	" "
	244	Serum, normal dog.	Egg albumin.	" "
	244	" " "	Serum, normal dog.	Symptoms. Dead about 30 hours.
	342	" " "	Egg albumin.	No symptoms.
	342	" " "	Normal dog serum.	Symptoms. Dead about 30 hours.
	670	" " "	Egg albumin.	No symptoms.
	793	" " "	Serum, normal dog.	Severe symptoms. Recovered.
	271	" " "	" " "	Severe symptoms. Recovered.
	288	" " "	" " "	Severe symptoms. Recovered.
	269	" " "	" " "	Moderate symptoms.
	796	" " "	" " "	Severe symptoms. Recovered.
	336	Egg albumin.	Egg albumin.	Dead, 20 minutes.
	286	" "	" "	Dead, 15 minutes.

In table No. III. given below is shown the results obtained by sensitizing pigs with serum drawn after varying periods following the placing of the egg albumin in the Thiry-Vella fistula.

TABLE III. — *Fistula dogs.*

No. of Dog.	No. of Guinea-Pig.	Fistula.	Substance used for Sensitizing Dose.	Substance used for Toxic Dose.	Results of Toxic Dose.
667 ..	612	8 inches from cecum.	Serum drawn 1 hour 40 minutes after introducing albumin in fistula.	Egg albumin.	Dead 45 minutes.
	656	8 inches from cecum.	(1) Washings of fistula after 1 hour.	" "	Severe symptoms.
675 ..	273	Middle of small intestine.	Serum drawn before albumin in fistula.	" "	No symptoms.
675 ..	337	Middle of small intestine.	Serum drawn before albumin in fistula.	" "	" "
675 ..	300	Middle of small intestine.	Serum drawn 1 hour 30 minutes after albumin in fistula.	" "	Severe symptoms.
675 ..	335	Middle of small intestine.	Serum drawn 1 hour 30 minutes after albumin in fistula.	" "	Severe symptoms.
646 ..	333	8 inches below duodeno jejunal fossa.	Serum drawn 3 hours after albumin in fistula.	" "	Severe symptoms. Recovered.

From an examination of the results it is evident that the egg albumin was rapidly absorbed and that it was present as such in the circulating blood.

In table No. IV. is shown the results obtained by sensitizing the pigs with urine passed during the first twenty-four hours after the introduction of the albumin into the fistula. In two instances the pigs were sensitized with the proteid obtained by half saturating the urine with ammonium sulphate, while in the other cases the whole urine was used for sensitizing purposes.

TABLE IV. — *Fistula dogs.*

No. of Dog.	No. of Guinea-pig.	Fistula.	Substance used for Sensitizing Dose.	Substance used for Toxic Dose.	Results of Toxic Dose.
	354	8 inches from cecum.	(2) Albumin from Urine.*	Egg albumin.	Marked symptoms.
	351	8 inches from cecum.	Albumin from urine.*	" "	Moderate symptoms.
	327	8 inches below duodeno-jejunal fossa.	Urine for 12 to 24 hours after albumin in fistula.	" "	Severe symptoms. Recovered.
	329	8 inches below duodeno-jejunal fossa.	Urine for 12 to 24 hours after albumin in fistula.	" "	Severe symptoms. Recovered.

* These pigs were sensitized by giving them the proteid obtained by one-half saturation of the urine with ammonium sulphate.

In this connection it may be of interest to report the following experiments with the urine from a case of alimentary albuminuria in a human subject.

The subject was a young man twenty-four years of age, who was in apparent good health with the exception that after partaking abundantly of some form of proteid food, particularly eggs and milk, he would excrete varying quantities of albumin in the urine, depending somewhat upon his activity and the amount of proteid eaten.

For the purposes of this experiment he was placed upon a low proteid diet, containing particularly no eggs or milk, for a period of five days. At the end of this time his urine was free from albumin by the ordinary chemical tests and when inoculated into guinea-pigs it did not sensitize them to either egg or milk proteid.

Instead of taking his usual breakfast one morning he took six raw eggs during a period of one and one-half hours. The urine collected during the subsequent twenty-four hours contained an abundant precipitate of albumin to the usual chemical tests. Four pigs were sensitized by intraperitoneal inoculation of five cubic centimeters of this urine. Fifteen days later two of the pigs were given the usual toxic dose of egg albumin with the result that they reacted in a characteristic manner with fatal outcome. The other two pigs were given an injection of two cubic centimeters inactivated human serum and gave a strong positive but not a fatal anaphylactic reaction. Two days later they were given the usual dose of egg albumin and reacted in the characteristic fatal manner.

The same plan of experiment was repeated on the same subject after a rest of ten days; in this instance milk was given instead of eggs, the subject drinking one and one-half quarts of milk during a period of one-half an hour.

The urine collected during the following period of twenty-four hours contained a small quantity of albumin as determined by the usual chemical tests. A quantitative estimate was not made in either the preceding experiment or this one, but from qualitative tests it would appear that not more than one-quarter as much proteid was excreted in the urine following the drinking of the milk as appeared during the egg experiment. The same plan of testing the urine was followed as in the previous case. Four pigs were sensitized by peritoneal inoculations of five cubic centimeters of the urine. Fifteen days later two of the pigs were given the usual toxic dose of egg albumin. They both reacted in a characteristic manner with fatal results. The remaining two pigs were given two cubic centimeters of inactivated human serum with the result that both had a strong positive reaction, one being fatal. One pig recovered after a time and four days later was given a toxic dose of egg albumin with fatal results. It must be concluded from these experiments that in this subject the taking of a large quantity of native albumin was followed by an absorption and excretion of a portion of it in an undigested form. Not all of the albumin appearing in the urine under these conditions is the foreign albumin taken in as food, for the urine of this experimental subject contained human albumin in sufficient quantity to sensitize pigs. The foreign albumin must have acted as a kidney irritant to a sufficient extent to permit some of the normal albumin to pass through the kidney filter.

In this connection it may be of interest to cite an experiment which proves practically the same point. A healthy dog was kept under observation for three days and the urine found to contain no albumin by the usual chemical tests. At the end of this period under morphine and cocaine anesthesia, the jugular vein was opened and a known quantity of egg albumin injected directly into the vein. The injection was made slowly during a period of fifteen minutes. The dog was then placed in the cage and the urine collected in periods of twenty-four hours. During the first twenty-four-hour period a large quantity of albumin appeared in the urine. During the second period of twenty-four hours a very little appeared in the urine.

The total quantity of proteid injected into the vein was .2130. The total quantity of proteid excreted during the first twenty-four-hour period was .3016. It is evident from these figures that a considerable portion of the proteid excreted in the urine must have been a serum proteid from the animal's own blood and it would appear that the passage of the serum albumin through the kidney had acted to some extent as a kidney irritant permitting the loss of serum albumin. Only one experiment of this type can be cited and it may be that this animal was in a peculiarly suitable condition, because of his kidney condition, to obtain these figures.

From the experiments above quoted, it must be concluded that egg albumin may pass through the epithelium of the intestine undigested and appear as such in the blood and be excreted in the urine. Further experiments will determine to what extent this may be a normal process. It may be that by such a process as this many pathological conditions may arise, and further investigation is being undertaken to elucidate some of these questions.

(Printed without author's corrections.)

THE

Journal of Medical Research.

(NEW SERIES, VOLUME XX.)

| VOL. XXV., No. 3. | FEBRUARY, 1912. | Whole No. 129. |

THE METABOLISM OF THE HYPOPHYSECTOMIZED DOG.[*]

FRANCIS G. BENEDICT AND JOHN HOMANS.

(*From the Nutrition Laboratory of the Carnegie Institution of Washington, Boston, Massachusetts, and Laboratory of Surgical Research, Harvard Medical School, Boston, Massachusetts.*)

SYNOPSIS.

INTRODUCTION.

PREVIOUS RESEARCHES ON THE RELATION OF THE HYPOPHYSIS TO METABOLISM.

CARBON–DIOXIDE PRODUCTION AS AN INDEX OF TOTAL METABOLISM.

APPARATUS USED IN THIS RESEARCH:
Respiration chamber.
Absorption system.
Tests of efficiency.
Absorbents.
Oxygen supply.
Arrangements for continuous observation.
Further use of the apparatus.

METHOD OF DETERMINING THE MUSCULAR ACTIVITY.

IMPORTANCE OF THE RECORD OF MUSCULAR ACTIVITY.

THE TECHNIC OF THE OBSERVATIONS:
Selection of animals.
The routine care and feeding.
Daily observations.
Use of the respiration chamber.
Use of the kymograph records.
The use of statistical data in establishing a base line.
Operative methods.

STATISTICS OF OBSERVATIONS:
Observation IX.
Observation X.
Observation XI.
Observation XIII.
Observation XIV.

DISCUSSION OF RESULTS.

SUMMARY.

BIBLIOGRAPHY.

[*] Received for publication Dec. 29, 1911.

Introduction. — The incentive to this study originally came from observations made by Crowe, Cushing, and Homans,[1] during experiments in the Hunterian Laboratory in Baltimore. In these experiments peculiar changes in the temperature of the animals were noted in the last stages of cachexia hypophyseopriva. Since the gradual failing of some of the animals after a period of apparently perfect health was accompanied by a fall of temperature before death to below 30° C., it seemed desirable to study by some accurate means the metabolism of the hypophysectomized dog.

Up to the present time, most of the experiments on the effect of hypophyseal secretion upon metabolism have been made by administering the extract, powdered preparation, or fresh gland of one animal to another by way of the blood vessels, peritoneal cavity, subcutaneous tissues, or the alimentary canal. The results of these investigations are of interest in connection with this research only in showing the effects upon metabolism of adding hypophysis to the organism as opposed to those due to its removal. In a general way, the results of such experiments follow two main lines: First, the immediate brief physiological effects upon the pulse rate, blood pressure, kidneys, and unstriped muscles in all parts of the body, of single doses of preparations containing the posterior lobe of the gland; second, the more permanent changes in the general metabolism of the animal under the influence of repeated doses of the entire gland or the anterior lobe.

That the gland contains substances capable of causing a rise in blood pressure and a slowing of the pulse rate was first demonstrated by Oliver and Schäfer[2] in 1895; later, this property was fitly ascribed by Howell[3] to the posterior lobe. Subsequently, kidney expansion and diuresis due to posterior lobe injections were produced by Schäfer and Magnus.[4] Recently v. Frankl-Hochwart and Fröhlich[5] have made a careful study of the effect of this same secretion upon the bladder and uterus, and prove it to have a direct stimulating action upon smooth muscle fibers. The refinement of the study of this secretion as, for instance, the possible presence of both pressor and depressor substances in the posterior lobe, are beyond the scope of this paper, but an excellent summary and discussion of the subject has recently been published by Wiggers.[6]

At the present time it is impossible to separate many of the effects of anterior lobe injection or feeding from the effects following the use of the whole gland. Apparently contradictory results may be explained by failure to separate properly the different parts of the gland, by differences in the dosage, and in the methods of preparing the extracts. It appears, however, that since the most sharply defined results have followed the prolonged use of the whole gland, it may be found that some supposed effects of anterior lobe injections are in reality due to the inclusion of the intermediate portion with the anterior. In large quantity the whole gland is evidently toxic. Mairet and Bosc[7] in 1896 found that considerable quantities of whole gland were somewhat toxic to both animals and men. Salvioli and Carraro[8] separated the posterior and anterior lobes, finding the former toxic and the latter not. Franchini[9] and Sandri[10] both found that the toxicity of the anterior lobe was due to the remains of pars intermedia left attached to it, while Crowe, Cushing, and Homans showed that puppies given repeated injections of whole gland extract gradually became cachectic and died. The latter investigators gave anterior lobe injections repeatedly without ill effects. On the other hand, Hamburger[11] has recently obtained a marked fall in blood pressure, sometimes accompanied by collapse and death, from the intravenous injection of anterior lobe extract. This result, which is contradictory to previous findings, may perhaps be explained by an imperfect separation of the different portions of the gland.

With the exception of these general poisonous effects, evidently due to over-doses of the extract, injections and feeding of the whole gland have produced fairly constant results. Malcolm[12] fed dogs upon dried extracts of both the anterior and the posterior lobe. Under each form of feeding, nitrogen was retained in the body, calcium and magnesium were increased in the excreta, while phosphorus was retained under anterior lobe feeding. When fresh gland was given — as much as twenty-five grams daily to a dog weighing twenty-one kilograms — the excretion of nitrogen and phosphorus was markedly increased, and that of calcium and magnesium to a lesser extent. Thompson and Johnston [13] obtained similar results, especially with the fresh gland of a calf. In addition, their subjects suffered a striking loss of weight, the effects of feeding the gland lasting for several days after its administration was discontinued. Cerletti [14] gave intraperitoneal injections to dogs and rabbits with a consequent loss of weight as compared with controls, a shortening of the long bones, and an occasional absolute increase in the width of the epiphyses, the quantities being sufficiently small to produce no obvious toxic symptoms. Falta, Rudinger, Bertelli, Bolaffeo, and Tedesco,[15] gave " Pituytrin " (Parke, Davis & Co.) to fasting dogs and brought about an increase in the excretion of nitrogen, phosphates, and salts. Somewhat similar results were obtained in experiments on man by Schiff [16] and by v. Moraczewski,[17] while Delille [18] and others have used the extract of whole gland medicinally in typhoid and other debilitating diseases with a reported rise in blood pressure and diuresis, and an improvement in the general condition of the patients. Finally, Exner,[19] making use of the fact that transplanted glands will often live for several days in the host before becoming absorbed, transplanted a number of rat hypophyses into young animals of the same species. While the process of absorption was going on and for a time afterwards, the hosts gained in weight over the controls. This increased growth in rats was also obtained in Schäfer's [20] recent experiments by the long-continued feeding of small amounts of the whole gland.

It may be inferred safely, therefore, that the effect of adding hypophyseal extract to an organism is to alter its metabolism. In large quantity the effect is toxic; in smaller quantities, there is generally an increased excretion of nitrogen, phosphorus, calcium, and magnesium, with some loss of weight. Under certain conditions there may be a gain in weight, and sometimes alterations in the growth of bone.

Previous researches on the relation of the hypophysis to metabolism. — The brilliant and thorough research of Paulesco[21] went far toward settling the vexed question whether or not the hypophysis was necessary to the existence of an animal. Previous researches, partly because of faulty methods of operating, and partly on account of a failure to show by a thorough microscopical examination after death whether or not the gland had been completely removed, had left this question open. Paulesco's findings that complete removal of the hypophysis was inconsistent with the life of a dog for more than a few days or weeks have been confirmed by investigators making use of equally thorough and careful methods. (On this subject, see articles by Reford and Cushing,[22] and also by Crowe, Cushing, and Homans,[1] in which Paulesco's results are confirmed and the previous literature of this subject summarized.) Apparent exceptions to this rule are probably due to the leaving behind at operation of microscopic fragments of the gland, or to epithelial rests in the sphenoid bone or even in the pharynx.

Removal of most of the gland, or in young animals of even the whole gland, is not, however, incompatible with life for considerable periods. Under these circumstances a study has been made of the effect upon the organism of the loss of the secretion. Aschner[23] removed the hypophysis by the buccal route in dogs with the following results: Young animals presented after operation a striking retardation of development as compared with controls. A failure of growth was observed, with hypoplasia of the genitals, shortening of the skull, anomalies in the growth of hair, a rich deposit of fat and, in general, an apathetic appearance and bearing suggestive of cretinism. As to the changes observed in older dogs, we quote the author's own words: " Bei den erwachsenen Tieren sind die erwähnten Störungen teils gut ausgeprägt, teils nur angedeutet, doch kann man bei ihnen regelmässig eine tiefgreifende Aenderung des Kohlehydratstoffwechsels konstatieren in eben dem Sinne, wie dies bei den thyreopriven Hunden in letzter Zeit gezeigt worden ist. So reagieren sie auf Adrenalin auffallend weniger mit Glykosurie und Sympathetikusreizerscheinungen als normale Hunde und es ist auch die alimentäre Glykosurie in ihren verschiedenen Formen herabgesetzt." In a later communication, Aschner reiterates these findings and adds that, in consequence of his researches, he finds that the hypophysis is not essential to life. Recently, in an informal communication to us, he reports that following operation there is in his dogs a falling off in the nitrogen metabolism and the total metabolism of the fasting animal. Aschner's observation that total hypophysectomy is compatible with prolonged existence in adult dogs may be explained on the grounds already stated; but one would be more ready to accept his evidence if he supported it by showing that when sacrificed his hypophysectomized dogs showed no remains of the gland on microscopic examination. This criticism, even if just, does not invalidate Aschner's other observations.

Similarly, under nearly complete hypophysectomy, Crowe, Cushing, and Homans [1] observed a failure of sexual development combined with pronounced adiposity in young animals — a phenomenon analogous to Fröhlich's syndrome [24] in man. Complete removal of the hypophysis they found, however, to be incompatible with the life of the dog. Later, the carbohydrate tolerance of hypophysectomized dogs was carefully studied by Goetsch, Cushing, and Jacobson,[25] with the result that an increased tolerance was ascribed to the loss of the posterior lobe secretion, an observation which at once confirms and refines Aschner's previous discovery.

The striking results of all the recent investigations upon hypophysectomized animals (most of which have been made to determine simply whether the hypophysis was essential to life) are the following: Animals dying as a result of

removal of the gland may make a sudden exit without any
definite illness and in a condition of apparently perfect
health; or they may, especially if they survive for weeks or
months, sink down in the course of a few days with the
appearance of cachexia, low temperature, and slow, feeble
pulse.   In the meantime they may preserve every appearance of health, or if operated upon when young, they may
become fat, asexual, and fail to complete their growth.
When the temperature of one of these animals has fallen
even as low as 27° C., it may be raised by artificial heat to
nearly normal, but without saving the life of the animal
(Crowe, Cushing, and Homans[1]).   These peculiarities
induced the writers to study the metabolism of dogs before
hypophysectomy, in the period of health following it, and
in the final decline.

We can only find three researches in the literature at all
resembling ours.   V. Narbut[26] and others in v. Bechterew's
laboratory made an observation on an animal for five days
preliminary to operating, determining the carbon-dioxide production on the last two days.   They then removed the gland
(the method is not stated) and kept the animal under observation until its death, which took place twenty-four days later,
with characteristic low temperature just before the end.
The carbon dioxide was determined for ten days following
the operation.   No mention is made of the diet or of the
control of the animal's movements, but there was observed a
diminution in the carbon-dioxide output after operation, a
slight diminution in the oxygen intake, an enormous loss of
water, and a marked increase in the excretion of nitrogen
and phosphoric acid.

Very recently Wolf and Sachs[27] have studied the carbon-
dioxide output of dogs which have been subjected to a
partial or complete removal of the hypophysis. They
obtained results similar to those from Bechterew's[26] laboratory with regard to the carbon-dioxide output in animals
partially or completely hypophysectomized, the falling off in
an animal which died forty-eight hours after operation being
particularly noticeable.   Whether or not allowance was made

for differences in the muscular movements of the animals before and after the operation, and for the effects of temperature changes, is not recorded in the brief report of their work.

Aschner's [20] researches in this direction, of which we have at present only a brief informal statement, have already been mentioned.

Carbon-dioxide production as an index of total metabolism. — The well-known relations of the internal secretions to metabolism are such that a disturbance of any of these relations may be considered naturally a precursor of a series of steps in the development of an abnormal metabolism. While for the most part these disturbances in metabolism are characterized by variations in the character and amount of the nitrogenous metabolism, certain internal secretions apparently take an important part in the general metabolism. When the nitrogen metabolism is to be studied, a long series of experiments is essential, in which careful determinations are made of the amounts and kinds of nitrogenous material in the intake as well as a determination of the total nitrogen outgo in the urine. These experiments not only usually cover a considerable period of time, but often the fluctuations in the nitrogenous metabolism are so masked by other uncontrollable factors that before deductions can be drawn a large number of experiments must be made. On the other hand, any effect on the general metabolism of the animal is shown with relative clearness by careful metabolism measurements, which need not extend over any considerable period of time. A disturbance in the metabolism involving a change in the kind and amount of materials metabolized in the body can only be adequately studied by means of elaborate respiration and calorimetric experiments, involving the determination of the total carbon-dioxide production, the oxygen consumption, and the heat production. The technic and apparatus necessary for such complete experiments are, however, very elaborate and are out of the question in most laboratories. The oxygen determination alone furnishes much valuable information with regard to the total metabolism, and is, indeed, a very accurate index of the total oxidative process; but the determination of oxygen is second in difficulty only to that of determining heat, and likewise requires very complicated apparatus and time-consuming experiments. Fortunately, the carbon-dioxide production of animals can be determined with relative ease, and it is this factor that is most commonly taken as a measure of the total metabolism under varying conditions.

The transformation of matter inside the body results in an absorption of oxygen from the atmosphere, and a production of carbon dioxide and water. As has been frequently pointed out, the carbon-dioxide production is not coincident with the energy production, inasmuch as the calorific equivalent of carbon dioxide, that is, the number of calories resulting from the production of one gram of carbon dioxide, varies considerably, depending upon whether the carbon dioxide is derived from the oxidation of protein, fat, or carbohydrate. On the other hand, the calorific value of oxygen is very much the same whether the oxygen be used to burn carbohydrate, fat, or protein; hence oxygen is a much more accurate index of the total heat production of the animal body. When apparatus for the measurement of heat production and for the determination of oxygen are not available, however, experiments can be made in which the carbon dioxide is accurately determined, thus obtaining results of much value.

In determining the carbon-dioxide production in small animals, one of two methods is used: either the animal is caused to breathe through a mask or through a tracheal fistula and the products of respiration analyzed, or the animal is placed in a chamber through which a ventilating current of air is passed, the air being analyzed before it enters and after it leaves the chamber. By the latter method, the carbon-dioxide production is computed from the results of the analyses and the data regarding the total ventilation of the chamber. To these two fundamental methods should be added the method of confinement first used by Kaufmann,[28] by which the animal is placed inside of a relatively large chamber and the carbon dioxide allowed to accumulate, frequent analyses of the air inside the chamber being made to indicate the change in composition. If the volume of air in the chamber is accurately known, these periodic analyses give information with regard to the carbon-dioxide production.

A careful study of the calorific values of carbon dioxide shows that although there are noticeable differences depending upon whether the carbon dioxide is derived from the combustion of protein, fat, or carbohydrate, nevertheless, if approximately constant dietetic conditions are maintained, the carbon dioxide may frequently be used as an excellent index of any change in metabolism incidental to an operation or to some other superimposed factor. The results are of very great value, approximating in importance those experiments in which oxygen and heat are also determined.

Apparatus used in this research. — When contemplating a study of the metabolism of small animals as influenced by the removal of the hypophysis we found that the pressure of other work in the laboratory precluded the use of a calorimeter for the purpose; but since a constant diet could be rigidly adhered to it was believed that for preliminary experimenting in such a research only the determination of the carbon-dioxide production was essential. An apparatus was therefore especially devised for the research by means of which the carbon-dioxide production could be determined.

With the earlier apparatus the method of confinement was used, *i.e.*, the animal was placed inside a respiration chamber of a known volume, and the carbon dioxide allowed to accumulate; samples of air were withdrawn from period to period and the content of carbon dioxide determined, usually on a Haldane gas analysis apparatus. The first chamber constructed contained one thousand liters of air, but it was soon found that for the size of the dog available and best adapted for the study under consideration (five kilograms or less) the chamber was too large; a second apparatus was therefore constructed with an air capacity of but two hundred and eighty liters. The efficiency of the apparatus was tested by burning known amounts of alcohol in the chamber, and determining the amount of carbon dioxide given off; it is probable that the experimental results were accurate to within three per cent. This apparatus, which has been described in detail in a previous paper,[29] was used in this research for the experiments in both Observations IX. and X., and also in Observations XI. and XIII. to May 11, 1911.

Laulanié [30] has shown that the metabolism is not materially altered in dogs kept in an atmosphere containing as high as six per cent of carbon dioxide. In none of our experiments was the percentage of carbon dioxide allowed to rise above three and one-half per cent; nevertheless, it was soon evident that while the results obtained were sufficiently accurate for preliminary experiments of relatively short duration, the apparatus could not be satisfactorily utilized for long-continued experiments such as are necessary for the adequate study of most metabolism problems. Fundamental alterations in the apparatus were therefore required before a more detailed study of the problem under investigation could be attempted.

As a prerequisite for this kind of experimenting, it was obvious that the carbon dioxide should be removed essentially as fast as it was produced. This could be done in one of two ways — by ventilating the chamber according to the well-known principle of Pettenkofer and Voit, and analyzing an aliquot sample of the air by the usual method of titration with barium hydroxide; or by using the principle of the large respiration calorimetric apparatus in this laboratory employed for studying the respiratory exchange of man.[31] By the latter method, which was selected for this research, the entire current of ventilating air is passed through absorption vessels containing soda lime, thus removing the carbon

dioxide.   The carbon-dioxide-free air is then returned to the
chamber, sufficient oxygen being supplied from a cylinder of
the highly-compressed gas to replace the amount used by
the subject.   With the large calorimetric apparatus for man,
this has been made a quantitative operation, both for carbon
dioxide and oxygen, and has long been in use not only in
this laboratory but previously in the laboratory of Wesleyan
University at Middletown, Conn.   More recently in an
apparatus used solely for respiration experiments,[32] and
hence lacking the calorimetric features, the expired air is
conducted from the mouth or nose of a man, passed through
purifying vessels in which the carbon dioxide is removed,
and then returned to the subject for inspiration, the deficiency in oxygen being made up by admitting oxygen from
a cylinder of the gas.

In adapting the apparatus for researches with animals, a
respiration chamber of a size suitable for experiments with a
dog was connected with the absorption system at the point
in the ventilating circuit ordinarily occupied by the nose-
pieces used in experiments with a man.   In this way the
vitiated air, containing water vapor and carbon dioxide, and
deficient in oxygen, was forced by a rotary blower through
a chain of absorbers in which the air was freed from carbon
dioxide and water vapor, and finally returned to the chamber.

Respiration chamber. — The apparatus is illustrated in the
accompanying figure.   The respiration chamber, which is
sixty centimeters on each side, is made of galvanized sheet
iron, with seams reinforced and soldered to make it perfectly
air-tight.   The top is removable and provided with a flange,
which, when the cover is in place, dips into a small trough,
H, H, firmly soldered to the upper part of the chamber.
When this trough is partly filled with water, a perfect seal is
provided.   In the center of the cover is a glass window, sixteen centimeters square, which gives an opportunity for
observing the movements of the dog.   A thermometer, T, is
shown passing through the cover.   On the right wall of the
chamber is attached a brass pipe supporting a shallow tin

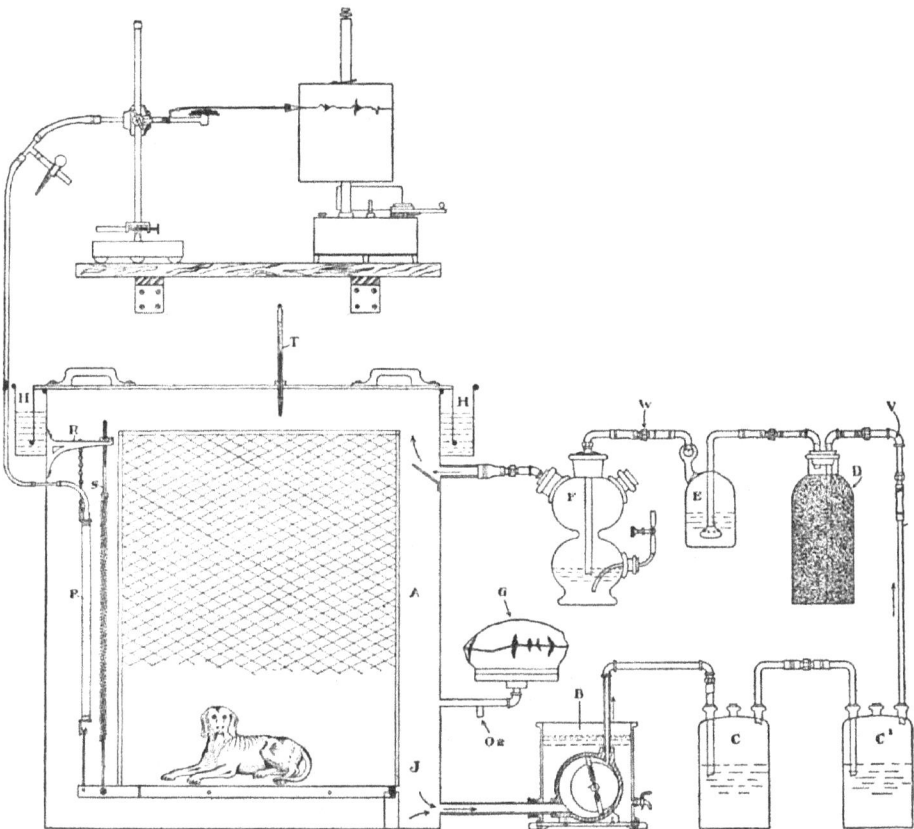

Fig. 1. — *Diagram of respiration apparatus and absorbing system.*

At the left side of the figure is the respiration chamber, with cover closed by a water seal, H, H. A thermometer, T, for recording the temperature of the air inside the chamber, is inserted through the top of the cover. One side of the cage, A, rests on a knife-edge bearing, J, the other side being supported by a brass spring, S, fastened to a support, R. The brass spring is paralleled by a pneumograph, P, by means of which any movement of the dog is transmitted to the tambour shown above the chamber, and recorded on the drum of the kymograph. A diagram of the absorbing system may be seen at the right of the chamber. The air first enters the blower, B, which forces it through the water absorbers, C and C¹; it then passes up through the carbon-dioxide absorber, D, and next into the drying vessel, E. Finally, moisture is added in the vessel, F, and the air is again returned to the chamber as indicated. The shallow pan at the right side of the chamber supports the diaphragm, G, by means of which fluctuations in the air content of the chamber are shown. Oxygen as required is supplied through the opening O_2.

pan covered by a rubber bathing cap, G, which acts as a diaphragm and allows for the expansion and contraction of the air inside the chamber. As the oxygen is consumed by the dog, any diminution in the volume of the air is at once shown by the falling of the diaphragm. To bring the volume again to constancy, oxygen may be introduced as needed by means of a petcock in the brass pipe at the point indicated by O_2. The total volume of the chamber is approximately two hundred and eighty liters, the full capacity of the rubber diaphragm when distended being an additional two and one-half or three liters.

Absorption system. — The absorption system is shown at the right in the figure. To save space the air current is represented as leaving and entering the chamber on the same side while, as a matter of fact, the entrance and exit points are opposite each other. As the air is withdrawn from the chamber it passes first into a rotary blower, B, which forces it into two Wolff bottles, C, C^1, containing sulphuric acid which removes the water vapor. From these vessels, the air then passes into a carbon-dioxide absorber, D, containing soda lime. Finally, in order to obtain the weight of the water which may have been taken up by the dry air as it passed over the moist soda lime, the ventilating current is carried through a drying vessel, E, filled with sulphuric acid. Although we have no reason to believe that the absence of sweat glands from the dog renders him susceptible to the cooling influence of extremely dry air, provision is made for supplying moisture to the current by passing it through the vessel, F. The air is then returned to the chamber, free from carbon dioxide, and partly saturated with moisture. The amount of carbon dioxide absorbed is determined by weighing together the carbon-dioxide absorber, D, and the drying vessel, E.

Formerly, a silver-plated brass can was used to contain the soda lime; recently, a less cumbersome and more economical container has been found by substituting an ordinary wide-mouth bottle which is fitted with tubes that conduct the gas to the bottom of the bottle and allow it to pass out at the top. In the original description of the absorbing system [32] it was stated that the moisture from the soda lime was collected in a sulphuric acid drying apparatus made from the bottom of a Kipp generator.   For this has been substituted a blown glass sulphuric acid absorber devised by H. B. Williams of the Department of Physiology of Columbia University.   When charged with the respective reagents the average combined weight of the soda lime bottle, D, and the sulphuric acid container, E, is not far from six to seven kilograms.   Inasmuch as with the Sauter balance employed, ten kilograms can be weighed to ten milligrams, it is readily seen that the weight of the carbon dioxide absorbed can be secured with great accuracy.

The rate of ventilation ordinarily used in experiments, namely, about thirty-five to forty liters per minute, is sufficient to maintain the carbon-dioxide content of the air inside the chamber at a very low value, and constant.   This has been repeatedly tested and found to be not far from .06 per cent of carbon dioxide.   With a ventilating rate of forty liters per minute it can easily be seen that with a cubical content of two hundred and eighty liters, theoretically the air may be completely changed once every seven minutes, this being amply sufficient to keep the percentage of carbon dioxide at a low level.   Accordingly, to know the carbon-dioxide production of the animal during a given experimental period, it is unnecessary to take note of possible changes in the carbon-dioxide residual inside the chamber, but only to determine the amount of carbon dioxide actually taken away by the ventilating air current.   This does away with the necessity for the sampling and analysis of air, as well as all observations regarding the volume of the chamber, and of temperature, pressure, and moisture conditions.

Tests of efficiency. — A test of the efficiency of this system can be made very readily by putting the cover on the respiration chamber and passing air through the system for half an hour.   During this time the soda lime bottle and its accompanying absorbers should not have materially increased in weight. If there is an increase in weight, it is due either to a slight amount of carbon dioxide residual in the chamber when the cover is put on, or more frequently to the fact that the air does not enter the soda lime bottle absolutely dry.

Absorbents. — The sulphuric acid used as an absorbent is the ordinary commercial sulphuric acid of 1.84 specific gravity.

The soda lime is made as follows: Seven hundred and fifty grams of lime and seven hundred and fifty grams of crude commercial sodium hydroxide are each weighed out separately. The caustic soda is dissolved in water in an iron stew-pot over a gas flame, using ordinarily four hundred and fifty cubic centimeters of water. When this is thoroughly dissolved the lime, which has been previously pulverized, is thrown into the stew-pot, the flame turned down, and the whole mass carefully stirred with a long-handled poker until slaking is finished. When it is complete, the material should be lumpy and slightly moist to the touch of the finger, but not so moist as to be sticky, nor so dry as to be easily powdered.

Oxygen supply. — While the air returned to the chamber is carbon-dioxide-free, the oxygen content is still below normal. Oxygen is therefore admitted from time to time through the petcock in the pipe leading to the diaphragm, but no attempt is made at present to measure quantitatively the amount introduced. (That this factor can readily be determined is already assured by preliminary experiments now in progress.) It has been shown by Durig[33] and others that the metabolism becomes abnormal if the percentage of oxygen in the air falls below eleven per cent. On the other hand, Benedict and Higgins[34] have shown that an increased percentage of oxygen in the air is without influence upon the metabolism of man, even when the proportion of oxygen is as high as ninety per cent. Accordingly, it is our custom to fill the apparatus with air slightly richer in oxygen than normal, so that during the progress of an experiment, the oxygen percentage can never fall to the eleven per cent value indicated by Durig as being inimical to normal metabolism.

Arrangements for continuous observation. — The noise of the electric motor and the rotary blower appears to have a soothing effect upon the dog. Since it has often been found that at the end of an experimental period when the absorber system was disconnected and a new set of absorbers put on, the stopping and starting of the blower would frequently cause a slight disturbance of the dog, it was planned so to arrange the system as to be able to deflect the air from one set of purifiers to another without interrupting the ventilating current of air or the sound of the blower and motor. This was easily secured by placing a two-way valve near the entrance end of the carbon-dioxide absorber (see V) and a tee with two wheel valves between the drying vessel and the air moistener (see W). Neither the two-way valve nor the wheel valves are shown in the diagram. After both sets of the absorber system are coupled up, the two rear valves are opened. Obviously, up to this point, the direction of the air current is not in any way altered. At the exact end of the experimental period, the two-way valve is suddenly deflected by hand, and the air is switched from one set of the absorbers to the other, free passage being provided through the rear valve, which has previously been opened. When the rear valve has been closed on the set previously used, the vessels can be removed and weighed without disturbance to the ventilating current in any way. This innovation has proved extremely satisfactory in experimenting with dogs and, indeed, with small infants, as the slight noise or purring sound of the blower is continuous throughout the whole experimental period.

Further use of the apparatus. — It should be stated that it is the plan of this laboratory to supply all of the numerous respiration apparatus here in use with two sets of connections, one for use with man under normal conditions and one for use with dog respiration chambers. As a matter of fact, by attaching to the apparatus a special form of chamber, the metabolism of infants has been recently studied in the same way with very great success. It is thus hoped that the respiration apparatus may be made so adaptable as to make possible accurate quantitative measurements of both oxygen and carbon dioxide with a man breathing through either a mouthpiece or a nosepiece, and also the accurate determination of carbon dioxide with animals or with infants in small chambers. Preliminary experiments have led to the belief that when used with chambers, slight modifications will also permit the quantitative measurements of oxygen; the fact should be emphasized, however, that the apparatus lacks calorimetric features.

Method of determining the muscular activity. — Not only has the method for determining the carbon dioxide in this research been fundamentally changed, but the method for indicating the muscular activity has also been materially altered, although in principle it remains the same. In the apparatus originally described,[*] the cage in which the dog was confined inside the respiration chamber was placed upon a platform, one end of which rested on a support with a knife-edge bearing, and the other on a rod, suspended from above by a stout spring. As the dog changed the center of gravity of his body there was a disturbance of the equilibrium between the weight upon the spring and the weight upon the knife edge; the spring, therefore, became either lengthened or shortened according to the alterations in weight. By means of a pointer upon the rod leading from the suspension point, the changes in the length of the spring were accurately indicated

[*] In the earlier description of this method for recording muscular activity we unfortunately neglected to mention the ingenious apparatus of Stewart (Am. Journ. Physiol., i, 1, 1898). The Stewart apparatus has recently been considerably modified and adapted for recording the activity of small mammals by Slonaker (Anat. Record, ii, 3, June, 1908, p. 116, and Journ. Comp. Neurol. and Psychol., xvii, 4, 1907, p. 342).

upon a smoked paper drum. This required an air-tight closure as it passed through the cover of the apparatus, which was provided by a movable cylinder and water seal. Obviously the record upon the drum was limited in size to the actual vertical movement produced by the shifting of the cage; no magnification of the movement under these conditions was permissible. Furthermore, the top of the box was encumbered with complicated apparatus, including the kymograph, pointer, rod, and water seal. In order to remove the lid of the box it was necessary to take these away and to disconnect the rod and pointer. In the description of the original apparatus it was pointed out that the method of recording the movements was undergoing modification and that a pneumograph was being substituted for the spiral spring. As a matter of fact this modified method is now perfected and has been in constant use for over a year with satisfactory results.

The new method for recording the muscular activity does not rely upon the movement of a vertical rod in any way. By reference to the figure illustrating the apparatus, it will be seen that, as in the former method, the cage is suspended on a platform which rests on a knife-edge bearing at J, the free end of the platform being suspended by a stout brass spring, S, which bears the greater part of the weight. Immediately beside this and holding a small percentage of the total weight, is the spring of a long-tube Fitz pneumograph, P, supplied by the Harvard Apparatus Company. The elongations of the brass spring are accompanied by elongations of the pneumograph tube, consequently with changes in the pressure of the air inside. By means of a brass tube passing through and soldered to the wall of the chamber, this air compression or alteration in pressure is transmitted through a rubber tube to a Marey tambour with a pointer, which registers directly upon the smoked drum the vertical movements of the cage.

By this modified method the cover of the chamber is in no way encumbered with apparatus, but a far more important advantage lies in the fact that the vertical movements of the cage may be magnified almost indefinitely by means of a tambour and pointer. The new apparatus is infinitely simpler and more sensitive than the earlier form, and can be readily applied to any respiration chamber. Not only has it been employed with the larger chamber containing one thousand liters, first constructed by us for dogs, but it has also been used for a small chamber in which experiments have been made upon infants, as well as with the bed calorimeter in this laboratory. By means of a tee placed in the transmission tube from the pneumograph to the tambour, provision is made for equalizing any pressure caused by sudden fluctuations or permanent changes in the body position of the animal resulting in a movement of the pointer and a permanent change in the center of gravity. By opening a pinchcock attached to a rubber tube on the glass tee the pressure is equalized, and the tambour is not kept permanently under tension or decreased pressure. With this appliance the records can be made exceedingly sensitive. Indeed, in a number of instances, the apparatus has been so adjusted that when the dog is otherwise lying absolutely still, the regular rhythmical respiration can be shown upon the curve. As the lungs are inflated with air, the center of gravity of the dog changes sufficiently to produce an alteration in the tension upon the spring and the pneumograph, with consequently an alteration in the line drawn upon the kymograph. Usually, such a degree of magnification of the movement is impracticable and is generally not employed; the possibility of such adjustment is cited primarily as an indication of the potential sensitiveness of the apparatus, rather than as a practical point to be recommended.

The kymograph records were obtained on a Porter kymograph, which was secured from the Harvard Apparatus Company and has been found to be extremely satisfactory. In experiments of short duration the time marking is usually recorded by means of a small Jaquet clock. In longer experiments, of an hour or several hours' duration, the slow-moving kymograph devised by Professor Porter and requiring about one hour for each revolution of the drum is employed with success.

Importance of the record of muscular activity. — Our experience with this form of apparatus clearly indicates that a metabolism experiment consists of two important parts: first, the chemical data with regard to the gaseous metabolism, and second, a graphic record of the muscular activity or degree of quietness during the period of experimentation. It is our firm belief that the chemical records of the gaseous metabolism are without value if unaccompanied by a record of the muscular activity, for various writers and experimenters may differ greatly in their impressions as to whether the dog is quiet, reasonably quiet, completely quiet, absolutely at rest, and other methods of expression of degrees of muscular activity. On the contrary, a graphic curve showing a straight line on the smoked paper kymograph can be interpreted only as showing that during the period referred to the dog did not move his leg, tail, or head, and consequently was absolutely quiet other than the movements incidental to respiration.

Knowing the enormous effects of muscular activity upon metabolism, it can be stated at the outset that in order to study properly the influence of a superimposed factor upon metabolism, it is necessary to have a base line to which the measurements subsequent to the imposition of the factor must be compared. This base line is usually obtained at least twenty hours after the last food is given to overcome the question of the specific dynamic action of foodstuffs and the active processes of digestion itself but, in addition to this, there comes into play the question of the muscular activity of the dog. Of the factors affecting this, wakefulness — perhaps the most important one — reaction to sound and to light, are all important; the importance of temperature regulation has also been emphasized by the striking series of experiments made by Rubner.[35] In our earlier paper we showed a series of curves indicating that this

temperature regulation was of very great importance, and
that not until temperatures of 25° C. to 26° C. were reached
was the dog lying absolutely quiet. Shivering was very
perceptible at 12° C., and minor muscular movements incidental to irregular respiration, shivering, or to involuntary
muscular contractions were noted as the temperature rose
gradually until 26° C. was reached. We wish to confirm
completely Rubner's view that a temperature must be
secured for experiments of this nature at which the dogs are
completely relaxed and muscular activity ceases. After an
examination of the curves in our first paper and of those
given here, no other deduction can be drawn than that for
accurate experiments on dogs, a temperature below 25° C.
is of little avail; furthermore, that periods to be compared
must indicate on the kymograph record the same degree of
muscular activity — preferably the minimum activity. It is,
however, rare to secure experimental periods in which the
dog had no muscular movements of any kind — periods
which would be indicated by a perfectly straight line. On
the other hand, it is perfectly feasible to compare periods
by noting the general activity and the general deviation
from the straight line as indicated by the pointer. An
examination of several hundred of these curves leads us to
believe that an inexperienced person can readily pick out a
series of curves from the general irregularities in the lines
that are reasonably comparable. Obviously, in any series
of experiments no attempt should be made to use for comparison purposes periods showing great irregularity in the
curves, for in metabolism experiments we can deal intelli-
gently only with periods showing but slight, if any, muscular
activity.

It is extremely unfortunate that at present no method for
registering or recording the mathematical values of these
different curves has yet occurred to us, since the expense
incidental to publishing a large series of curves is in most
instances prohibitive. On the other hand, it is believed that
when other experimenters become familiar with the method,
the value of these curves as a guide in the selection of

suitable periods for comparison will be at once evident. We have already found this to be true in several researches and in consequence all experiments in this laboratory, whether with men, infants, or animals, are at present accompanied by some form of graphic record showing the degree of restlessness of the subject during the experimental period. Formerly this was obtained with men by placing one or two pneumographs about the chest, this method being found of great value in an extensive research on diabetes. More recently an apparatus similar to that described herewith has been attached to the bed in the bed calorimeter, and another upon this principle has been used for several months in experiments with infants.

The technic of the observations. Selection of animals. — It has been found that young animals ·who have not yet reached maturity bear removal of the hypophysis much better than full-grown dogs, and are more likely to remain alive for a considerable period. Moreover, the operation of hypophysectomy is much more easily performed on the young animal. Accordingly for this work we have selected young animals which have nearly attained full growth, but who are as yet sexually undeveloped. The animals have been ordinarily six to eight months old, and in all cases well and lively.

The routine care and feeding. — As the object of our experiments has been to superimpose the effect of hypophysectomy upon a definite, constant metabolism, it has been necessary to obtain for our dogs surroundings in which such factors as temperature, noise, excitement, etc., are as nearly uniform as possible. The animals have been kept in metabolism cages in a large, well-lighted laboratory where they see and become accustomed to a considerable number of people. Their surroundings are, therefore, constant, and the animals are easily observed. It is hardly necessary to say that the dogs are kept free from fleas and guarded as far as possible from distemper.

The urine and feces are collected every morning before the dogs are taken to the respiration chamber, food being given once a day after the conclusion of the day's experiment. In order to secure uniform conditions of diet, the following plan has been adhered to:

Each dog is given a uniform ration of dry white bread soaked in a known quantity of water (in Observations IX. and in the early part of XI. dog bread was given); with this is mixed a weighed amount of fat-free cooked meat. As all of our animals have been of practically the same weight, we have evolved, after a certain amount of preliminary experimentation, a constant ration as follows: one hundred and fifty grams of white bread and fifty grams of fat-free meat in three hundred cubic centimeters of water. This the animals eat with relish for long periods of time with very few intestinal disturbances. If it becomes necessary to change this diet, one hundred and thirty grams of cracker dust are substituted for the bread. In addition, the dog is given five hundred cubic centimeters of water daily, and allowed to drink as much as he pleases, the amount left each morning being measured. By taking occasional samples of this diet for analysis, it has been possible to calculate the amount of nitrogen in the food, and to know that all the elements of the diet are in a relatively constant amount and proportion. This being so, it is only necessary to insure the performance of each day's experiment upon a fasting animal to obtain the constant conditions of nutrition which are demanded. It is the rule, therefore, though this is rarely necessary, to take away the pan containing the daily ration if the animal has not eaten the whole of the food within half an hour after it has been given him. The experiments of course are always performed in the morning, before feeding.

Daily observations. — Upon being taken from the cage, the dog is weighed and then allowed to lie still in the observer's lap while the pulse, respiration, and temperature are being taken. This requires considerable time, since the

animal must become completely relaxed in order that satisfactory records of the pulse and respiration may be
obtained.   Sometimes this relaxed condition is not reached
for fifteen to twenty minutes, for even after the animal appears
quiet, the pulse rate may continue to fall for several minutes.
In other words, one must train a dog in order to secure
accurate observations.   The slightest shivering or spasmodic
breathing increases the pulse and respiration rates materially; in order to do away with this source of error, it is
necessary that the room temperature should be at least
24° C. — a temperature which has been found comfortable
by our dogs.

Use of the respiration chamber. — On the experimental
days, the records of the pulse, respiration, and temperature
are obtained at about 9 A.M., or sixteen to twenty hours after
the feeding of the animal on the day before.   As soon as
the records have been made, the dog is placed in the respiration chamber, the cover closed, enough oxygen forced in
to distend the rubber cap upon the diaphragm, and the
motor started.   During the preliminary period, in which the
carbon dioxide in the chamber is reduced to a constant
level, the temperature is regulated to as nearly 25° C. as
possible.   This is done entirely by raising or lowering the
temperature of the room and not through any mechanism in
the chamber itself.   Meanwhile the pointer attached to the
Marey tambour is made to record upon the drum of the
kymograph, and the dog's movements are accurately noted.
As soon after the first fifteen minutes as the dog is seen to
have become absolutely quiet, the current of air, which has
hitherto been passing through the chamber and in which the
carbon dioxide is reduced to a minimum, is deflected into a
fresh set of absorbers which have been previously weighed,
the time is noted, a mark is made upon the smoked paper
record, and the experiment is begun.

The length of the experiment now becomes a matter of
chance; that is, if the dog remains quiet, it is unnecessary to
keep him in the chamber longer than to obtain two or three
periods of approximately half an hour.   Accordingly, if

after the experiment has begun, the animal remains quiet for twenty minutes, the next period is not begun until he shows signs of restlessness.   By changing from one set of absorbers to another and weighing the absorbers at each change, we can determine the amount of carbon dioxide given off by the dog in a number of periods, in some of which he will have been absolutely quiet, while in others he may have moved more or less.   When we have for comparison a sufficient number of periods the experiment is concluded; the animal is then given a short walk, and returned to his cage to be fed.

Use of the kymograph records. — At the present time, the method of recording the animal's movements while in the chamber is so delicate that it is almost always possible to count the respirations upon the drum.   It is, therefore, obvious that even very slight movements are recorded, but these are of considerable importance, especially such movements as are caused by shivering from cold.   Other movements if they are slight and not constantly repeated are of little moment, for under the influence of the steady hum of the blower and if undisturbed by noises in the room, the dog almost invariably remains in one place throughout the experiment after he has quieted down.   When studying the carbon-dioxide output for each day, only those periods are selected for comparison in which the movements of the dog have been negligible.   By comparing several of these periods, if there are more than one, we can be sure that our carbon-dioxide output has been constant, or the one quiet period may be compared with several controls in which there have been appreciable movements.   Having then established a standard, which is, in reality, the minimum carbon-dioxide output for each day, the carbon-dioxide output of successive days may be compared, and a base line for future work established, omitting from such comparison all experiments in which there are no actually quiet periods.

The method of using these kymograph records for determining the influence of the muscular activity upon the production of the carbon dioxide is shown in Figs. 2 and 3, which give specimen curves, and Tables 1 and 2, which show the carbon-dioxide production for the same periods.

TABLE 1.

Carbon-dioxide production in metabolism experiment with dog.
(Observation XIII., Experiment 51, June 6, 1911.)

Number of Period.	Duration of Period.	Temperature of Apparatus.	Carbon Dioxide.	
			Measured.	Per Hour.
	Minutes.	°C.	Grams.	Grams.
I.	15	27.0–28.1	1.00	4.00
II.	30	28.1–29.7	1.98	3.96
III.	11	29.7–30.0	.68	3.71
IV.	20	30.0–31.0	1.09	3.27
V.	30	31.0–31.0	1.66	3.32

TABLE 2.

Carbon-dioxide production in metabolism experiments with dog.
(Observation XIV., Sept. 18, 1911.)

Experiment Number.	Body Temperature Preceding Experiment.	Number of Period.	Duration of Period.	Temperature of Apparatus.	Carbon Dioxide.	
					Measured.	Per Hour.
	°C.		Minutes.	°C.	Grams.	Grams.
24 (A.M.)...	32.0	I.	30	27.6–28.0	1.03	2.06
		II.	30	28.0–28.5	.96	1.92
		III.	30	28.5–28.5	1.00	2.00
		IV.	30	28.5–29.0	1.06	2.12
		V.	30	29.0–	.96	1.92
25 (P.M.)...	36.2	I.	30	30.2–30.3	1.73	3.46
		II.	30	30.3–30.2	1.43	2.86
		III.	30	30.2–30.2	1.33	2.66
		IV.	30	30.2–30.2	1.34	2.68

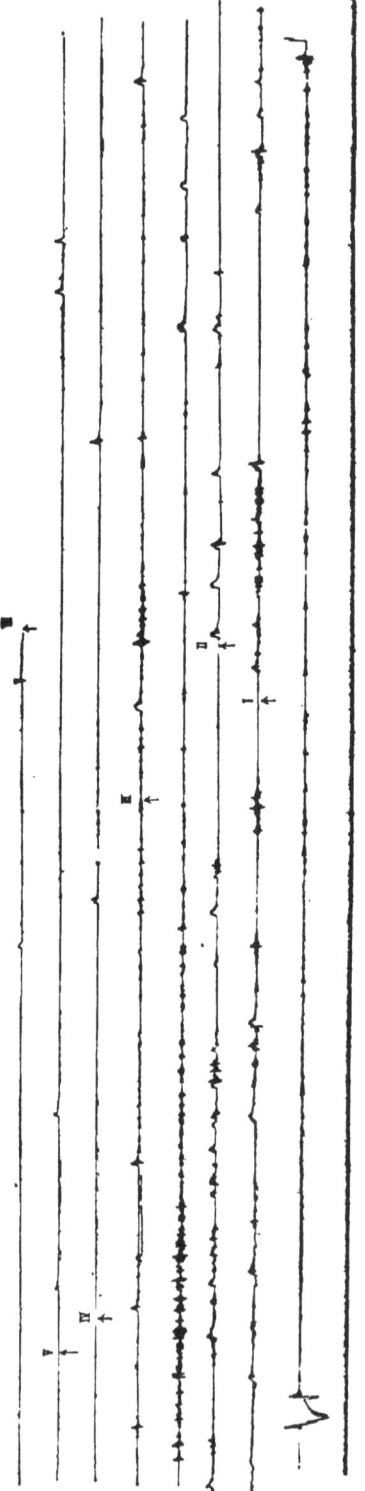

FIG. 2. — *Kymograph record of muscular activity of dog in Observation XIII, Experiment 51, June 6, 1911.*
The five periods of the experiment were of varying length, the beginning of each being represented by the Roman numerals I., II., III., etc., the lowest line showing the record for the preliminary period. The line below the activity curve gives the time record in seconds, showing the rate of movement of the kymograph. The irregularities in the first three periods show the activity of shivering due to the low temperature of the chamber in contrast with the comparative quiet of the last two periods, IV. and V., when the temperature of the chamber was higher.

ents 24 and 25, Sept. 18, 1911.

The fi each period is indicated by the Roman numerals I., II., III., etc., the lowest I

The first series of tracings, which should be read from below up, are taken from Observation XIII., Experiment 51, June 6, 1911.   At this time it was found impossible to obtain perfect relaxation on the part of the dog in the respiration chamber until a temperature of about 30° C. was reached.   The tracings taken from periods I., II., and III., at a temperature of 27 to 29° C., all present a considerable number of constant slight irregularities.   By inspection through the window of the chamber we have learned that these irregularities are due to shivering on the part of the animal.   Table 1 gives the temperature of the apparatus and the amount of carbon dioxide produced by the animal during these periods.   In the first three periods the amount of carbon dioxide is successively 4 grams, 3.96 grams, and 3.71 grams per hour.   At a temperature of 30° C +, in the last two periods when the tracings show that the animal was quiet, the amount falls off to 3.27 grams and 3.32 grams respectively, and these two periods, one of twenty and one of thirty minutes, are taken as the basis of the carbon-dioxide output for that day.   The considerable difference in the amount of carbon dioxide in the periods in which the dog was shivering and in which he was quiet is very striking.

The other two sets of tracings, Observation XIV., Experiments 24 and 25, were taken during the last few hours of the life of this animal.   Experiment 24 was made in the morning when the rectal temperature of the dog was 32° C.; Experiment 25 was made late in the afternoon of the same day when his temperature had been raised to 36.2° C.   One-half hour after the conclusion of the second experiment the animal died.   Both series of tracings, which again are to be read from below, show that at all times the animal was quiet and relaxed, the number of irregularities in the tracings being negligible.   It will be seen by consulting Table 2, that the carbon-dioxide output per hour was approximately the same during the five thirty-minute periods of Experiment 24, while the temperature of the apparatus during this time was sufficiently high to prevent any muscular effort.   It will be noted, also, that the rate of carbon-dioxide production was extremely low, very much lower than was obtained in any other experiment and considerably lower than that obtained in the succeeding experiment. In Experiment 25, in which the animal's temperature had been raised to 36.2° C., the carbon-dioxide output rose considerably, in fact, nearly as high as some of the lowest observations made while the animal was apparently in a state of health. This is a further evidence that our efforts to secure a minimum metabolism have been consistently successful.

The use of statistical data in establishing a base line. — By the use of the methods outlined, therefore, it will be seen that we have a series of experiments made upon a fasting animal whose surroundings, exercise, and diet are approximately constant. When the dog has become thoroughly accustomed to the chamber, which may require several weeks, the operation is performed and the experiments are continued. As the control of the animal's movements in the chamber is as nearly absolute as possible, we can safely compare the carbon-dioxide output of successive days before and after the operation, using the established level at the beginning of the series as a base line. Moreover, as a check upon our results, we have the pulse rate, which experience has shown to be an important indication of the metabolism.

While the pulse rate is falling we have made it a rule not to consider our base line as established. In other words, there is a time at the beginning of each observation when for more than a week, while the dog is becoming used to the routine, there may be a successively lower pulse rate from day to day. It may happen, perhaps because of the darkness and comfort of the chamber, that the minimum carbon-dioxide output in the respiration apparatus is obtained before the pulse rate has reached constancy, but until the dog has been so trained that the pulse rate has become constant, he is not considered ready for the operation. The uniformity of the changes in the pulse rate and carbon-dioxide output are clearly shown in the curves and tables. It has not been possible, however, to obtain accurate blood-pressure observations for our routine work.

Operative methods. — The method of removing the hypo-physis is that devised by Cushing, that is, the temporal route successfully used by Paulesco [21] with the addition by Cushing of the double-sided opening of the head and the resection of the zygoma. The technic, which differs in no way from that described by Crowe, Cushing, and Homans,[1] is as follows:

The animal is given a moderate dose of morphine, placed upon its abdomen on the operating table, and etherized. The head is shaved, scrubbed with soap and water, corrosive sublimate, and alcohol. The skin is then incised in the median line from the prominence above and between the eyes well back upon the neck. It is turned up enough on each side to allow both temporal muscles to be incised along their upper edges and turned back from the skull. After the posterior half of the zygoma has been resected to allow free access on the side on which entrance is to be made, both sides of the skull are extensively opened; on the side opposite to which entrance is to be made, the dura is widely split to allow the brain to be partly dislocated through this opening during the manipulations made necessary by the removal of the gland. From this time on a head light is necessary.

On the side of entrance a specially-formed curved retractor is passed into the temporal fossa outside the dura until it is stopped by the adhesion of the dura to a ridge on the floor of the fossa. Here the dura is incised with a fine hook knife and the retractor is passed through the slit beneath the temporal lobe toward the shallow sella turcica. The position of the hypophysis is easily identified by its relation to the carotid artery in front and the third nerve which slopes downward and forward at the side. When the fine vessels which pass into the posterior lobe from behind have been torn away, the gland can be seized with forceps and removed. The muscles and skin are then replaced. The operation need not and does not, if gently done, injure the brain, which is dislocated to the oppo-site side by the retraction necessary to disclose the gland. The details of the absolute amount of gland removed and of the treatment of the differ-ent animals will be given under the history of each case.

Statistics of observations. — In a series of fourteen obser-vations only five have been brought to completion, Nos. IX., X., XI., XIII., and XIV. The first few experiments made upon cats were operative failures and several upon dogs were purely tentative or were brought to a conclusion by the death of the animals from distemper.

The following histories and tables describe as briefly as possible the condition of the animals during the experiments, and the changes brought about by a removal of the hypophysis. The earlier form of apparatus in which the carbon dioxide was determined by analysis was used for the experiments in Observations IX. and X., and also for the experiments in Observations XI. and XIII. up to May 11, 1911, but beginning with this date, the experiments in these observations were carried out with the later form of apparatus, in which the carbon dioxide was absorbed and weighed. This apparatus was also used for all of the experiments in Observation XIV. The modified method for determining the muscular activity was used in Observation XI., beginning with Feb. 7, 1911, and for all experiments in Observations XIII. and XIV.

OBSERVATION IX. *Description of subject.* — Short-haired mongrel bull. Female. Weight, 6.2 kilograms. Age, about eight months. Nearly full grown. In good condition but not fat. Not sexually mature.

Before operation. — Beginning Feb. 3, 1910, the animal was kept in a metabolism cage in an airy, well-lighted laboratory having a temperature of about 19° C. The diet consisted of meat and dog biscuit (see remarks, Table 3). For the first twenty days of the observation, the dog was made to become accustomed to handling, and to lying in the respiration chamber. During this time, records of the temperature and pulse were frequently taken, but owing to the rather cold atmosphere in which the dog lived and the constant state of shivering which resulted, the respirations were almost impossible to obtain. The first four experiences in the respiration chamber were unsatisfactory, owing to the failure of the animal to remain quiet. In the succeeding four, however, the carbon-dioxide output was established.

Operation, nearly complete hypophysectomy, Feb. 23, 1910. — Operation at 5 P.M. The usual approach was made. A little laceration of the temporal lobe. The sella turcica was unusually deep, making the removal of the gland somewhat difficult. The hypophysis was removed in pieces and a few minute fragments were left adherent to the floor of the third ventricle.

After the operation. — The day following the operation, the dog drank water, and in the evening ate meat. She was rather dull and indifferent but showed no paralysis. The weight was 5.6 kilograms, a loss of nearly a kilogram.

From February 25 until March 11, the subject gained steadily in weight to seven kilograms, and seemed well and lively. At this time a note was made that she seemed to be "getting fat." Soon after this her maximum weight of 7.3 kilograms was reached and at almost the same time her pulse, which had averaged about 110 before the operation, had sunk to 60-70. She seemed more quiet than formerly.

From March 22 to April 1 the dog suffered from diarrhea; her weight fell off to 6.8 kilograms, and her pulse sank to 50–60. During this period and for the rest of the animal's life, the temperature which, previous to the operation, had been 39° C., began to fall below 38° C., and as low as 37.4° C. at times.

From April 1 to April 24, when the animal died, her condition was unchanged until just before the end. She suffered from slight diarrhea at times and seemed rather foolish and aimless in her behavior, although she was to all appearances well. On April 23 a bloody diarrhea set in; she ate nothing, and there was a slight discharge from one eye. These appearances led to a suspicion of distemper. She died on the evening of the 24th, two months after the operation.

Autopsy, twelve hours post-mortem. — General appearances. — Subcutaneous fat upon chest and abdomen, one centimeter thick. Mucous membrane of mouth and nose appeared normal.

Head. — No wound infection. No evidence of hemorrhage or infection at the base of the brain. The sella turcica was unusually deep and contained a pit in its center, evidently the remains of the embryonic connection with the pharynx. Microscopic serial sections of the base of the brain showed a few minute fragments of anterior lobe tissue connected directly with a few fine tabs of pars intermedia.

Chest. — No free fluid. Lungs unusually firm. No evidence of pneumonia.

Abdomen. — Unusual amount of sub-peritoneal fat. All abdominal organs appeared normal except the intestines. These presented on the surface possibly half a dozen thickened patches of lymphoid tissue, pale in color, slightly raised, and varying in size from two to six centimeters in length. The mucous membrane was evidently the seat of hemorrhage in several places, but there were no gross ulcerations except in the stomach, where there was one definite ulcer. On microscopic examination the patches in the intestine appeared to be composed of hypertrophied lymphoid tissues with beginning ulceration and infection.

Microscopic examination of the liver, pancreas, and adrenal not remarkable.

The ovary contained many normal-looking follicles.

Immediate cause of death, distemper.

SUMMARY OF OBSERVATION IX.

Almost complete hypophysectomy upon a nearly adult female dog. Duration of life, two months, accompanied by a fall of pulse and temperature, and a moderate laying-on of fat. No cachexia. Death from distemper.

TABLE 3.

Carbon-dioxide production — Observation IX.

Experiment No.	Date.	Body Temperature.[1] °C.	Pulse.[1]	Respiration.[1]	Body Weight. Kilos.	Activity.[3]	Temperature of Apparatus. °C.	Duration of Experiment. Minutes.	Carbon Dioxide Produced. Per Hour. Grams.	Carbon Dioxide Produced. Per Kilogram Per Hour. Grams.	Remarks.
5 ..	1910. Feb. 11	39.0	106	....	6.3	Slightly active.	15.3–18.3	150	7.59	1.205	Dog given daily ration of about 50 grams meat and 1¼ dog biscuit, *i.e.,* 149 grams. On experimental days, the food was given after the experiment. Nitrogen equivalent of the food about 6.8 grams.
6 ..	" 14	....	....	....	6.6	Slightly active.	15.0–18.2	180	7.36	1.115	
7 ..	" 16	39.0	120	30	6.8	Slightly active.	17.4–21.0	180	7.17	1.054	
8 ..	" 18	39.0	112	24	6.8	Fairly quiet.	13.6–18.0	180	8.25	1.213	

Operation, Feb. 23, 1910.

No.	Date											Remarks
9..	Feb. 24	34.9	132	15	5.6	Very quiet	17.0–19.8	180	3.41	0.609	No food on the 23d, and refused it the morning of the 24th.
10..	" 25	37.0	150	18	5.7	"	13.0–17.7	180	5.24	0.919	Had eaten only the 50 grams meat on February 24, but that at night. Seems better.
11..	" 26	38.0	140	15	5.7	"	14.9–19.0	180	4.78	0.839	Had probably eaten little more than the 50 grams of meat on February 25. Seems bright.
12..	" 28	38.2	86	24	6.0	Quiet.	16.2–18.7	180	5.85	0.975	Lively. Ate all the food on February 27.
13..	March 1	37.7	70	15	6.2	Very quiet.	16.8–18.0	180	4.49	0.724	Ate meat but only a small amount of dog bread on February 28.
14..	" 2	38.4	80	......	6.3	Quiet.	15.4–16.6	180	6.36	1.010	
15..	" 4	39.0	75	......	6.5	"	16.2–19.1	180	5.97	0.918	Reported lively and well on March 3 and eating practically all food.
16..	" 8	38.0	76	......	6.7	Slightly active.	16.8–19.0	180	5.64	0.842	
17..	" 14	38.5	70	16	7.2	Active.	15.0–18.2	180	6.41	0.890	Noted on March 12 that dog seemed to be getting fat.
18..	" 22	38.2	55.	......	6.9	"	17.1–19.8	180	5.02	0.728	Ate small part of food on March 19 and vomited after eating. Ate nothing on March 21. Seems rather more quiet than formerly.

TABLE 3. — *Continued.*

Experiment No.	Date.	Body Temperature.[1] °C.	Pulse.[1]	Respiration.[1]	Body Weight. Kilos.	Activity.[2]	Temperature of Apparatus. °C.	Duration of Experiment. Minutes.	Carbon Dioxide Produced. Per Hour. Grams.	Per Kilogram Per Hour. Grams.	Remarks.
19..	1910. March 24	38.2	65	......	6.7	Active.	19.6–21.1	180	4.56	0.681	Bowels loose.
20..	" 28	39.0	60	......	6.7	Slightly active.	19.0–21.1	180	5.46	0.815	Probably ate during the night, as food had been left in the metabolism cage.
21..	" 30	39.0	56	......	6.7	Fairly quiet.	21.7–22.7	180	4.14	0.618	Looseness of the bowels on the 28th, with diarrhea on the 30th, this condition persisting to a slight degree on March 31 and April 1.
22..	April 2	38.0	60	......	6.9	Slightly active.	19.0–20.8	180	4.97	0.720	On April 2, bowels were constipated.

[1] Body temperature, and pulse and respiration rates were taken before the dog entered the apparatus unless otherwise noted.

[2] Estimated from the examination of the kymograph records for the entire duration of each experiment. See discussion later of Observations IX. and X., and compare with method used in Observations XI., XIII., and XIV.

CHART 1 .

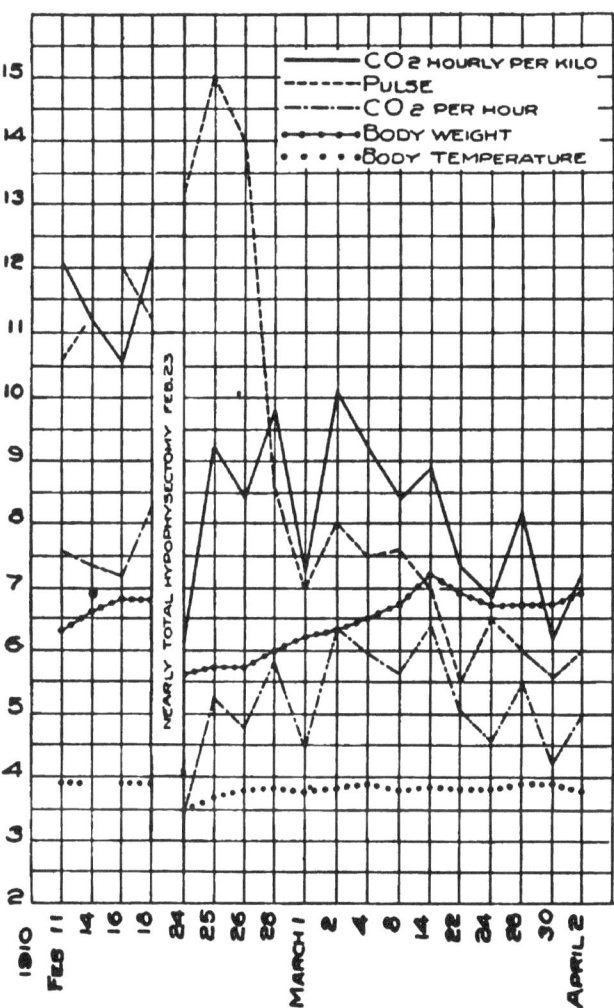

CHART I. Observation IX. — The curves record the carbon dioxide per hour per kilogram of body weight, the pulse rate, the total carbon dioxide per hour, the body weight in kilograms, and the body temperature. The experiments to determine the carbon-dioxide output were of uniform length, and the activity of the animal in the respiration chamber was estimated from the kymograph records.   (See Table 3.)

OBSERVATION X. *Description of subject.* — Short-haired mongrel. Female. Weight, 6.1 kilograms. Age, about eight months. In good condition and rather thin.

Before operation. — The animal was placed on a regular diet on March 9, 1910, and was kept under observation in the same laboratory with Observation IX. from this date until March 17, when the operation was performed. During this time the pulse ranged from 84 to 100, and the temperature from 39° to 38.2° C. As in Observation IX. it was found impossible to get a satisfactory examination of the respirations owing to the cold.

Operation, March 17, 1910. — The usual procedure with the insertion of the retractor beneath the temporal lobe was carried out, except that the gland itself was not disturbed. The time of the operation was made to conform to the usual time taken for the complete operation with excision of the gland.

After the operation. — Following the operation, the weight fell to 5.9 kilograms, and then came back to 6.2 kilograms in the course of five days. There was no appreciable change in the pulse rate except for a slight rise on the day following the operation. It was noticeable that the animal remained more quiet than before, especially in the respiration chamber. It was sacrificed under chloroform on March 19, and the wound of operation was found to be normal.

SUMMARY OF OBSERVATION X.

A nearly adult female dog, observed for a short time before and after operation to determine the effect of trauma. Except for making the animal more tractable in the respiration chamber, no especial effect was noticed, the operation producing no change in metabolism.

Experiment No.	Date.	Body Temperature.[1] °C.	Pulse.[1]	Respiration.[1]	Body Weight. Kilos.	Activity.[2]	Temperature of Apparatus. °C.	Duration of Experiment. Minutes.	Carbon Dioxide Produced. Per Hour. Grams.	Per Kilogram Per Hour. Grams.	Remarks.
1	1910. March 10	39.0	108	......	6.0	Slightly active.	17.4–19.4	180	5.93	0.988	Dog reported on March 9 as active and restless. To be trained if possible. Given daily ration of about 50 grams meat and 149 grams dog biscuit. Nitrogen equivalent of food about 6.8 grams.
3 ...	" 12	39.0	100	......	6.1	Active.	16.3–18.3	180	8.27	1.356	
4	" 15	38.2	96	20	6.3	"	15.0–19.9	180	7.65	1.214	

Operation, March 17, 1910.

Experiment No.	Date.	Body Temperature.[1] °C.	Pulse.[1]	Respiration.[1]	Body Weight. Kilos.	Activity.[2]	Temperature of Apparatus. °C.	Duration of Experiment. Minutes.	Carbon Dioxide Produced. Per Hour. Grams.	Per Kilogram Per Hour. Grams.	Remarks.
5 ...	March 18	39.4	114	......	5.9	Quiet.	15.3–18.1	80	5.62	0.953	In good condition.
6 ...	" 19	39.0	100	20	5.8	"	15.0–18.7	180	5.69	0.981	
7 ...	" 21	38.8	90	......	6.2	"	17.5–20.0	180	6.21	1.002	Reported in good condition on March 20 and eating all food.

[1] Body temperature, and pulse and respiration rates were taken before the dog entered the apparatus unless otherwise noted.

[2] Estimated from the examination of the kymograph records for the entire duration of each experiment. See discussion later of Observations IX. and X., and compare with method used in Observations XI., XIII., and XIV.

CHART 2.

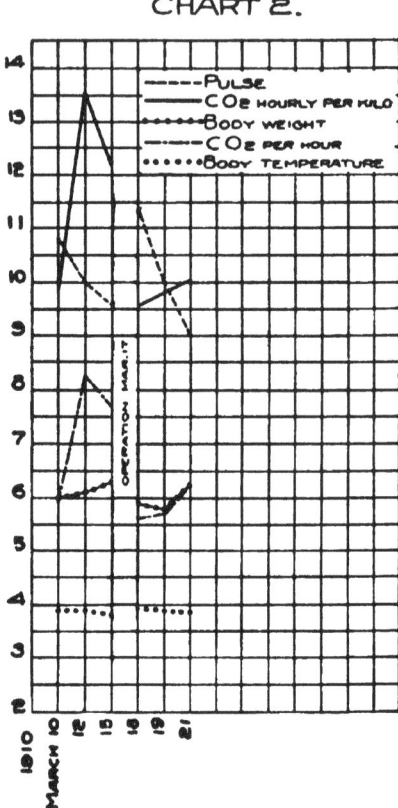

CHART II. Observation X. — The curves record the pulse rate, the carbon dioxide per hour per kilogram of body weight, the body weight in kilograms, the total carbon dioxide per hour, and the body temperature. The experiments to determine the carbon-dioxide output were of uniform length and the activity of the animal in the respiration chamber was estimated from the kymograph records. (See Table 4.)

OBSERVATION XI. *Description of subject.* — Short-haired mongrel. Female. Weight, four kilograms. Age, six months, or about two-thirds grown. In good condition and rather thin. Not sexually mature.

Before operation. — Previous to the operation, the animal was kept under observation from May 4, 1910, to May 16, 1910, the conditions being the same as those in Observations IX. and X. She was very tractable and rapidly learned to lie quiet for the recording of her pulse, respiration, and temperature, as well as in the respiration chamber. During the preliminary period, the pulse, which was 90 when the dog was first examined, rose several times and then remained between 82 and 88 from May 10 to May 16. The temperature was usually registered as 39° C., never lower and once as high as 40° C. The weight hardly varied from four kilograms. The respirations were from 10 to 15. The diet consisted of meat and dog biscuit, this being changed later to meat and bread, although no alteration in the amount of meat was made. The nitrogen equivalent of the food is given under remarks in Table 5.

Operation, May 16, 1910. — The usual approach was made. There was considerable tearing of the temporal lobe with moderate venous oozing, which was checked by packing. The gland was removed in three pieces. The floor of the sella turcica was scraped, but a few minute fragments of anterior lobe were left adherent to the floor of the third ventricle. The field of the operation was left dry. On the evening of the day of operation, the condition of the animal was good; she moved about without difficulty, and responded to the voice. The first urine contained no sugar.

After operation. — On May 17, the day after operation, the animal was bright and lively; the weight was 3.6 kilograms; the pulse rate was 124. There were both polyuria and polydipsia. From this date until May 28 the general condition of the animal remained about the same. On only one day were there signs of cachexia, this being shown in a slight arching of the back. Doubtless owing to the steady diet of dog biscuit, the animal suffered considerably from diarrhea, so that on May 28 the ration was changed to meat and dry bread instead of meat and dog biscuit. As this change produced no improvement the dog was sent to the animal farm on June 3. When the diarrhea was at its worst there was considerable cough, but the general health and appearance of the dog seemed to be sufficiently good. The weight at this time was four kilograms.

During the period from May 16 to June 3, there were marked changes in the pulse rate, and a slight fall in the carbon-dioxide output. On the two days following the operation, the pulse rate was 124 and 120 respectively, but on the third day (May 19) it fell to 84, which was approximately the pulse rate previous to the operation. On May 20 it fell to 60, and with an occasional rise and one fall to 52, this rate was maintained until the animal was sent to the farm on June 3. The fall in the carbon-dioxide output was much less striking; when the diarrhea was at its worst this factor even rose to the preoperative level.

When the dog returned from the farm on June 13 its weight had fallen to 3.8 kilograms; the general condition was good and the appetite was voracious. The animal appeared quite somnolent and lethargic, especially when not disturbed, though she still was active when roused. The pulse rate remained in the neighborhood of 60, but the carbon-dioxide output fell off sharply to a considerably lower level.

On June 15 a subcutaneous injection was given of .05 gram of boiled powdered anterior lobe (Armour & Co.), with a temperature reaction from 38.8° to 39.2° C. There was no increase in the carbon-dioxide output on this day, but a perceptible increase in the two succeeding days. A part of the injected material remained unabsorbed in the tissue several days later, though there was no obvious infection.

During the month of June the animal began to put on fat, the weight rapidly increasing to 4.6 kilograms; in other respects there was no change.

In July the dog's weight increased from 4.6 kilograms to 5.2 kilograms, with a marked increase in fat. At the same time the carbon-dioxide output per kilogram fell off markedly, although the absolute amount was only slightly lower. The average pulse rate was approximately the same as in June, while the temperature fell slightly. In this period the dog, though she had never been in heat, showed signs of sexual excitement, a peculiarity which remained as one of her characteristics until her death. In August there was a further gain in weight. An injection of .10 gram of anterior lobe extract produced the same temperature reaction as before, but without a subsequent rise in the carbon-dioxide output. During the following months of September, October, November, and December, 1910, the animal continued to gain in weight and its general condition remained unchanged. As the weather grew colder, it was found impossible to prevent shivering in the respiration chamber. Even under these circumstances, however, the carbon-dioxide output usually remained low and the pulse did not vary materially. An examination of the kymograph records at this time shows that the method of recording muscular movements then in use was not accurate enough to give a satisfactory idea of the animal's fine muscular tremor. This error was not rectified until Feb. 7, 1911, when the pneumograph system described on page 427 was employed. During this interval many of the experiments probably gave too high a value to the carbon-dioxide output. Attention is called to this error in the margin on Table 5, under the heading of remarks, and also on the chart.

Beginning December 11, 0.5 gram of powdered anterior lobe was given with the food daily, and on December 19 this was increased to one gram. No obvious change was produced by this feeding, though it was kept up for many months, but the animal continued to react to the subcutaneous administration of the extract.

In January, 1911, the maximum weight of six kilograms was attained, and the dog was fat and well. Treatment for tapeworm, which was discovered at the end of this month, resulted, however, in considerable loss of weight. In February the animal was taken to the farm for the second time. While there she exercised considerably more than when confined in the metabolism cage. The regular diet was discontinued with the exception of the one gram of anterior lobe extract which was continued until her return.

On April 28, 1911, the dog was brought from the farm and the experiments continued. At this time she weighed only five kilograms, the pulse rate was nearer 70 than 60, and the carbon-dioxide output per kilogram was considerably higher than for some months. With the regular diet and confinement, the weight increased from five kilograms to 5.7 kilograms, while the carbon dioxide, both absolute and per kilogram of body weight, fell off markedly. After June 8, 1911, no more observations were made.

From Aug. 1, 1911, to Sept. 27, 1911, the dog was allowed to run loose about the animal house. As a result, she grew comparatively thin, but seemed well. She was sacrificed under chloroform on Sept. 27, 1911.

Autopsy, immediate. — Weight, 4.8 kilograms. Subcutaneous fat one centimeter thick.

Head. — Wound of operation cleanly healed. No evidence of past hemorrhage or infection at base of brain. Serial sections of the base of the third ventricle and the dura below it which were removed in one piece showed a definite fragment of anterior lobe tissue of microscopic size buried in the thickened dura.

Abdomen. — Considerable fat in omentum. Abdominal organs appeared normal.

Liver. — Microscopic examination showed nothing remarkable except a moderate increase of connective tissue about the portal vessels.

Pancreas. — Not remarkable.

Ovary. — Microscopic examination showed only a few undeveloped follicles which appeared incomplete and rudimentary.

Adrenal. — Not remarkable.

Thyroid. — The acini were small and very numerous. The cells were low and cuboidal. There were a great many groups of cells not distinctly formed into acini. The gland seemed to be in an inactive state.

SUMMARY OF OBSERVATION XI.

Almost complete hypophysectomy upon a nearly full-grown female dog. Immediate fall of pulse rate. More gradual fall in carbon-dioxide production. Some cachexia in the first few months following operation. Considerable laying on of fat with preservation of infantile characteristics and failure of sexual development. No return to normal under the influence of prolonged feeding of dried hypophyseal anterior lobe extract. Duration of life, sixteen months following operation, after which the animal was sacrificed under chloroform. Age at the time of death nearly two years.

TABLE 5.

Carbon-dioxide production. — Observation XI.

Experiment No.	Date.	Body Temperature.[1] °C.	Pulse.[1]	Respiration.[1]	Body Weight. Kilos.	Temperature of Apparatus During Quiet Periods. °C.	Carbon Dioxide Produced During Minimum Activity.			Remarks.
							Duration of Quiet Periods. Minutes.	Per Hour. Grams.	Per Kilogram Per Hour. Grams.	
1....	1910. May 5	40.0	120	14	4.2	20.2–20.0	90	4.10	0.976	Dog given daily ration of 50 grams meat, 49.5 grams dog biscuit soaked in water. On experimental days food is given after the experiment. The nitrogen equivalent of food was 4 grams.
2....	" 6	39.0	100	11	4.0	22.0–22.6	135	3.70	0.925	
3....	" 9	39.2	100	15	4.1	20.5–23.1²	90	3.92	0.956	
4....	" 10	39.0	86	13	4.2	22.0–23.0³	90	3.74	0.890	
5....	" 11	39.3	88	14	4.1	21.0–23.1³	45	3.36	0.820	
6....	" 14	39.0	82	12	4.0	³	90	3.38	0.845	Dog thin but otherwise well. On May 12 the portion of dog biscuit was increased to 99 grams, the nitrogen equivalent of the food now being 5.4 grams.

Operation, May 16, 1910.

	Date								Remarks
7...	May 18	39.4	120	10	3.8	°	90	3.34	0.879	Good recovery from ether. Moves without rotating; responds well. Urine at 6 P.M., May 16, showed no sugar. Reported as bright and lively on May 17. Wound in good condition. Given usual food with a little milk. Ate all. No sugar in urine.
8...	" 19	39.0	84	10	3.8	°	45	3.40	0.895	Moderate diarrhea; otherwise well and eats all food. A slight cough beginning on May 17 persists. No sign of cerebral injury.
9...	" 20	39.0	60	9	3.9	°	90	3.14	0.805	Feces solid.
10...	" 21	38.8	64	9	3.9	°	135	3.24	0.831	Moderate diarrhea. Slight arching of back.
11...	" 23	38.6	60	9	3.9	20.9–23.1	135	3.18	0.815	Bowels in good condition.
12...	" 26	39.0	60	10	4.0	24.0–25.7	90	3.15	0.788	Slight cough reported on previous days still persists. Bowels loose. Seems bright.
13...	" 28	39.0	64	9	4.0	°	90	3.23	0.868	Severe diarrhea and cough, making it necessary to change diet. This condition continued on the 29th and 30th. Stale bread, 150 grams, substituted for the dog biscuit on March 28, the nitrogen equivalent of the food now being 4.9 grams.
14...	" 31	38.9	70	10	4.0	°	90	3.73	0.932	Diarrhea worse. Coughs and sneezes considerably, but no discharge from nose.

TABLE 5. — *Continued.*

Experiment No.	Date.	Body Temperature.[1] °C.	Pulse.[1]	Respiration.[1]	Body Weight. Kilos.	Temperature of Apparatus During Quiet Periods. °C.	Carbon Dioxide Produced During Minimum Activity.			Remarks.
							Duration of Quiet Periods. Minutes.	Per Hour. Grams.	Per Kilogram Per Hour. Grams.	
15...	1910. June 2	39.0	68	12	4.0	?	90	3.59	0.898	Movements solid. Diarrhea improved on June 1, but grew worse again on the 3d. Cough. Dog was sent away to the farm on June 3.
16...	" 13	38.8	56	8	3.8	?	90	2.58	0.679	Returns from the farm in excellent condition.
17...	" 14	39.2	58	10	3.9	?	90	2.98	0.764	
18...	" 15	38.8[4]	62[4]	11[4]	4.1	?	90	2.84	0.693	Injection of 0.05 gram anterior lobe at 9.20 A.M., about one hour before beginning the experiment.
19...	" 16	39.0	72	10	4.2	?	135	3.24	0.771	Voracious appetite. Defecation every other day. In excellent shape. Since operation, tendency to be quiet for long periods if undisturbed, otherwise playful.
20...	" 17	39.0	70	9	4.2	?	90	3.69	0.879	
21...	" 18	39.0	68	10	4.24	?	90	3.44	0.811	Bowels in good condition.

										Remarks
22...	20	39.0	64	12	4.4	26.9-27.6	45	3.37	0.766	Seems rather sleepy most of the time since returning from the farm; lies still for long periods. Lively when roused. Small abscess at site of injection opened.
23...	22	39.0	68	16	4.5	29.3-30.3	90	3.95	0.878	Very hot on this day and the day before.
24...	24	38.9	64	14	4.56	27.0-27.8	45	3.76	0.825	Bowels loose again on the 23d. Seemed to be putting on flesh. Respiration more rapid, similar to panting and seemingly due to hot weather.
25...	" 27	38.0	58	16	4.4	24.0-24.8	90	3.04	0.691	
26...	" 28	38.6	65	10	45	4.5-25.3	90	3.58	0.796	
27...	" 29	38.3	60	13	4.6	24.0-.2	45	3.45	0.750	Apparatus leaked a little. Results only approximate.
28...	July 21	38.9	76	15	5.1	24.0-25.0	90	3.58	0.702	Reported on July 2 that dog was evidently putting on fat and seemed to be growing. Some sexual excitement also noted. Rapid respiration continued to be noted.
29...	" 23	38.3	64	16	5.16	25.7-26.3	45	2.91	0.564	Considerable sexual excitement noted before the experiment.
30...	" 26	38.3	56	13	5.1	26.2-27.1	90	3.25	0.637	In first rate condition, lively and well, but on being made to lie still, goes to sleep easily. Less sexual excitement.
31...	" 27	38.3	54	9	5.1	26.0-26.7	45	2.62	0.514	Less sexual excitement.
32...	" 28	38.4	62	11	5.14	24.4-26.3	90	3.32	0.646	

TABLE 5. — *Continued.*

Experiment No.	Date.	Body Temperature.	Pulse.	Respiration.	Body Weight.	Temperature of Apparatus During Quiet Periods. °C.	Duration of Quiet Periods. Minutes.	Carbon Dioxide Produced During Minimum Activity. Per Hour. Grams.	Per Kilogram Per Hour. Grams.	Remarks.
	1910.	°C.			Kilos.					
33...	July 29	38.5	56	12	5.0	26.4–27.4	90	2.65	0.530	Has cough or sneeze, something like that after the operation. Seems perfectly well otherwise.
34...	Aug. 2	38.7	60	12	5.1	24.6–25.6	90	2.85	0.559	Was not fed on July 31, so had 50 grams meat extra on August 1. Well and lively.
35...	" 3	38.3	58	12	5.0	25.3–26.3	90	2.73	0.546	
36...	" 4	38.7	66	12	5.2	26.5–27.3	45	3.03	0.583	
37...	" 5	38.5	62	14	5.3	24.3–25.2	90	3.34	0.630	
38...	" 8	38.7	60	12	5.2	24.5–25.5	90	3.10	0.596	
39...	" 9	39.3[b]	62[a]	12[a]	5.3	26.0–26.2	90	3.54	0.668	Injection of 0.10 gram anterior lobe one-half hour before experiment.
40...	" 10	38.7	76	13	5.4	24.6–25.7	90	3.53	0.654	
41...	" 12	38.4	62	12	5.3	23.9–24.9	60	3.31	0.625	
42...	" 16	38.7	68	12	5.3	24.2–24.6	90	3.32	0.626	

										Remarks
43...	" 17	38.7	68	14	5.4	23.8–24.6	90	3.58	0.663	Vomited part of food night before.
44...	" 18	38.6	58	13	5.4	23.4–24.7	60	3.64	0.674	Bowels rather loose, but material is digested. Shivering considerably in respiration apparatus.
45...	" 22	38.8	58	13	5.3	23.7–24.4	45	3.18	0.600	Good condition. No diarrhea.
46...	" 24	38.5 [7]	58	12	5.4	26.2–26.7	45	3.06	0.567	Feces solid.
47...	" 29	38.9	76	15	5.5	23.6–24.7	90	3.38	0.615	
48...	" 31	38.6	68	13	5.54	21.7–22.6	45	3.80	0.686	Has shivering type of respiration as soon as he is made to relax. Weather colder. Pulse varies, 68 being lowest count.
49...	Sept. 3	39.0	92	19	5.56	22.6–23.1	45	4.54	0.817	Had been exercising considerably before preparing for the experiment.
50...	6	38.8	68	.14	5.5	23.8–24.6	60	3.70	0.673	
51...	10	38.5	60	9	5.56	22.2–23.3	60	3.61	0.649	Hard to get pulse and respiration steady after the experiment as dog began to have usual trembling, shivering respiration on being kept quiet.
52...	17	38.8	88	16	5.6	21.3–22.0	45	3.81	0.680	Reported on September 15 that dog had been having rapid pulse and higher temperature on the two days preceding. Weather distinctly cooler. Skin red from application for fleas. Fat evidently increasing. Shivering constantly in the respiration apparatus on September 17.

TABLE 5. — *Continued.*

Experiment No.	Date.	Body Temperature.[1] °C.	Pulse.[1]	Respiration.[1]	Body Weight. Kilos.	Temperature of Apparatus During Quiet Periods. °C.	Carbon Dioxide Produced During Minimum Activity.			Remarks.
							Duration of Quiet Periods. Minutes.	Per Hour. Grams.	Per Kilogram Per Hour. Grams.	
53...	1910. Sept. 19	39.0	70	10	5.6	21.0–21.3	45	3.82	0.682	Respiration a little less spasmodic and shivering when taken before the experiment.
54...	" 22	38.9	80	14	5.74	19.8–21.3	80	5.18	0.902	Animal shivering; evidently abnormal results in consequence.
56a...	Oct. 6	38.5	70	14	5.7	28.3–28.2	60	3.02	0.530	Animal shivering while in the respiration apparatus, but not while the records of the pulse and respiration were taken beforehand.
57...	" 8	38.2	60	10	5.5	22.8–23.7	60	3.38	0.615	
58...	" 11	38.5	70	12	5.4	24.2–24.0	30	3.19	0.591	Animal shivering in respiration apparatus.
59...	" 15	38.5	70	13	5.6	21.6–21.4	45	4.22	0.754	Shivering respiration at all times in respiration apparatus.
60...	" 17	38.4	64	15	5.56	24.5–24.6	30	3.43	0.617	
61...	" 19	38.4	56	11	5.5	26.0–25.9	60	3.40	0.618	
62...	" . 28	38.5	66	15	5.54	22.3–22.8	30	4.39	0.792	On October 23 was given 1 gram anterior lobe extract with the food, which, beginning with October 22, was soaked with 300 cubic centimeters of water. Shivering a little in apparatus on October 28.

	Date									Remarks
63...	" 29	38.5	63	12	5.5	23.0–23.8	60	3.98	0.724	Shivering a little while in respiration apparatus.
64...	Nov. 4	38.7	66	14	5.64	22.4–22.8	30	3.51	0.622	Shivered a little while in the apparatus. Trembled violently in colder air of the room after the experiment.
65...	" 9	38.5	66	13	5.7	24.1–24.2	60	3.12	0.547	Had exercised more than usual before preparing for the experiment. No shivering while in the apparatus.
66...	" 14	38.5	58	13	5.6	24.3–23.8	60	3.49	0.623	
67...	" 16	38.4	63	12	5.66	24.7–24.5	60	3.49	0.617	
68...	Dec. 5	38.5	62	12	5.7	23.3–24.0	30	2.86	0.502	Beginning on November 19, dog was sick, vomiting on November 19 and troubled for some days with severe diarrhea. It was necessary during this time to vary the diet somewhat. Reported free of diarrhea and eating well on December 3.
69...	" 8	38.6	62	12	5.8	24.8–24.0	30	2.63	0.453	
70...	" 12	38.5	64	15	5.8	23.8–24.2	30	3.15	0.543	Given 0.5 gram anterior lobe extract on December 11 and each following day until amount was increased to 1 gram on December 19.
71...	" 14	38.8	66	13	5.74	24.0–25.0	60	3.61	0.629	
72...	" 15	38.6	66	13	5.9	24.4–24.7	30	3.33	0.564	
73...	" 17	38.6	64	12	5.8	25.8–25.2	90	3.27	0.564	

TABLE 5. — *Continued.*

Experiment No.	Date.	Body Temperature.[1] °C.	Pulse.[1]	Respiration.[1]	Body Weight. Kilos.	Temperature of Apparatus During Quiet Periods. °C.	Carbon Dioxide Produced During Minimum Activity.			Remarks.
							Duration of Quiet Periods. Minutes.	Per Hour. Grams.	Per Kilogram Per Hour. Grams.	
74...	1910. Dec. 20	38.6	70	11	5.8	24.7–25.3	90	3.18	0.548	Anterior lobe extract increased to 1 gram on December 19 and that amount was given each day till dog was sent to the farm in February, 1911.
75...	" 23	38.5	72	12	5.9	25.2–25.1	90	3.52	0.597	
76...	" 24	38.8	74	14	5.9	24.7–24.3	60	3.46	0.586	
77...	" 27	38.7	68	14	5.9	25.1–25.3	30	3.45	0.585	Reported as spilling most of food on December 25, and vomiting small part of food eaten.
78...	1911. Feb. 7	38.3[9]	62[9]	14[9]	5.4	25.0–25.3	30	2.86	0.530	Treatment was begun on the dog on Jan. 19, 1911, for tapeworm, two segments of which had appeared daily. Vomiting and diarrhea resulted. Reported free of diarrhea on February 4. Diet was changed on February 5, 90 grams of bread being given end of 150 grams; total nitrogen in the food 4.1 gms. All food was eaten. The modified tard for determining the muscular activity was used for the first rime in this spent, and for all succeeding experiments.

										Remarks
79...	11	38.3	58	12	5.4	26.2–26.0	60	2.87	0.531	Diet changed again on February 11 so that dog was again getting 150 grams bread; nitrogen equivalent of the food 4.9 grams.
80...	13	38.5	57	12	5.4	24.8–25.5	60	3.21	0.594	
81...	17	39.1[10]	60[11]	12[11]	5.5	24.8–25.4 / 25.4–26.0	90 / 90	4.33 / 3.58	0.787 / 0.651	Given subcutaneous injection of 0.1 gram boiled anterior lobe about one hour before the experiment began.
82...	18	39.4	82	15	5.48	25.9–26.1	60	3.28	0.599	Dog sensitive on whole right side. Abscesses formed at the site of the injections were noted February 20 and treated with success. Dog was kept in good condition.
83...	April 28	38.8	67	16	5.0	25.7–26.8	92	3.84	0.768	Back from the farm 3 days before. In good condition. Less subcutaneous fat.
84[13].	May 11	....	......	......	5.3	25.1–26.8	120	3.90	0.736	
85...	" 17	38.8	68	18	5.5	25.8–26.1	60	3.62	0.658	
86...	" 19	38.9	72	22	5.6	27.0–27.0	30	3.52	0.629	
87...	" 23	38.8	66	26	5.6	26.2–26.5	60	3.46	0.618	
88...	" 31	38.8	70	24	5.7	25.3–27.0	60	3.45	0.605	During experiment, one record of respiration, 13.

TABLE 5.— *Concluded.*

Experiment No.	Date.	Body Temperature.[1] °C	Pulse.[1]	Respiration.[1]	Body Weight. Kilos.	Temperature of Apparatus During Quiet Periods. C.	Carbon Dioxide Produced During Minimum Activity. Duration of Quiet Periods. Minutes.	Per Hour. Grams.	Per Kilogram Per Hour. Grams.	Remarks.
89...	1911. June 8	38.7	64	20	5.7	26.0–27.0	90	2.86	0.502	Average respiration during the experiment, 10. Dog was let out to run on August 1. Reported on Sept. 27, 1911, as seeming well and weighing 4.8 kilos.

[1] Body ⋯⋯ we taken before the dog entered the ⋯⋯ unless otherwise noted.

[2] ⋯⋯ nge of ⋯⋯ ⋯⋯ it; the ⋯⋯ for the ⋯⋯ ⋯⋯ pods are ⋯t ⋯ ⋯il ⋯le.

[3] The ⋯ bk ⋯⋯ ⋯g the ⋯ ⋯⋯ts May 5–June 18 was lost and ⋯ be ⋯⋯ ⋯ of the apparatus t ⋯ ⋯ps are not ⋯vailable. For days where no apparatus ⋯⋯ ⋯es are set ⋯⋯n in the table, ⋯ ⋯, it is known definitely that the t ⋯ ⋯ pe was ⋯r below 20° C. and ⋯r ⋯e 25° C.

[4] Records ⋯⋯ ⋯ em at 9.30 A.M. after ⋯⋯ ⋯⋯n. Records taken before ⋯⋯ ⋯on were : Body ten ⋯⋯ ⋯pe, 38.8° C.; pulse, 54; ⋯ration, 9.

[5] Record taken at 9.48 A.M. after the ⋯⋯ ⋯ion which was at 9.30 A.M. Body ⋯⋯ ⋯ tal em ⋯ ⋯e the ⋯ ⋯in was 38.7° C.

[6] Records of the ⋯ ⋯se and ⋯⋯ ⋯in were ⋯⋯ aken ⋯e the i ⋯⋯ ⋯in; ⋯y could ⋯t be ⋯ ⋯ken after the ⋯ ⋯ ⋯ ⋯e the dog was restless and hard to ⋯ ⋯t. Records at 1.30 P.M., after the experiment, were : Body temperat ⋯e, 39.1° C.; p⋯ ⋯e, 76; ⋯iration, 18.

[7] Record taken after the ⋯ ⋯t.

[8] Experiment 55 ⋯t ⋯d; apparatus ⋯king.

[9] ⋯e ⋯⋯ we taken at the end of the ⋯ ⋯ ent.

[10] Body temperature ⋯ ed taken at 9.55 A.M. after the inj ⋯ ⋯ion which was at 9.35 A.M. Record before the injection was 38.3° C. Record before the injection was 38.3° C.

[11] ⋯is taken ⋯e i ⋯⋯ ⋯n.

[12] The first ⋯ ⋯ ⋯nt in which the carbon ⋯e was ⋯ ⋯d. Experiments 85–89 were carried out by the same method.

CHART 3.

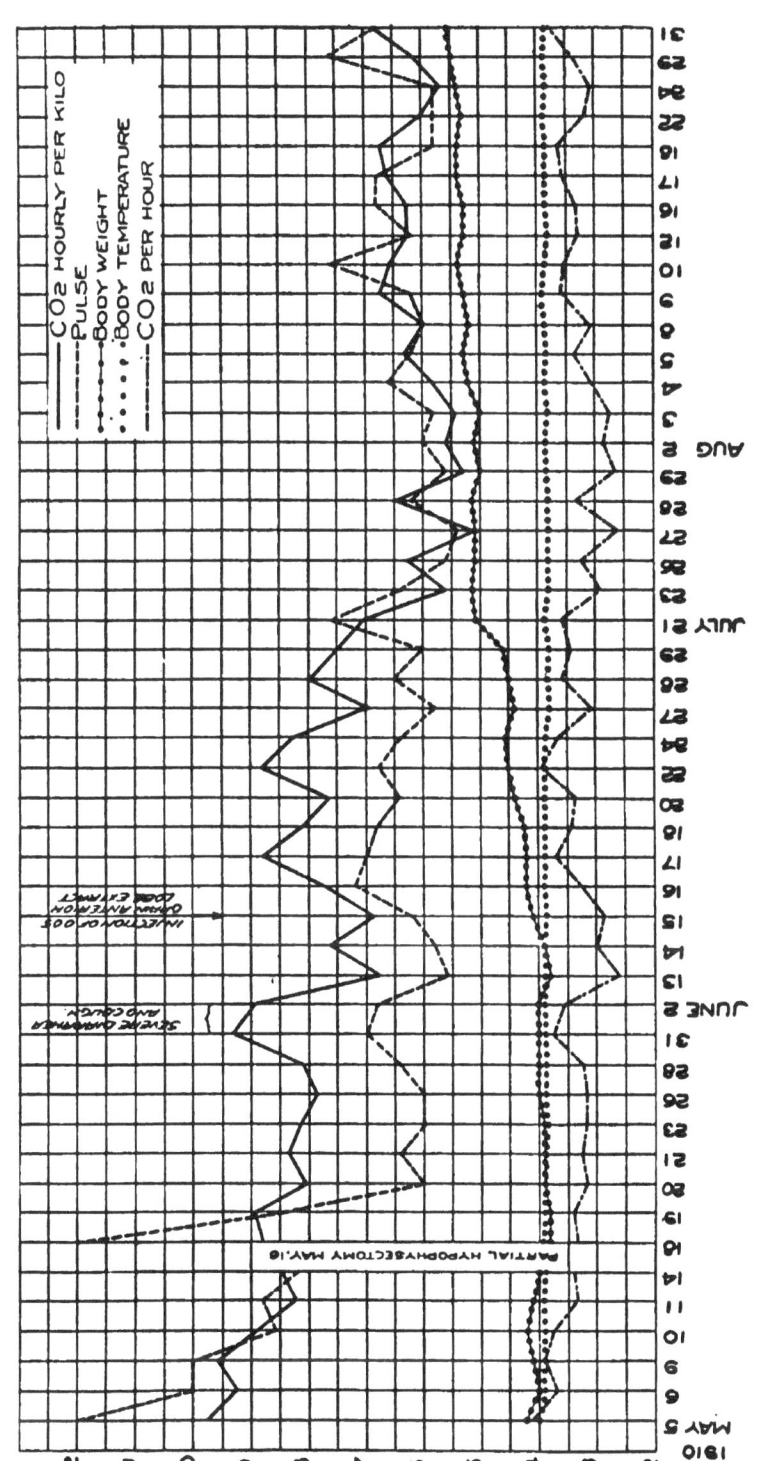

CHART III. Observation XI. — The curves record the carbon dioxide per hour per kilogram of body weight, the pulse rate, the body weight in kilograms, the body temperature, and the carbon dioxide per hour. The experiments to determine the carbon-dioxide output were not of uniform length. Only periods in which the muscular activity was at a minimum were chosen for record. (See Table 5.)

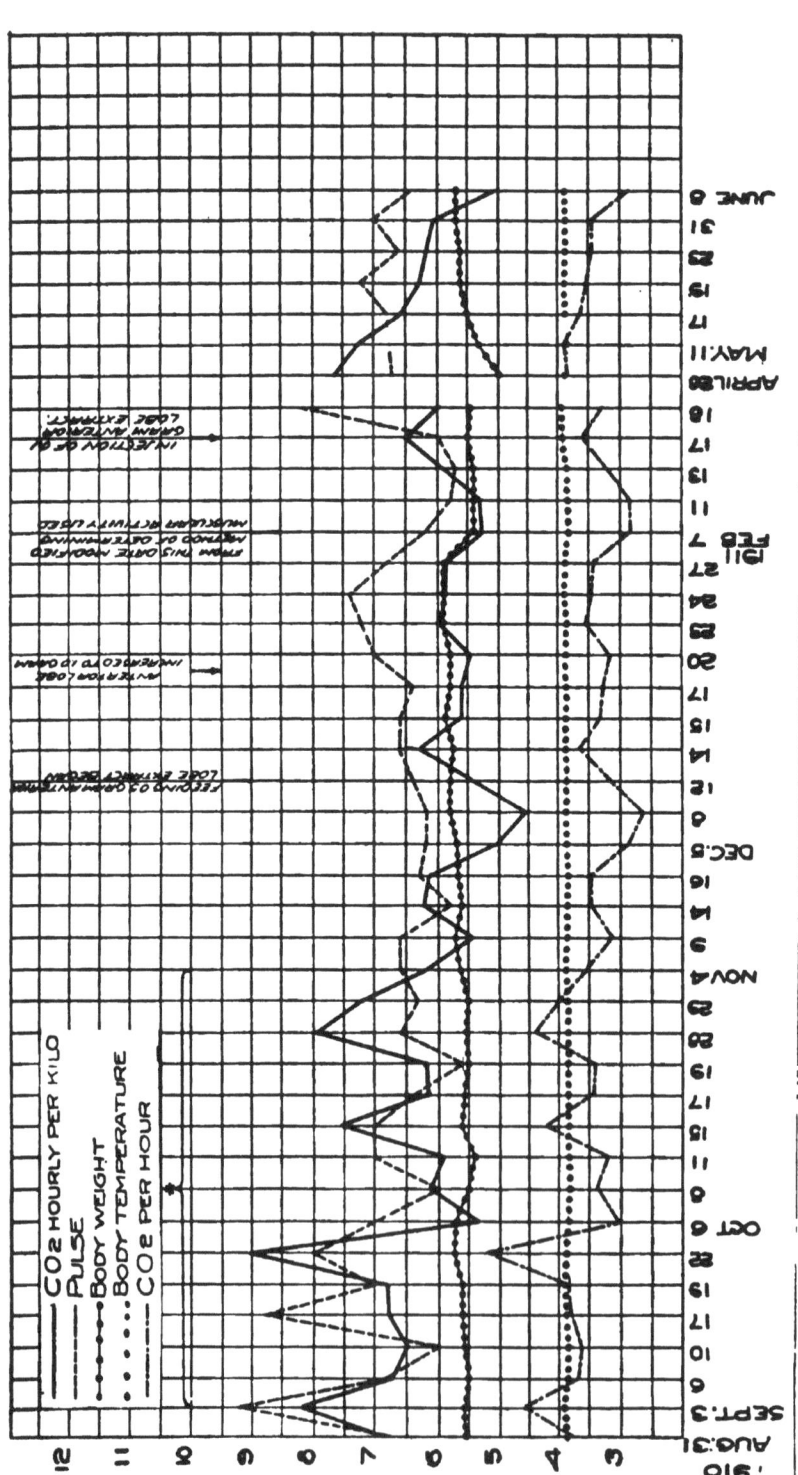

CHART 3.

* During the months of September and October the animal shivered at times in the respiration chamber owing to the low temperature. While the registering apparatus then in use was not sufficiently sensitive to clearly record this activity, it is probable that the carbon-dioxide production was in general increased by it.

OBSERVATION XIII. *Description of subject.* — Fox terrier. Male. Weight, 4.5 kilograms. Probably more than eight months old and nearly full grown, though not sexually mature. In good condition and rather thin.

Before the operation. — The dog was under observation previous to the operation from Feb. 25, 1911, to April 21, 1911, in the same surroundings as his predecessors. During this time he was kept in the metabolism cage and placed upon a constant diet of meat and bread and, when this seemed to disagree with him, on meat and cracker dust. He was not, however, a very regular eater and occasionally refused part of his bread or cracker diet, but never the meat.

The weight rose from 4.5 kilograms at the beginning to 4.9 kilograms upon the day of the operation. The pulse was at all times unusually slow for a dog of his size. During the first few weeks it ranged from 50 to 66, and on one or two occasions it was found to be between 45 and 50. The respirations ranged usually from 7 to 11, averaging about 9. The body temperature was quite variable, though it was generally above 38.5° C. and below 39° C. The muscular movements in this observation were recorded by the modified pneumograph method, previously described, which supplied a very accurate index of the activity, including that of shivering. On March 30, 1911, a subcutaneous injection of .1 gram of boiled powdered anterior lobe was given. The temperature at 9.15 A.M., when the injection was given, was 38.5° C. At 1.30 P.M. it was 39.5° C., and at 5 P.M. it was still 39.5° C. The dog was very sensitive all over and yelped when he was touched.

On April 3 the dog was etherized for the purpose of opening the abscess which had appeared at the site of injection. Like the abscess following injections in other observations, this contained a peculiar gelatinous brownish material, without an inflammatory wall. A section of the right testicle was also taken for microscopical examination, and as a control to the effect of subsequent procedures.

Operation, April 21, 1911. — Morphia and ether. The approach to the gland was free from bleeding. No tearing of the temporal lobe. The gland was crushed close to the stalk so as to leave a small fringe of anterior lobe adherent to it, and was removed in one piece without hemorrhage. The wound was closed as usual. The dog came out of the ether well, walked about on the same afternoon, and ate his usual ration of food. The first urine was free from sugar.

After the operation. — During the next few days the animal ate the usual amount of food. The weight remained constant at 4.8 kilograms. The only untoward symptoms were a slight exophthalmus of the left eye, with some reddening of the eyeball.

No observations of the carbon-dioxide production were made on the two days following the operation (April 22 and 23), as the pulse and temperature were both slightly elevated. On the 24th, when the carbon-dioxide observations were resumed, the pulse rate was 68, the respiration rate 12, the temperature 39.1° C., and the weight five kilograms. On the 26th and 27th, the fifth and sixth days after operation, two suppurating areas were found in the scalp wound, but these healed immediately upon being drained. The fall in pulse rate following the operation was less noticeable with this animal than in the two previous observations, evidently being interrupted during the wound infection. The same is true of the carbon-dioxide output which fell, however, in the course of two weeks to considerably lower levels than had hitherto been observed. During the remainder of the time in which this factor was observed it returned practically to the preoperative level, though the pulse rate remained comparatively low.

On June 10, 1911, when the experiments in the respiration chamber were brought to an end, the weight of the dog was 5.5 kilograms, and he was in excellent condition. Both previous and subsequent to the operation he had seemed to be in rather an erotic state and continued so up to the last two months of his life.

On July 6, 1911, he was in fine condition, with a weight of 6.1 kilograms. On August 1 he was let out to run. During the last months he became very shy and hard to handle, and this condition increased until his death. He was no longer in the erotic state noted before. On Sept. 21, 1911, the animal was sacrificed under chloroform.

Autopsy, immediate. — The subcutaneous fat on the chest and abdomen was about one centimeter thick; otherwise there was nothing remarkable about the general appearance.

Head. — Operative wound cleanly healed. No evidence at the base of the brain of previous bleeding or infection. A thin layer of scar tissue was adherent to the base of the third ventricle; this was removed with the brain, cut in serial section, and examined microscopically. No discrete fragments of anterior lobe tissue were found in this examination, but in many sections a number of eosinophile staining cells, resembling those of the anterior lobe, were seen adherent to the remains of the pars intermedia which fringed the infundibulum. Very deep in the main scar tissue mass there were also a few small scattered collections of cells, which may or may not have belonged to the anterior lobe.

Chest. — Not remarkable.

Abdomen. — Not remarkable. The liver appeared normal, as did the pancreas and adrenals. Microscopical examination of these organs showed that while the liver was rather fatty the others were apparently normal.

Thyroid gland. — Only one thyroid could be found. This appeared to be of normal size. On microscopical examination this showed an apparently normal gland with rather an excess of colloid.

The genitalia appeared infantile. Both testicles were unusually small, especially the left one, from which a specimen had already been removed. On microscopic examination the section removed at the operation showed that the basal cells of the tubules were moderately active. A number of spermatids could be seen. The interstitial cells were fairly numerous.

The testicle removed at autopsy differed very markedly in showing a distinct atrophy and mucoid degeneration of the tubules in a disappearance of most of the basal cells and an evident absence of spermatids of all that were left.   Although there was a considerable increase of fibrous tissue about the degenerated tubules, there was no obvious alteration in the interstitial cells.

SUMMARY OF OBSERVATION XIII.

Almost complete hypophysectomy in a nearly adult male dog. No marked change in general appearance and only a slight increase in fat. Moderate fall in pulse rate, respiration rate, and in the carbon-dioxide output; the latter fell noticeably at first, but later returned to practically its preoperative level.   The sexual development which, at operation, was nearly complete, was slowly but noticeably altered.   The animal was sacrificed seven months after the beginning of the observation and when he was about fifteen months old.

TABLE 6.

Carbon-dioxide production — Observation XIII.

Experiment No.	Date	Body Temperature.[1] °C.	Pulse.[1]	Respiration.[1]	Body Weight. Kilos.	Temperature of Apparatus During Quiet Periods. °C.	Carbon Dioxide Produced During Minimum Activity.			Remarks.
							Duration of Quiet Periods. Minutes.	Per Hour. Grams.	Per Kilogram Per Hour. Grams.	
1....	1911. Feb. 28	39.2	60	7	4.5	27.2–28.2	30	3.09	0.687	Animal given 150 grams bread soaked in 300 cubic centimeters of water and 50 grams meat each day. Fed after the experiment on experimental days. Nitrogen equivalent of food, 4.9 grams.
2....	March 2	38.2[2]	58	8	4.58	25.8–25.5	30	3.76	0.821	Record shows that dog was never quiet. Whining and moving in preliminary period.
3....	3	38.7	54	8	4.6	25.8–27.0	30	3.17	0.689	Whined and moved a good deal when placed in apparatus. Became so excited in second half-hour period that experiment was stopped.
4....	6	9. 0	60	11	4.63	25.0–25.1	30	3.52	0.760	
5....	7	38.8	50	7	4.5	26.4–26.7	60	3.32	0.738	
6....	8	39.0	60	10	4.6	26.1–25.8	90	3.57	0.776	

										Remarks
7...	10	38.8	56	8	4.63	24.5–26.0	90	3.48	0.752	Found asleep at the end of the experiment, and had to be awakened.
8...	11	39.0	57	11	4.67	27.3–26.8	30	3.56	0.762	
9...	13	38.8	60	10	4.7	25.6–26.0	60	3.10	0.660	Dog hard to quiet before the experiment; yawned a good deal and would not lie still for any length of time; was quiet during the experiment.
10...	" 15	39.0	66	11	4.8	26.2–27.3	90	3.84	0.800	Dog was restless and hard to quiet before the experiment.
11...	" 16	39.0	60	14	4.8	27.2–26.7	60	3.40	0.708	
12...	" 17	38.7	60	10	4.73	26.7–27.0	60	3.58	0.757	Vomited food on night of March 17.
13...	" 22	38.7	50	8	4.6	27.0–26.7	63	2.93	0.637	Reported on March 20 as not eating well.
14...	" 23	38.7	54	10	4.6	27.4–26.8	60	3.03	0.659	
15...	" 25	39.0	65	10	4.6	26.3–26.7	103	3.03	0.659	
17³...	" 28	38.4	49	8	4.7	25.7–26.7[4]	60	3.77	0.802	
18...	" 29	38.4	46	8	4.65	26.0–26.4	60	3.54	0.761	
19...	" 30	38.5[6]	50[5]	8[6]	4.7	25.?–25.8[8]	60	4.55[7]	0.968[7]	Injection of about .1 gram of boiled anterior lobe extract about one hour before the experiment. Very sensitive after the experiment and yelped on being touched. Sensitiveness had disappeared on April 2.

TABLE 6. — Continued.

Experiment No.	Date.	Body Temperature.[1] °C.	Pulse.[1]	Respiration.[1]	Body Weight. Kilos.	Temperature of Apparatus During Quiet Periods. °C.	Carbon Dioxide Produced During Minimum Activity.			Remarks.
							Duration of Quiet Periods. Minutes.	Per Hour. Grams.	Per Kilogram Per Hour. Grams.	
21⁸...	1911. April 4	39.2	64	11	4.6	25.8–25.0	90	4.03	0.876	On April 1, 130 gms of aft er meal was ted for the bread in the diet, the nitrogen equivalent of 6d being 4.8 grams. tied on ,pril 3, a non-inflammatory bes nd at site of jin. Animal te ll Section of right tide tal en. thr quiet on on April 3, but nd morning of April 4. Was lhd vell on April 4 and April 5.
22...	6	38.8	56	10	4.8	25.4–25.0	60	3.85	0.802	
23...	7	38.4	48	8	4.8	25.0–27.0	150	4.53	0.944	Possibly food was eaten within a few hours of the experiment. Animal lay still and did not want to move when chamber was opened. Sat half up as if stiff. Seemed rather sleepy.
24...	8	39.3	66	14	4.84	27.0–26.0	60	5.03	1.039	Vomited a little on April 7. Food eaten within a few hours of the experiment. Respirations recorded during first half hour were: 14, 18, 20, 22.

									Remarks
25...	" 11	38.6	58	12	4.8	27.0–26.8	70	4.02	0.838	ossibly ood was eaten within a few hours of the ...t. Average respiration during ...t, 10.
26...	" 12	38.7	53	10	4.78	26.6–27.0	90	3.65	0.764	During ...riment, average respiration 8.
27...	" 14	38.4	52	9	4.7	27.3–27.0	120	3.69	0.785	During experiment, average respiration 10.
28...	" 18	38.8	50	10	4.8	25.8–26.0	22	3.66	0.763	On ...ril 16, dog again given 150 grams bread in the diet, nitrogen equivalent of ...d being 4.9 grams. Ate well after the change. During recorded period average ...ion 8.

Operation, 10.30 A.M., April 21, 1911.

									Remarks
29...	April 24⁰	39.1	68	12	5.0	28.0–27.4	60	3.82	0.764	Given ¼ gram morphine at 10 A.M. Rather dull at 3.30 P.M. Walked about. Urinated. Ate all bread and meat (about the usual amount). Drank much water on April 22. Wound looked well. Pupil of left eye dilated; a little exophthalmus and eye reddened. / Expiration rather spasmodic; otherwise normal condition. Ate all food on April 23. During experiment, average respiration 12.
30...	" 25	38.9	45	8	4.9	26.8–26.3	90	3.02	0.616	Ate meat but only about 50 grams bread on April 24. During experiment, average respiration 8.

TABLE 6. — *Continued.*

Experiment No.	Date	Body Temperature.[1] °C.	Pulse.[1]	Respiration.[1]	Body Weight. Kilos.	Temperature Apparatus During Quiet Periods. C.	Carbon Dioxide Produced During Minimum Activity.			Remarks.
							Duration of Quiet Periods. Minutes.	Per Hour. Grams.	Per Kilogram Per Hour. Grams.	
31...	1911. April 26	38.8	62	11	5.1	27.2–26.6	90	3.53	0.692	Two small suppurating areas in wound, one on top of head, one at posterior part over neck. During experiment, average respiration 10.
32...	" 27	38.8	58	10	5.1	26.5–26.5	90	3.84	0.753	Wound still a little infected. During experiment, average respiration 8.
33...	" 29	38.2	42	8	5.14	26.3–27.0	120	3.37	0.656	During recorded periods, average respiration 8.
34...	May 1	38.2	36	7	5.3	26.0–26.3	60	3.50	0.660	During recorded periods, average respiration 7.
35...	" 2	38.7	45	10	5.3	25.7–25.9	60	3.21	0.606	During experiment, average respiration 7.
36...	" 4	38.7	48	8	5.2	27.0–26.8	30	2.97	0.571	During recorded period, average respiration 7.
37...	" 5	38.6	54	9	5.2	26.3–26.1	60	3.02	.581	During recorded periods, average respiration 7.
38...	" 6	38.5	50	10	5.2	27.1–26.9	60	2.74	0.527	Respiration in last half hour was 7.
39...	" 9	38.4	44	8	5.3	25.8–26.9	90	3.50	0.660	During recorded periods, average respiration 7.
40...	" 10	38.3	45	9	5.3	25.1–27.4	120	3.61	0.681	During recorded periods, average respiration 7.

										Remarks
41[10]	" 12	38.3	45	10	5.4	27.0–27.3	120	3.88	0.719	Average respiration in first half hour of the experiment was 8.
42	" 13	38.5	42	10	5.4	25.8[11]—	120	3.93	0.728	Recorded respiration during first and third half hours of the experiment, 8.
43	" 15	38.3	43	8	5.4	26.0–27.0	120	3.45	0.639	
45[12]	" 18	38.4	50	10	5.5	26.2–27.0	60	3.68	0.669	During first half hour of experiment, respiration was 7. During latter part of the experiment there was shivering.
46	" 20	38.7	48	10	5.5	27.0–28.0	120	3.73	0.678	During first half hour of experiment, average respiration 8. Shivered before the experiment and was hard to quiet at temperature of 25° C.
47	" 22	38.7	52	11	5.5	28.2–28.4[14]	90	3.72	0.676	During last half hour of experiment, average respiration 10.
49[13]	" 26	38.5	44	10	5.5	29.3–29.1	56	3.64	0.662	Dog shivered some before the experiment at temperature of 26° C.
50	June 2	38.7	47	8	5.45	31.2–31.0	30	3.64	0.668	The recorded period was about two hours after the observation of temperature, etc. Dog was very quiet at first at temperature of 28° C. Sudden shivering which continued till temperature was at 30° C. Throughout the recorded period, the respiration was 8.
51	6	38.8	44	7	5.4	30.0–31.0	50	3.30	0.611	

TABLE 6. — *Concluded.*

Experiment No.	Date.	Body Temperature.[1]	Pulse.[1]	Respiration.[1]	Body Weight.	Temperature of Apparatus During Quiet Periods.	Carbon Dioxide Produced During Minimum Activity.			Remarks.
							Duration of Quiet Periods.	Per Hour.	Per Kilogram Per Hour.	
		°C.			Kilos.	°C.	Minutes.	Grams.	Grams.	
52...	1911. June 10	38.5	46	10	5.5	29.7–29.3	90	3 45	0.627	Average respiration during experiment, 8. On July 6, the dog was reported in good condition and weighing 6.1 kilograms. Let out to run on August 1. Weighed 6.1 kilograms on September 21.

[1] By ... ts ... we taken before the dog entered the apparatus unless otherwise noted.
[2] Body ... rte tal en ... for the ... int.

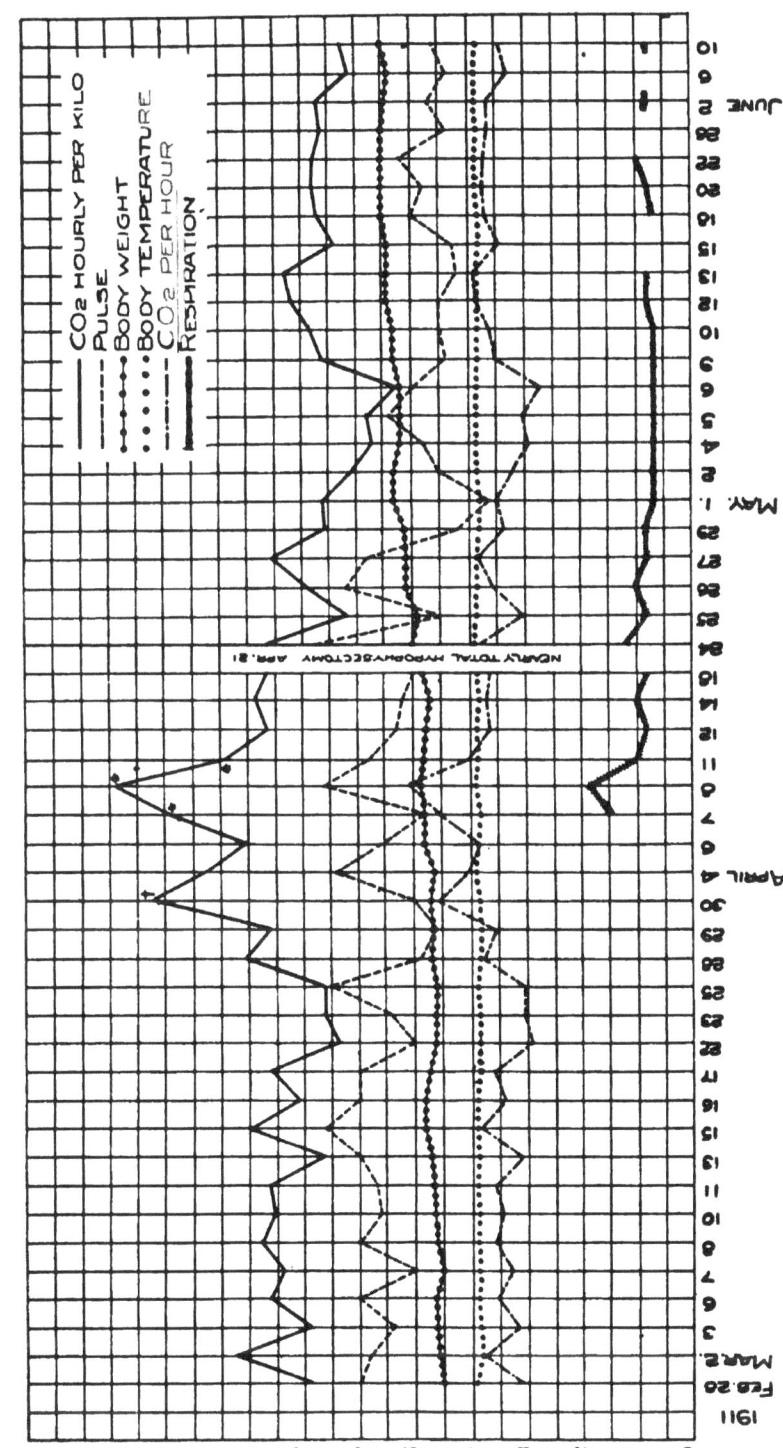

CHART 4

CHART IV. Observation XIII. — The curves record the carbon dioxide per hour per kilogram of body weight, the pulse rate, the body weight in kilograms, the body temperature, the total carbon dioxide per hour, and the respiration rate in the respiration chamber. The experiments to determine the carbon-dioxide output were not of uniform length. Only periods in which the

OBSERVATION XIV. *Description of subject.* — Boston terrier puppy. Female. Weight, 3.4 kilograms. Age, about four months. Rather fat; very active.

Before the operation. — This animal was kept under observation previous to the operation from May 25, 1911, to June 21, 1911. The first few days were devoted to training the dog to lie still, which was found to be somewhat difficult. She soon became accustomed, however, to the respiration chamber. During the preliminary period, the appetite was voracious. At first a ration was given of thirty grams of meat and one hundred grams of bread, but two weeks before operation the amount of bread and meat was increased and the same ration was given as for the somewhat larger animals. As the dog was kept in a large cage and not in the metabolism cage, it exercised rather more than the previous subjects. The weight increased from 3.4 kilograms to 4.2 kilograms. The pulse ranged from 104 to 140, and the respirations from 22 to 40. The body temperature was always 39° C. or over.

Operation, June 21, 1911. — Total removal of the gland without hemorrhage by the usual procedure. The animal made a good recovery from ether and on the following day seemed lively and well, except for considerable polyuria and polydipsia.

After operation. — On June 23, the second day after the operation, the weight had fallen to 3.8 kilograms, although the animal had eaten as usual. From this time on, the weight steadily rose to 5.2 kilograms, this being the weight at the time of death.

The fall in the pulse and respiration rates, temperature, and carbon-dioxide output was almost immediate and very marked. After the first week in which the changes occurred, there was practically no further alteration in these factors until just before death three months later.

In the latter part of July, the animal, which had then grown very fat and unwieldy, began to suffer from a thickening and scaling of the skin and falling out of the hair. This condition persisted in spite of treatment with sulphur ointment. Aside from this, the dog seemed well and as active and playful as ever. On account of its fat and prematurely old appearance, its playfulness seemed particularly silly, yet its intelligence was apparently normal. From the early part of July until the end of the experiment .5 gram of dried powdered extract of the whole gland was given daily with the food, but without apparent effect.

During the entire experiment, the weather was unusually hot, a condition which may have agreed with the animal for, with the first advent of colder weather, certain very marked changes occurred. The 13th and 14th of September were both cold days (about 0° C.), and the laboratory became considerably chilled. On September 15, the animal seemed very feeble and did not eat. On September 16, she seemed rather stronger and made attempts to jump about as usual. The rectal temperature was found to be 33.6° C., and the pulse 68; the dog was shivering and could not be properly quieted. On September 17, there was no change. The body temperature was 32° C.

On September 18, the rectal temperature in the morning was 32° C., but was raised by heating the room to 36.2° C. Both the pulse and respiration rose slightly with the rise of temperature, and the carbon-dioxide output rose considerably; nevertheless, the animal died within one-half hour of the conclusion of the experiment. During the last four days of life the dog drank water, but ate no food. There was no loss of weight, which remained 5.2 kilograms.

Autopsy, twelve hours post-mortem. — The animal presented the appearance of an old, fat dog. The hair was coarse and thin. The skin was thickened and scaly and hung in folds about the face, neck, and the various joints of the body. The subcutaneous fat on the abdomen and chest was two centimeters thick, and the intra-abdominal fat was excessive in amount.

Head. — Wound of operation cleanly healed. Base of the brain showed no evidence of blood clot or infection except for an encapsulated collection of leucocytes, microscopic in size, which was discovered in the serial sections of the base of the brain. There were no visible remains of the hypophysis, but the dural covering of the sella turcica was dissected and removed with the brain for microscopic examination. Serial sections of this tissue cut through the dura and the base of the third ventricle showed no definite fragments of anterior lobe. There were found, however, in association with the fringe of pars intermedia surrounding the infundibulum, a few eosinophile staining cells resembling those of the anterior lobe.

Chest. — Not remarkable.

Abdomen. — Liver " nutmeg." The pancreas and adrenals appeared normal and showed nothing striking on macroscopic examination. Sections of the adrenals and pancreas gave evidence of no tangible alteration. There was moderate passive congestion of the liver.

Thyroids. — By mistake, these were not saved.

Genitalia, undeveloped. Ovaries very small and soft. On microscopical examination there were found a few small undeveloped follicles. These consisted of simple groups of cells without characteristic arrangement.

SUMMARY OF OBSERVATION XIV.

Female puppy, four months old. Complete removal of the hypophysis, followed by a striking fall in pulse and respiration rates, temperature, and carbon-dioxide production. Change in two months to the appearance of an old, fat dog, but without alteration in the disposition of the animal. During the last three days of life the temperature of the animal was 5° to 7° below normal. During this period the metabolism was very low, with a rise just before death corresponding to the artificially-raised temperature, — practically the existence of a cold-blooded animal. The duration of life after operation was about three months and the age at death eight months. No cause of death aside from the absence of the hypophysis was discovered.

TABLE 7.

Carbon-dioxide production — Observation XIV.

Experiment No.	Date.	Body Temperature.[1] °C.	Pulse.[1]	Respiration.[1]	Body Weight. Kilos.	Temperature of Apparatus During Quiet Periods. °C.	Carbon Dioxide Produced During Minimum Activity.			Remarks.
							Duration of Quiet Periods. Minutes.	Per Hour. Grams.	Per Kilogram Per Hour. Grams.	
3...	1911. June 1	39.4	120	30	3.6	26.9–28.6	60	3.27	0.908	Diet ... ed of 100 grams ... 200 gms of ..., nitrogen being 3.1 grams; ... on experimental days. ... Did ... or ... 23 ... period, 24.
4...	" 3	39.4	130	40	3.6	27.3–27.5	75	3.49	0.969	Average respiration in first half hour, 26. Respiration in preliminary period, 30–35. Cover was not put on apparatus till 21 minutes after animal entered.
5...	" 5	39.3	120	26	3.6	26.4–27.2	90	3.21	0.892	Average respiration in first period, 22; in second period, 20.
6...	" 7	39.2	116	24	3.7	25.8–26.4	30	3.58	0.968	Bread in diet was increased on June 6 to 150 grams and water to 300 cubic centimeters, nitrogen equivalent of food now being 3.8 grams. Average respiration 21.

7...	" 9	39.0	120	24	3 75	26.2-26.4	30	3.58	0.955	Meat in diet increased to 50 grams on June 8, nitrogen equivalent of food now being 4.9 grams.
8...	" 12	39.3	110	30	3.8	25.5-26.7	50	3.52	0.926	Was fed at 7 A.M., June 11; not fed again until after this experiment.
9...	" 13	39.4	110	24	3.8	26.3-26.0	30	3.46	0.911	
10...	" 15	39.4	120	26	4.0	25.5[8]	44	3.94	0.985	Average respiration first half hour, 21.
11...	" 16	39.3	120	22	4.1	26.0[3]	28	3.64	0.888	Average respiration 19.
12...	" 20	39.4	104	24	4.2	27.9-28.3	30	3.74	0.890	Respiration during period, 20.

Operation, June 21, 1911.

13...	June 23	39.5	110	14	3.8	27.0-27.0	30	3.32	0.874	Dog reported in good condition on June 22, body weight, 4.05 kilos; excess urine on night of June 21. Ate all food on June 22 and drank all water. Very pronounced polydypsia and polyuria. Well and lively on June 23. Average respiration 14.
14...	" 24	39.3	116	25	4.4	27.2-27.8	90	4.61	1.048	Drank large quantity of water in preceding 24 hours. Ate food on June 23. Abdomen looks full. Respirations rather violent, i.e., forced respiration. Average respiration in first half hour measured was 23.
15...	" 26	38 8	80	16	4.3	26.0-27.0	84	3.55	0.826	Good condition. Seems perhaps a little less lively. Average respiration 13.

TABLE 7. — *Continued.*

Experiment No.	Date.	Body Temperature.[1] °C.	Pulse.[1]	Respiration.[1]	Body Weight. Kilos.	Temperature of Apparatus During Quiet Periods. °C.	Carbon Dioxide Produced During Minimum Activity. Duration of Quiet Periods. Minutes.	Per Hour. Grams.	Per Kilogram Per Hour. Grams.	Remarks.
16...	1911. June 27	38.6	75	17	4.4	26.0–26.0	60	3.56	0.809	Average respiration, first period, 14; last two periods, 13. Wound clean. For the last few days has been evidently thirsty. Keeps licking lips.
17...	" 28	38.4	68	16	4.5	25.8–27.5	81	3.30	0.733	In good condition. Lively. Average respiration in first hour, 14.
18...	" 29	38.4	68	15	4.5	27.5–27.7	47	3.36	0.747	In fine condition. Average respiration, 1st period, 14; second period, 12.
19...	" 30	38.6	65	13	4.6	27.2–27.3	60	3.17	0.689	
20...	July 8	39.0	72	19	4.9	28.0–29.1	90	3.93	0.802	Respiration in first half hour, 15. Reported in fine condition on July 5, and weighing 5 kilos. Is obviously getting fat and seems to be growing larger also.
21...	" 14	38.5	66	15	5.1	29.2–29.6	90	2.98	0.584	Dog obviously much fatter.
22...	" 29	38.3	70	17[4]	5.2	28.5–28.5	30	3.72	0.715	Given whole gland, 0.5 gram since July 17. Has had eczema for last week. Is getting very fat. Has general appearance of an old, fat dog. Treated with sulphur ointment with considerable improvement. Weighed 4.9 kilos.

23...	Sept. 16	33.6	68	12[5]	5.2	27.0–27.7	30	2.76[6]	0.531	Did not eat on September 15; seemed very feeble, coincident with cold weather. Responded to petting and tried rather heavily to jump about.
24...	" 18 A.M.	32.0	52	8	5.2	27.6–29.0	150	2.00	0.385	Body [illegible] on Sept[illegible] 17 was 32° C. Drank water [illegible] of September 18 [illegible] entering [illegible]. Body of dog [illegible] the [illegible]; [illegible] in age. [illegible] barely [illegible]. Several [illegible] of the [illegible] at 1.55 [illegible] gave 15. He was [illegible] and [illegible]. No [illegible], no shi[illegible]ing. Temperature of [illegible] was 30° C.
25...	" 18 P.M.	36.2	60	14	5.2	30.3–30 2	120	2.92	0.562	Quiet period began at 3.21 P.M. and experiment ended at 5.21 P.M. Body temperature when dog left the apparatus was 36.8° C.; pulse, 96; respiration, 22. On being placed on side on the floor, motions with his legs as though running. Would not drink. Died within half hour after leaving apparatus.

[1] Body [illegible], [illegible] and [illegible] ates were taken before the dog entered the apparatus unless otherwise noted.

[2] [illegible] at [illegible] of [illegible].

[3] [illegible] at [illegible] of [illegible].

[4] [illegible] the l[illegible] of the [illegible], *i.e.*, at 10.45 A.M.

[5] [illegible] as [illegible] in the last [illegible] of the [illegible].

[6] [illegible] in the [illegible] was 1.70 [illegible] 1.70 grams respectively or 3.40 grams per hour and .654 gram [illegible] kilogram of [illegible] weight [illegible]. [illegible] not [illegible], owing to shivering. The temperature of the apparatus in these two [illegible] was 25.1–27.0° C.

CHART 5.

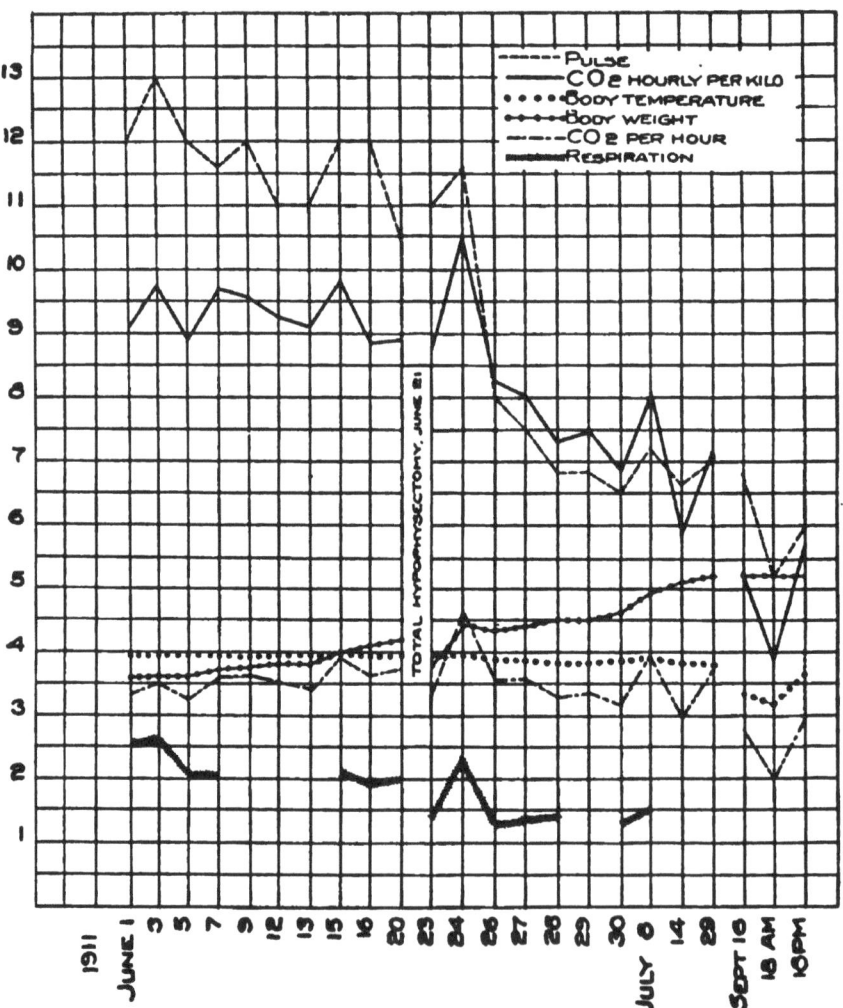

CHART V. Observation XIV. — The curves record the pulse rate, the carbon dioxide per hour per kilogram of body weight, the body temperature, the body weight in kilograms, the total carbon dioxide per hour, and the respiration rate in the respiration chamber. The experiments to determine the carbon-dioxide output were not of uniform length. Only periods in which the muscular activity of the animal was at a minimum were chosen for record. (See Table 7.)

DISCUSSION OF RESULTS.

The main purpose of this investigation was to determine the influence of hypophysectomy on total metabolism; as the investigation progressed, however, and each subject was carefully studied, it was soon seen that in any discussion of the results it was better to consider the observations individually than to discuss generally the fundamental principles involved.

We arrived at this conclusion after considering a number of factors. In the first place, this investigation covered a period of approximately two years; during this time there were many unavoidable and essential alterations in both the technic and the care of the animal, due to a larger insight into the general problem of experimental technic along these lines. Among these changes, one of the most important was the regulation of the temperature of the respiration chamber. In the earlier experiments, the need for the careful regulation of the temperature of the environment was not sufficiently realized, but later this point was emphasized. Furthermore, we were at first inclined to take too much for granted in regard to the metabolism of the dogs in the preoperative period, and adequate controls of the earlier experiments are unfortunately absent. Finally, alterations in the principle of determining the carbon-dioxide production, as well as in the measurement of the muscular activity, were made about the middle of the observation period. Although these latter alterations were fundamental, nevertheless, by thorough test, it was shown that the advantages lay in increased facility of operation rather than in increased accuracy.

The animals selected were of varying ages and hence we might reasonably expect differences in the effects following the removal of the hypophysis. Until the autopsies were performed and serial sections made, however, we were never sure of the completeness of removal, and it soon became apparent that animals differed markedly in reaction after the removal of the hypophysis. Taking these points into consideration, therefore, it is clear that, first, each case should be discussed by itself, and finally, a general summary should be made of the fundamental points upon which light is definitely thrown by the investigations.

Observation IX. — The series of experiments on Observation IX. must be considered imperfect for two reasons: first, because the experiments were made at a temperature at which the dog was constantly shivering and, second, because of the faulty method of interpreting the animal's activity in the respiration chamber; these two difficulties must be considered together. Almost all of the experiments were made at a temperature below 19° C., some as low as 13° and 14° C. At this temperature the animal's muscular activity was far from being at a minimum. Furthermore, instead of the carbon-dioxide output of the animal being determined in short periods in which the exact muscular activity was noted, it was determined, in all but one case, in experiments covering a period of three hours, and the muscular activity was interpreted for the whole period as slightly active, fairly quiet, etc. As a result of these defects, the only inferences which may properly be drawn from this observation are the following:

The effect of nearly complete removal of the hypophysis in this case was to cause the pulse rate to fall from an average of possibly 110 before operation to an average of nearly 60 after operation. At the same time the respiration rate underwent almost as great a fall. In the course of about three weeks the carbon-dioxide output was diminished by the removal of the gland to a considerably lower level than that established before operation, but unfortunately the number of experiments in the respiration chamber before operation are so few that they hardly seem sufficient to establish a proper base line for this factor. This discrepancy is due to the fact that the first few experiments in the respiration chamber were considered failures on account of the dog's evident annoyance and restlessness while in the apparatus. On the other hand, the pulse rate may be said to have been obtained with reasonable accuracy, and a number of observations which are not here recorded confirm the level which the accompanying table establishes.

It will be noted from an examination of the chart that for several days following operation there was a rise in pulse rate, together with a considerable loss in weight. It was only some five or six days after removal of the gland that the pulse rate fell below the preoperative level. The falling off in the carbon-dioxide production was more tardy. On the day following operation it was very low, then rose on the next few days to a level which, if we take into consideration the animal's diminished muscular activity in the respiration chamber, was about the same as before operation. It was not until some two weeks later that it was decidedly and permanently lowered. At this time, with the exception of the change in pulse rate and in the carbon-dioxide output, the animal was in practically the same condition as before operation.

Aside from the initial loss in weight following operation, the dog remained until the end of the observation of essentially the same body weight as at the beginning; as this dog was practically full grown we may fairly conclude that we are dealing with essentially the same number of cells, particularly of active protoplasmic tissue. In view of this fact the observations in regard to the carbon-dioxide production are especially significant and interesting. Toward the end of the observation the carbon-dioxide production fell to sixty or seventy per cent of its original value. Indeed, on March thirtieth it was exactly one-half the output on February eighteenth, five days before operation. This noticeable fall is in accord with the belief that the metabolism is essentially proportional to the pulse rate since we find that this fall in metabolism paralleled the fall in the pulse rate. As the dog did not materially alter in weight, we find, likewise, that the carbon-dioxide production per kilogram of body weight per hour underwent the same relative alteration. Only once after operation does the carbon-dioxide production per kilogram of body weight even approximate that before operation, i.e., on the second of March, when it was

1.01 grams per kilogram per hour. After that there was a regular, persistent fall.

The only abnormality on the part of the animal was a more or less severe diarrhea, apparently due to the steady diet of dog biscuit upon which the animal was kept, and which finally became so annoying that it was necessary to abandon the experiments. It is possible that this condition may have had some effect on the general metabolism, though the steadily increasing weight of the animal in the first few weeks following the operation would tend to contradict this suggestion.

An examination of the temperature of the respiration apparatus shows that, in general, the temperatures during the preoperative period were slightly lower than those in the post-operative period. From Rubner's [35] experiments one would expect to find a somewhat higher metabolism with a low temperature, but the differences in the temperature of the apparatus are by no means sufficient to account for the enormous variations in the metabolism. Thus on February sixteenth, when the dog was slightly active, the temperature of the apparatus was 17.4° C. to 21.0° C., and the carbon dioxide was 7.17 grams per hour. On March twenty-second, when the dog was designated as being active, the temperature was 17.1° C. and 19.8° C., and the actual carbon-dioxide output was only 5.02 grams per hour. While, therefore, the somewhat higher metabolism during the preoperative period may properly be considered with some reserve, we believe that the numerous individual experiments, together with records of activity, indicate that, in general, the two periods can fairly be compared so far as the temperature of the apparatus is concerned, and that the actual fall in carbon-dioxide production is a true measure of the effect of the nearly complete removal of the hypophysis.

Observation X. — The results of this observation, which was made as a control, are given in Table 4. In all, seven experiments were made; here again, while the temperature of the apparatus was not maintained at the high point which

we have found most desirable, nevertheless, in the preoperative and post-operative periods the temperatures were essentially alike, and hence they can fairly be compared.   Unfortunately, the selection of individual periods showing minimum
muscular activity was not made, and we have only the
averages of three-hour experiments with an estimation of
the activity of the dog.   The records show that during the
preoperative period the dog was distinctly more active than
during the post-operative period.   There was no pronounced
effect upon the body temperature, respiration, or pulse rate
as shown by the few records made.   The body weight at the
end of the experiment was essentially the same as at the
beginning.   The considerable variations in the carbon-dioxide production per hour, both before and after operation, are
all easily explained by the variations in activity observed in
the dog, and while the observation has not the extent that
one could wish, nevertheless it seems reasonable to conclude
that the operation as such had no material effect upon the
metabolism of the animal.

Since no important structures were removed or injured,
the operation can be looked upon as neither more nor less
than an ordinary surgical opening from which dogs usually
recover without the slightest disturbance in metabolism or
general health; hence, the observation is fully in line with
the experience of practically all observers in which a simple
operation is carried out.   It merely demonstrates that a dog
of characteristics similar to the one described under Observation IX. suffers no obvious change either in the pulse rate
or in the carbon-dioxide production from exposure to an
operation in which all of the manipulation required for the
removal of the hypophysis without removal of the gland
itself was carried out.

Observation XI. — The earlier portion of this observation
suffered, though in a less degree than Observation IX., from
the effect upon the muscular activity of the animal of the
low temperature of the respiration chamber.   An attempt
was made to keep the temperature not far from 20° C., and,

save in the excessively warm weather in June and later in the summer, the temperature was always between 20° and 25° C. Beginning with the early fall, care was taken to keep the temperature not far from 25° C., this being usually the optimum temperature for dogs. Nevertheless, it will be noticed that in the column headed remarks in Table 5, and also in Chart 3, attention is called to this abnormal factor during the months of September and October. Subsequently, however, the experiments are free from this possible source of error. In selecting the quiet periods for comparison, the inspection of the kymograph records was very thorough, and only those periods in which there were no movements, or very slight movements, are here included. It may conservatively be stated that the number of periods used for comparison represents only about one-third of those actually obtained, the others being discarded on account of irregularities in the muscular movements of the animal. In interpreting the values shown in Table 5, therefore, we may consider that the variations in muscular activity may be excluded.

The subject of this observation was probably somewhat younger than the subject of Observation IX. and in many ways was a very different animal. She was thin and nervous, in constant activity, and, on the whole, hard to train. In all, eighty-nine experiments were made with this subject. As the operation was performed just prior to the seventh experiment, it will be seen that the animal, after a nearly complete removal of the hypophysis, was under careful observation for over one year.

An examination of the body temperature records shows no material alteration. Before operation the temperature was either 39° C. or above; after operation, it rarely exceeded or, indeed, reached 39° C., and never went below 38° C. The temperature limits, therefore, cannot be interpreted in any way as indicating that the operation produced an abnormal effect upon the temperature regulation of the body with this particular animal.

The high pulse rate in the initial experiment, which was

undoubtedly due to the fact that it was the first time the dog had been in the respiration chamber, immediately fell in the two following experiments to 100, and thereafter remained quite constant at 86 to 82. After operation the high initial pulse rate fell rapidly to 60, and thereafter, until the end of the experiment, the pulse rate hovered between this point and 80, exceeding the latter only when, owing to low temperature or other exciting cause, the dog had been under muscular stimulus. Accordingly, we have here the characteristic fall in pulse rate after hypophysectomy.

The maximum fall in the pulse rate occurred a very few days after operation, but the corresponding fall in the carbon-dioxide output did not occur for nearly a month. This discrepancy, which, however, is noticed in a lesser degree in other observations, may perhaps be accounted for by the fact that the dog was in a particularly abnormal condition in the first few experiments following operation. The bowels were constantly loose, and there was considerable cough; in fact the dog's condition became so alarming two weeks after operation that it was considered prudent to discontinue the experiments temporarily and send the animal to the country for a couple of weeks. Upon her return in excellent condition and with a weight almost exactly the same as the preoperative weight, it was found that the very low carbon-dioxide output had become a constant phenomenon.

The general appearance and condition of the animal remained essentially the same throughout the whole experiment, except that as it grew older it appeared more infantile and silly on account of its rather prematurely old face in combination with its very puppy-like behavior. Probably the dog gained very little in active protoplasmic tissue from the time of operation till its death. Although no measurements were made of its size, its only obvious change was the laying on of fat. It would perhaps be fairer to assume that in actual flesh outside of fat the dog gained possibly one-half kilogram. This statement is borne out by the fact that when the observation had been continued for about a year, and the animal had returned from a trip to the farm where she had been running loose for several months, the weight was five kilograms as compared with four and two-tenths kilograms at the very beginning of the experiment, and the animal, to the eye at least, was thin. Similarly it can be seen that when the dog was sacrificed at the end of the observation, her weight after several months of activity outside of the metabolism cage had fallen to four and eight-tenths kilograms, thus implying a gain in weight, other than fat, of six hundred grams. It is fair to assume, therefore, that although the animal had gained somewhat in muscle and bone, yet the amount of increase, which at one time had been as high as two kilograms with a maximum body weight of approximately six kilograms, was principally fat. The chart shows that the carbon-dioxide output per kilogram of body weight per hour, which fell off markedly within six weeks of hypophysectomy, remained at approximately the same level during the rest of the time the dog was under observation. The carbon-dioxide production per hour, on the other hand, although it fell off considerably at first, afterwards returned to approximately the preoperative level and remained there through the rest of the experiment. Admittedly the carbon-dioxide production for this animal is complicated somewhat by the fact that in the preoperative period the temperature of the chamber was somewhat lower than it was during the later experiments in the post-operative period. Nevertheless the temperature differences are not so great that a reasonable comparison cannot be made between the preoperative period and the post-operative period for several months, especially as during this period no shivering was observed, such as was noticed subsequent to operation.

It is apparent, therefore, that the greater part of the diminution in carbon-dioxide output is due in this observation to the laying on of inert fat; on the other hand, when we remember that the dog, which was only six months old at the beginning of the experiment and nearly two years old at the end, had probably. gained an appreciable amount of protoplasmic tissue, there is a distinct fall in the carbon-dioxide production to be accounted for. If the new tissue had participated in the metabolic processes, the carbon-dioxide production per hour ought also to have increased. On the contrary, it will be noted that at the end of the experiment it appears, if anything, to be a little lower than at the beginning. On April twenty-eighth, 1911, the animal was returned from the farm in good condition, but obviously thin, the weight having fallen from nearly six kilograms to five kilograms. At this time it was found that the carbon-dioxide production, both total and per kilogram of body weight per hour, had risen considerably above the average post-operative level and practically to the preoperative level. In the course of the next month, as the animal gained fat, the carbon-dioxide production fell off to the level maintained before the animal had lost its fat; this is very strikingly shown in the curve.

The experiments conducted in this month will therefore give a very good idea of the effect upon the carbon-dioxide production of the animal of the laying on of inert fat. We have, then, in Observation XI., two obvious changes in metabolism: first, the same, though less striking, fall in pulse rate and carbon-dioxide production as was seen in Observation IX.; and, second, a secondary falling off in the carbon-dioxide production (unaccompanied by any further diminution in pulse rate), which was probably due to the laying on of inert fat. It is hardly proper, therefore, to ascribe the later changes in the animal's metabolism to such conditions as increasing age and training, for, as we have already stated, the carbon-dioxide production per hour never rose above the level established before operation, though a tendency to rise above that level might be expected from the slight increase in weight of the animal. Since during the whole of the post-operative observation, the pulse rate only once rose above the level established within a few days of operation, it is further evident that the animal's metabolism was lower, for we have found in all of our experiments that when properly controlled, the pulse rate and the carbon-dioxide production follow each other very closely. Finally, we should hardly expect to find in this observation changes as striking as in Observation IX., for the autopsy in this case showed that a very definite but microscopical fragment of the anterior lobe of the hypophysis remained attached to the base of the brain.

Experiments 86, 87, 88, and 89 of Observation XI. were made toward the end of our experimental season, and hence may rightly be assumed as having the advantage of all the improved technic. These experiments show certain apparent anomalies which require further consideration. An examination of the pulse rate shows that it remained essentially constant from the twenty-eighth of April until the eighth of June. Nevertheless, during this time there was a falling off in the total carbon-dioxide production per hour which culminated in a very striking fall on the last day. From the well known relationship between the pulse rate and metabolism, one expects to find a fall from the thirty-first of May until the eighth of June, accompanying the drop in pulse from 70 to 64; but if the records for May 23 are examined, we find that the pulse rate was approximately the same as on June 8, namely, 66 as compared with 64; yet there was a fall in the carbon-dioxide production of .6 gram or about seventeen per cent; a perceptible fall in the respiration rate was also noticed. While we wish to reassert the importance of the relationship between the pulse rate and metabolism, it is necessary to take into consideration that the pulse rates that are obtained on animals are at best somewhat uncertain. The greatest care is taken to record those pulse rates prior to the experiment when the animal is completely relaxed and quiet, but from the intermittent records of respiration, which were counted while the animal was inside the chamber, it can be seen that there were, in most instances, enormous falls in the respiration rate after the animal entered the chamber. Thus, on May 31, while the respiration rate prior to the experiment was 24, one of the records obtained during the experiment was 13; again, on June 8, the respiration rate before the experiment was 20, while the average respiration recorded inside the chamber was 10. Although we have no direct observations of the pulse rate inside the chamber, owing to the difficulties of technic, it is reasonable to suppose that there are approximately corresponding drops in the pulse rate. Attempts to check this by records of pulse rate after the animal left the chamber were almost always futile, as the excitement incidental to closing the experiment and removing the animal from the chamber precluded careful observation. While, therefore, at first sight it would appear that these results are somewhat anomalous, nevertheless an inspection of these respiration rates prior to and during the experiment leads one to the conclusion that the pulse rates may only be taken as rough indices of the general trend of the katabolism.

An examination of these experiments shows, therefore, the futility of attempting to draw deductions from any single experiment, for if the values for the last experiment are taken by themselves, one could easily be led into very erroneous conclusions. Accordingly, we can make no further deduction from this series of experiments than that there was a noticeable, general fall in metabolism, both in the amount per hour and in the amount per kilogram per hour, from the twenty-eighth of April until June 8.

Injection and feeding of hypophyseal extract. — The results obtained with this animal by the subcutaneous injection and feeding of hypophyseal extract are not noteworthy in any respect. About one month after operation the animal was given subcutaneously .05 gram of powdered anterior lobe which had previously been boiled with a little water, this being followed by a rise in body temperature some hours later of .4° C. While it does not appear that the carbon-dioxide production was altered in the experiment immediately following the injection, there is seen in the next few days a distinct rise for which no other known factor can be held accountable. Several succeeding injections of double the amount produced a distinct temperature reaction, but no obvious change in the general metabolism. Six months after operation, the feeding of the hypophyseal extract was begun. Dry powdered anterior lobe was given in amounts of .5 gram a day for a week, after which the amount was increased to one gram a day. No obvious change in the animal's condition occurred, nor was there any alteration in the pulse rate or in the carbon-dioxide output. The feeding of this preparation was kept up until the animal was sacrificed nearly a year later.

We must conclude, therefore, that although the subcutaneous injection of extract of the anterior lobe caused a distinct temperature rise, neither the injection nor long continued feeding in such amounts as we used with the possible exception of the first injection previously referred to can be said to have had any appreciable effect upon the metabolism of the hypophysectomized animal. Although the tests of feeding and ingestion were incidental to the larger study of hypophysectomy, it is to be regretted that we did not make use of a fresh gland for this purpose, as, according to the results of Malcolm,[12] and of Thompson and Johnston,[13] a fresh preparation of the gland has a much more noticeable effect upon the metabolism than a dried preparation.

Observation XIII. — This observation differed from the previous ones in having a long preliminary period in which both the pulse rate and the carbon-dioxide output were very carefully established. It also differed in that the animal was a male and very nearly full grown. In fact, on account of the full development of this animal, we feared that the nearly complete removal of the gland would be fatal. This was not the case, however, and in many respects fewer changes followed the removal of the hypophysis than were apparent with the other dogs. In this observation, moreover, the arrangements with regard to the temperature of the environment and the movements of the animal in the respiration chamber were accurately controlled, so that no disturbing influences mar the accuracy of the results.

The early intermediate changes in the pulse rate and in the carbon-dioxide output are less striking in this case than in any of the other observations. The pulse rate was unusually slow before operation, ranging from 50 to 60, generally in the neighborhood of 50, and sometimes even a little slower. Following operation the pulse rate, after a preliminary high period, ultimately fell to a point considerably lower than that before operation. The respiration rate on the whole was also somewhat lower after operation. During the whole post-operative period, there was a progressive rise in body weight, amounting to approximately half a kilogram, or ten per cent of the body weight. The total metabolism, as indicated by the carbon dioxide per hour, did not undergo any material alterations, although the metabolism per kilogram of body weight per hour was distinctly lower on the average during the post-operative period than before operation. It is interesting to note that the characteristic fall in carbon-dioxide production, occurring usually about two weeks after operation, was here observed.

The most noticeable effects of hypophysectomy in this observation were the characteristic change in the pulse rate, a slight increase in body weight, no alteration in the carbon-dioxide production per hour, and a slight decrease in the carbon dioxide per kilogram of body weight. This latter decrease was by no means so large that it cannot be properly ascribed to the laying on by the animal of inert body fat. The sharply definite effect of hypophysectomy observed in this dog is that of a noticeable change in the pulse rate.

One peculiarity of the observation is the comparatively slight effect which a practically complete hypophysectomy produced upon a nearly full-grown animal. The fact that the dog showed fewer signs of the lack of hypophyseal secretion than any of the other subjects may be accounted for by the fact that there may in any animal be hypophyseal rests in the sphenoid bone and in the vault of the pharynx, which can hardly be detected without the most painstaking examination; or it may be true, as cited by Aschner,[23] that grown animals, provided they survive the removal of the hypophysis, present fewer changes than do animals hypophysectomized in infancy and early youth.

The changes in the sexual activity of this animal were particularly interesting. Before operation the dog appeared to be in rather an erotic state, and a section of one of the testicles was taken as a control of subsequent procedures. After operation the animal's condition remained about the same until within a month or two of his death, when he became normal or probably more backward than usual in sexual activity. At autopsy, indeed, he presented every appearance of an infantile condition. Examination of the testicle which had not been touched previous to operation showed very extensive fatty degeneration in the tubules with a disappearance of most of the spermatogenous cells and an increase of fibrous tissue between the tubules. It was evident that the testicle had undergone degenerative and retrogressive changes during the five months subsequent to the hypophysectomy.

Injection of hypophyseal extract. — In one respect this observation was very unsatisfactory. Previous to operation an injection of .1 gram of powdered anterior lobe was given; a pronounced temperature reaction was noted which we had supposed would appear only in hypophysectomized animals. The dog was made very sensitive by the injection and difficult to handle. When he was placed in the respiration chamber it was evident that not only did he profoundly feel the effect of the injection, but that his carbon-dioxide output was materially raised. The temperature reaction differed in no respect from that seen in other animals of the series, notably the subject of Observation XL, to which subsequent to hypophysectomy we had given subcutaneous injections of this extract. It would therefore appear that the preparation which we used was in some way toxic or else that the animal was perhaps abnormal from the beginning. This point deserves further study.

Observation XIV. — This observation is somewhat more complete than any other of the series. The subject was a female puppy only four months old, younger than any of the dogs previously used. She was kept under observation for about a month previous to operation, during which time the pulse rate and the carbon-dioxide output were very thoroughly established. The pulse rate was unusually high and the rectal temperature and respiration rates were also somewhat higher than with the other dogs. During this preoperative period, the animal gained .6 of a kilogram in weight; at the same time there was a moderate but steady increase in the carbon-dioxide production per hour, although the carbon-dioxide production per kilogram of body weight remained remarkably constant. This observation is therefore peculiarly valuable as indicating the effects of the steady growth of the animal under constant surroundings upon the carbon-dioxide production per hour, the curve as shown on the chart being parallel to that indicating the body weight. It will also be noted upon the chart that the fluctuations of the pulse rate and the carbon-dioxide production per kilogram of body weight per hour were approximately parallel and that the body temperature remained at a very definite level.

Following operation, which a subsequent autopsy showed to have been a complete removal of the gland, the change in the metabolism of the animal was particularly striking. Within five days the pulse had fallen from an average of 110 to 80, and within two more days, or a week after the operation, the pulse had fallen to 68, at which level it remained for the greater part of the time. This was the first animal in our series which showed a material fall in temperature after operation, the fall remaining not far from 1° C. with but slight changes until just before the end of the experiment. At this time, the characteristic fall incidental to cachexia hypophyseopriva was observed, the rectal temperature on two days being 33.6° and 32° C., respectively.

With the alteration in the apparatus for recording muscular activity, it was found that the tambour arrangement provided sufficient sensitiveness to give a graphic record of the respiration rate. Consequently, in this observation we have an admirable opportunity for comparing the respiration rate as determined before the experiment while the dog was held in the observer's lap, with the respiration rate during the experiment inside the respiration chamber. In the earlier experiments, we have no such control of the rates recorded by the observer outside the chamber, but in this observation we can speak of the respiration rate with a considerable degree of confidence. Although a number of extraneous factors, more or less uncontrollable, may influence decidedly the respiration rate of the animal before entering the chamber, the records of the respiration determined while the animal was inside the chamber have a peculiar value and significance; accordingly these records are plotted on the chart for this observation. A general tendency is shown toward a definite falling off in the respiration rate inside the chamber over that before entering the apparatus (see Table 7, page 478). While no definite percentage decrease can properly be computed, in the preoperative period we find the respiration rates as determined inside the chamber were approximately four to five per minute lower than outside. After operation, the differences between the rates inside and outside the chamber are, if anything, slightly less. The effect of the operation was to cause a noticeable fall in the respiration rate both inside and outside the chamber, the rate after operation being about two-thirds that prior to operation. This fall in respiration rate is reasonably proportional to the fall in pulse rate.

The carbon-dioxide production per kilogram of body weight showed a similar fall, although the total carbon-dioxide production remained approximately the same as immediately before the operation. In other words, up to the few days immediately preceding the death of the dog the total carbon-dioxide production remained at about the same or a little lower level than before operation and ceased to show the steady rise which was apparent before operation.

The experiments upon this animal after operation were begun during June, and the hot weather of July prevented us from making as frequent experiments as hitherto. During the few observations in July there was no appreciable change in the condition of the dog, although the weight steadily increased, until, at the end of the observation, it was over one kilogram, or twenty-five per cent more than on the day before operation.

As previously stated, this animal soon developed a prematurely old, fat appearance and in September, or about three months after operation, the animal showed definite changes in its general condition, and very acute changes in its reaction to its surroundings. With the advent of somewhat colder weather, the rectal temperature fell to 32° C.; this fall in body temperature was accompanied by a further falling off in the pulse rate and a very striking fall in the carbon-dioxide output. Although the rectal temperature of the animal was about 32° C. for several days, it was finally possible, by heating the surrounding atmosphere, to increase it from 32° C. to 36° C. Observations in the respiration chamber obtained at both of these temperatures show that there was a distinct increase in the carbon-dioxide output with the rise in the temperature of the animal. It is most singular that the reaction to the warm environment did not have as great an effect upon the pulse rate, although it had a profound effect upon the total katabolism, and likewise produced an enormous percentage increase in the respiration rate.

Since all of the experiments were made with a new and improved form of apparatus and at a temperature that had been found normal for experiments with dogs, and since only periods with the minimum activity are included, it can be seen that the observations before and after operation can be compared with the greatest degree of exactness. No animal that we have thus far operated upon has shown so noticeably the effect of the removal of the hypophysis as did this dog, — the lowered body temperature, pulse rate, respiration rate, carbon-dioxide production per hour, and carbon-dioxide production per kilogram of body weight per hour, all indicating a profound effect in retarding the metabolism of the dog. This effect reached its greatest moment on the eighteenth of September, when the body temperature of the dog fell to 32° C.

Of almost equal importance with regard to the effect upon the metabolism of the removal of the hypophysis is the observation made upon the animal when the temperature was sub-normal, and of the reaction when the environment was artificially warmed. While it has been repeatedly observed that cold-blooded animals vary in the production of carbon dioxide when the temperature of the environment is increased, we believe that this is the first instance that the metabolism of a warm-blooded animal during a period when the body temperature was 6 to 7° below normal has been compared with its metabolism when the temperature has been artificially raised to approximately normal. As a matter of fact, it can be seen that when the body temperature of this animal was artificially increased, the metabolism rose nearly to that shown two months previous, *i.e.*, on July fourteenth, when the total carbon-dioxide production per hour was 2.98 grams, and per hour per kilogram of body weight, .584 gram, while on September eighteenth, P.M., the total carbon-dioxide production was 2.92 grams, and per hour per kilogram of body weight, .562 gram. Apparently the slight, almost insignificant differences in the pulse rate, as recorded on September eighteenth in the morning and in the afternoon, are such as to militate strongly against the belief that the pulse rate is a general index of the total metabolism or muscular tone. It should be stated, however, that had it been possible to determine the pulse pressure by careful measurement of blood pressure, noticeable differences might reasonably have been expected between the period of low temperature and of high temperature. Unfortunately, all of our efforts to determine the blood pressure with dogs have been without avail, as the usual procedure of employing an artery is obviously precluded in observations of this kind. From these results, therefore, it is evident that hypophysectomy here has produced a condition of thermal regulation fully comparable to the poikilothermic animal.

It is unfortunate that the rapid post-mortem or possibly ante-mortem changes which occurred on account of the peculiar condition of this animal for three days preceding its death prevented an accurate, histological examination of its various organs. It is to be hoped that subsequently a more adequate study of such conditions can be made.

Feeding of extract of whole gland. — Beginning July 17, about a month after operation, and continuing until the death of the animal, .5 gram dried powdered whole gland was fed to the dog daily. No appreciable effect was observed, and this feeding certainly did not prevent the animal from dying with symptoms of lack of hypophyseal secretion. It is unfortunate, however, that in this observation, as in Observation XI., we did not make use of larger doses or possibly of fresh gland preparation, and we can only conclude that as far as our work goes it shows that the ingestion of hypophyseal extract of the kind and quantity we used has no appreciable effect upon the metabolism of the hypophysectomized dog.

SUMMARY.
General effects of hypophysectomy.

On the body weight. — A tendency to retard the normal growth of the animal. Gain in weight is principally due to the deposition of fat.

On body temperature. — A tendency to slightly lowered body temperature, which may result ultimately in a disturbance of the heat regulation sufficiently profound to produce a very marked drop in body temperature immediately prior to death.

On pulse. — A marked fall in pulse rate occurring a few days after operation and remaining at essentially the same level throughout the life of the animal.

On respiration. — A fall in the respiration rate approximately parallel and similar to that of the pulse rate.

On total metabolism. — A marked fall after operation of the total metabolism as measured by the carbon-dioxide production. The fall in carbon-dioxide production per kilogram of body weight per hour is still more noticeable owing to the deposition of inert body fat.

On development and general appearance. — The growth of young animals is checked and their infantile characteristics are preserved. The sexual activity, if not already developed, never develops. If nearly or quite established, it is profoundly affected. The tendency to the deposition of an excessive amount of body fat is sometimes accompanied by thickening of the skin and falling of the hair analogous to changes noticed after thyroidectomy. The change in appearance of older animals surviving a nearly complete removal of the hypophysis is hardly noticeable.

BIBLIOGRAPHY.

1. Crowe, Cushing, and Homans. Experimental hypophysectomy. Johns Hopkins Hosp. Bull., xxi, May, 1910, 127–169.

2. Oliver and Schaefer. On the physiological action of extracts of the pituitary body, etc. Journ. of Physiol., 1895, xviii, 277–279.

3. Howell, W. W. The physiological effects of extracts of the hypophysis cerebri and infundibular body. Journ. Exp. Med., 1898, iii, 245–258.

4. Schaefer and Magnus. On the action of pituitary extracts upon the kidney. Proc. Physiol. Soc., July 20, 1901, pp. ix–x in Journ. of Physiol., 1901–2, xxvii.

5. V. Frankl-Hochwart und Fröhlich. Zur Kenntnis die Wirkung des "Hypophysins" auf das sympathetic und autonome Nervensystem. Archiv. f. exp. Pathol., 1910, lxiii, 347–356.

6. Wiggers, C. J. The physiology of the pituitary gland and the actions of its extracts. Am. Journ. Med. Sc., 1911, cxli, 502–515.

7. Mairet and Bosc. Recherches sur les Effets de la Glande Pituitaire. Arch. de Physiol., 1896, 5ᵐᵉ série, viii, 600–613.

8. Salvioli and Carraro. Archivio per le Scienze Mediche, 1907, xxxi, 242–294. Cit. Delille.

9. Franchini, G. Die Funktion der Hypophyse und die Wirkungen der Injektion ihres Extraktes bei tieren. Berliner klin. Woch., 1910, xlvii, 613–617, 670–673, 719–723.

10. Sandri, O. Contribution à l'anatomie et à la physiologie de l'hypophyse. Arch. Ital. de Biol., 1909, li, 337–348; and Rivista di patologica nervosa e mentale, 1908, xiii, 518–550.

11. Hamburger, W. W. Action of extracts of the anterior lobe of the pituitary gland upon the blood pressure. Am. Journ. Physiol., 1910, xxvi, 178–180.

12. Malcolm, John. On the influence of pituitary gland substance on metabolism. Journ. of Physiol., 1903–4, xxx, 270–280.

13. Thompson and Johnston. Note on the effects of pituitary feeding. Journ. of Physiol., 1905–6, xxxiii, 189–197.

14. Cerletti, U. Effets des injections de suc d'hypophyse sur l'accroissement somatique. Arch. Ital. de Biol., 1907, xlvii, 123–134; and Rend. della R. Accad. dei Lincei, 1906, xv, 2ᵉ sem., série V, 142–151; 213–216.

15. Falta, Rudinger, Bertelli, Bolaffeo, and Tedesco. Ueber Beziehungen der inneren Sekretion zum Salzstoffwechsel. Verhand. des Cong. f. Inn. Med., 1909, xxvi, 138–149.

16. Schiff, A. Hypophysis und Thyroidea in ihrer Einwirkung auf den menschlichen Stoffwechsel. Wiener Klin. Woch., 1897, x, 277–285.

17. V. Moraczewski, W. D. Stoffwechsel bei Akromegalie unter der Behandlung mit Sauerstoff, Phosphor, etc. Zeitschr. f. Klin. Med., 1901, xliii, 336–360.

18. Delille, Arthur. L'Hypophyse et la Medication Hypophysaire. Paris, G. Steinheil, 1909.

19. Exner, A. Ueber Hypophysentransplantationen und die Wirkung dieser experimentellen Hypersekretion. Deutsch. Zeitschr. f. Chir., 1910, cvii, 172–181.

20. Schaefer, E. A. The functions of the pituitary body. Proc. Royal Soc., 1909, Series B, lxxxi, 442–468.

21. Paulesco. L'Hypophyse du Cerveau. Vigot Frères, Eds., Paris, 1908.

22. Reford and Cushing. Is the pituitary gland essential to the maintenance of life? Johns Hopkins Hosp. Bull., 1909, xx, 105.

23. Aschner, Bernard. Demonstration von Hunden nach Exstirpation der Hypophyse. Wiener Klin. Woch., 1909, xxii, 1730–1731. Ueber die Folgeerscheinungen nach Exstirpation der Hypophyse. Verhand. der Deutsch. Gesell. f. Chir., 1910, xxxix, 46–49.

24. Fröhlich, A. Ein Fall von Tumor der Hypophysis cerebri ohne Akromegalie. Wiener Klin. Rundschau, 1901, xv, 883–886.

25. Goetsch, Cushing, and Jacobson. Carbohydrate tolerance and the posterior lobe of the hypophysis. Johns Hopkins Hosp. Bull., 1911, xxii, 165–190.

26. V. Narbut. Inaug. Dissertation, St. Petersburg. Cit. v. Bechterew, Die Funktionen der Nervencentra, Jena, G. Fischer, 1909, 1215–1220.

27. Wolf and Sachs. Metabolism after hypophysectomy. Proc. Soc. Exp. Biol. and Med., 1910, viii, 36.

28. Kaufmann. Arch. de physiol., 1896, viii, 329.

29. Benedict and Homans. A respiration apparatus for the determination of the carbon dioxide produced by small animals. Am. Journ. Physiol., 1911, xxviii, 29–48.

30. Laulanié, F. Arch. de physiol., 1894, vi, 845.

31. Benedict and Carpenter. Respiration calorimeters for studying the respiratory exchange and energy transformations of man. Carnegie Institution of Washington Pub. No. 123, 1910.

32. Benedict, F. G. An apparatus for studying the respiratory exchange. Am. Journ. Physiol., 1909, xxiv, 345–374.

33. Durig, A. Ueber Aufnahme und Verbrauch von Sauerstoff bei Aenderung seines Partiardruckes in der Alveolarluft. Archiv. f. Anat. u. Physiol., Physiol. Abth., Suppl , 1903, 209.

34. Benedict and Higgins. Effects on men at rest of breathing oxygen-rich gas mixtures. Am. Journ. Physiol., 1911, xxviii, 1–28.

35. Rubner, M. Die Gesetze des Energieverbrauchs bei der Ernährung. Franz Deuticke, Leipzig u. Wien., 1902.

Respiration apparatus, with absorbing system, for determining the carbon-dioxide production of small animals.

The respiration chamber is in the center, with the cover removed and resting on the floor at the left. On the left wall of the chamber is the shallow pan supporting the rubber diaphragm which indicates the variations in air content. At the right of the chamber is the table on which is placed the absorbing system. On the lower shelf of the table are the motor and the two water absorbers, while on the upper shelf is the double set of carbon-dioxide absorbers and drying vessels, with the two-way valve and wheel valves for deflecting the current of air from one set of absorbers to the other. On the table at the left of the chamber is the apparatus for recording the muscular activity, including a Jaquet clock and tambour with pointer for recording respectively the time and the activity on the smoked drum of the kymograph. The oxygen cylinder with reduction valve is in the rear.

ON THE RELATIVE LOCAL INFLUENCE OF COEXISTING TUBERCULOUS INFLAMMATION AND CANCER IN THE LUNG.*

HORST OERTEL, M.D.

(*From the Russell Sage Institute of Pathology, New York.*)

A considerable literature has accumulated which deals with the coexistence of tuberculosis and cancer in the same individual and in the same organ. While this combination is no longer considered one of great rarity, nevertheless it must be regarded as of infrequent occurrence. Of much practical importance has become the fact that glands in the neighborhood of tumors may not uncommonly show only evidences of tuberculosis — a fact which makes the clinical evidence of enlarged glands of doubtful value in the determination of malignancy. This is, in my experience, particularly the case in glands of the jaw and neck which occur in connection with tumors of the mouth and pharynx, particularly in ulcerating growths. Bastedo cites similar instances in other parts of the body. The discussion of this combination amongst pathologists has centered around their etiological relation. It seems well established that long continued tuberculous lesions may pave the way for the development of cancer. This has been demonstrated satisfactorily in the relationship which lupus has to cancer of the skin. There exist also a few instances in which this seems to have occurred in the lung (Wolf, Friedlander, Lubarsch, Schwalbe), in the larynx (Crone, Pepper and Edsall), and possibly in the gut — the latter, however, not with absolute certainty (Naegeli). More frequently are those cases in which a cancer develops either in an individual with old tuberculosis or in which a tuberculous inflammation is engrafted upon a flourishing carcinomatous disease.

Lubarsch, whose extensive study is generally quoted in this connection, found that the majority of reported cases of

* Received for publication Dec. 4, 1911.

cancer and tuberculosis belong to the first of these two groups. This is then an accidental complication, just as in case of combination with any other disease. The instances of the second group — cancer plus a fresh tuberculous infection — he regards as rare. These develop either from latent tubercle bacilli or constitute an entirely new infection; either instance can be determined only with great difficulty without reliable clinical data.

In the first large group, in which an apparently accidental association of cancer and tuberculosis exists and in distant parts of the body, the reported instances do not show any characteristic influence of one disease upon the other. But even in the cases where they have occurred side by side, or locally associated, detailed accounts of their relative influence and fate are meager.

Lubarsch describes the coexistence in the lung as follows: The patient, a man of twenty-nine years, had suffered from a carcinoma of the testicle. After operation symptoms occurred which were regarded as due to cancer metastases in the lung, but in the sputum were found tubercle bacilli. Autopsy showed numerous cancerous nodules in the lung and pleuræ and fresh cheesy peribronchial and bronchopneumonic foci and even cavities. It was interesting to observe in one and the same microscopic slide and indeed in one field cheesy tubercles with giant cells, and bacilli as well as cancer alveoli. In this case Lubarsch believes that the tuberculous development was undoubtedly favored by the cancerous cachexia   He emphasizes the rarity of such an occurrence.

More definite in this regard is Naegeli in his description of a case of coexisting primary carcinoma of the stomach and primary carcinoma of the ileum, the latter associated with a tuberculous inflammation of that part of the gut. He states: Most interesting were the pictures in places where carcinoma and tuberculosis grew through one another. It showed conditions which so far have not been described. There could be no doubt that the cancer progressed victoriously against the slower developing tuberculous tissue. It advanced in columns into the stroma between the tubercles, avoided constantly the necrotic center which usually contained giant cells, and therefore grew in ring fashion around the tubercle. Behind this latter the cancer columns met, almost like two bounding waves. In later stages the carcinoma had assumed greater dimensions in the form of increasing columns within the stroma between the tubercles, while the tuberculous center remained within cancerous tissue. Not infrequently typical giant cells were still lying between carcinomatous columns; but finally even these disappeared and the whole tubercle had

been replaced by a laminated cancerous tissue.   In analogy to mineralogi-
cal terminology and usage Naegeli speaks of this peculiar process and
picture as " pseudomorphism of cancer after tuberculosis." Its development indicates the resistance which a necrosing tissue offers to an invading
carcinoma where the cancer finds unfavorable conditions for its growth.
Naegeli believes this a constant state of affairs for, in a case of cancer of
the esophagus, a tuberculous bronchial gland, closely attached to the
esophagus, showed an advance of cancer columns into its interior, with
very similar microscopic pictures.

Schwalbe, in describing the development of a carcinoma from the wall
of a tuberculous cavity, saw alveolar cancer within a poorly-staining
necrotic tissue which contained tubercle bacilli.   The boundary between
cancer and tuberculous tissue was not formed by connective tissue.   On
the contrary, cancer was embedded in the tuberculous necrotic parts.
Some cell groups appeared to be torn from the main portion and were at
a distance from the principal focus.   Nowhere, however, could be seen
any changes which pointed towards any necrotic disintegration of the
cancer cells.   Their nuclei always retained good staining quality.   Thus
resulted a more or less intimate association of cancer and tubercle cells.
In the case of Baumgarten (cancer of larynx with a recent but not cheesy
tuberculous granulation) existed also a thorough mixture of cancer and
tuberculous tissue, a " symbiosis " as Baumgarten terms it.   Giant cells
occurred between cancer nests, sometimes with, sometimes without epithelial cells.   This was especially prominent in the portion between the lobes
of the thyroid gland.   The case evidently did not allow very definite conclusions as to the effect of one lesion upon the other, but Baumgarten
considers two possibilities: the tuberculous infection may have restrained
the rapidity of the cancer growth, on the other hand the absence of
destructive and cheesy changes in the tubercle allows one to suspect a
restraining influence of the cancer on the growth of the tuberculous
inflammation.

Warthin in a case of cancer of the breast observed invasion of the
normal acini by the tuberculous process, particularly by giant cells.   The
infiltration surrounded the acini, which were in some places enclosed by a
thickened basement membrane, in others free with collapsed lumen and
closely packed cells.   The giant cells invaded the epithelial cells and by
phagocytic action disposed of them.   But in some of these free epithelial
cells existed a tendency to proliferate, forming large cells with large
nuclei.   These were usually surrounded by tuberculous tissue, but in some
instances it appeared that the former growth had exceeded the latter and
broken through into the gland structure.   Similar pictures were presented
in the ducts.   In a second similar case, reported by Warthin, both lesions
appeared distinctly separated.   The tuberculous process was mostly at
the outside and surrounded the carcinomatous tissue.   But t he cancer
appeared here also more rapid in its growth.

Moack has described several cases of combination of tuberculosis with
cancer.   Of these the third case is of interest in this connection.   A

tuberculous gland became invaded by a scirrhus cancer metastases. This grew between the different tuberculous areas and through them into caseous homogenous centers, forming alveoli. Its ultimate fate there is not given. The tuberculous lesion was regarded as quiescent. In his fifth case existed an adenocarcinoma of the prostate with extensive metastases; ascending infection from bladder; tuberculosis of apices of lung with fresh tuberculous pneumonia; tuberculosis of bronchial glands, and kidneys. Concurrence of tuberculosis with cancer in lungs, liver, spleen and adrenals, bronchial gland and retroperitoneal lymph gland. The microscopic evidence was perivascular metastases in the lung; tubercles lie adjoining the new growth and at times in actual contact. Similar were the conditions in the other organs.

Borst saw in a metastatic cancerous nodule in tuberculosis of the lung the cancer advance into the tuberculous tissue, the cancer progressed to the periphery of the cheesy masses, where it stopped. A similar observation was reported by Nehrkorn in a metastatic epithelioma in a tuberculous lymph gland. The cancer advanced into the tuberculous nodule. Giant cells showed greater resistance and sometimes were entirely surrounded by cancer cells.

In Crone's case (epithelioma of the larynx with tuberculosis, the epithelioma developed in his opinion on a basis of the tuberculous infection) the epithelial proliferation never reached or invaded the deeper tuberculous lesion, but consisted of squamous epithelial cells with isolated nests in the submucosa. The tuberculous inflammation remained below these.

Pepper and Edsall in their report of a case of tuberculous obstruction of the esophagus with what they regard as secondary cancerous development, saw the cancerous change in the portions near the mucous membrane, the base was tuberculous. But there existed partial infiltration of cancer nests by round cells, or there were few epithelial cells within round cells. It is, however, not apparent whether cancer or tubercle successfully invaded or replaced the other.

A different state of affairs from those so far described existed in one of the cases described by Clement of a pylorus carcinoma associated with and entirely overshadowed by an extensive cheesy tuberculosis, with nodular tubercles in the lower and periportal and axillary lymph glands. The pyloric tumor consisted of a mucoid cylindrical cell carcinoma with very cellular stroma. The inflammatory infiltration extended to the muscular and serous coats irrespective of the presence of cancer. At the base of the tumor preponderates necrosis and cellular infiltration, so that little can be seen of any cancer. But even in the submucosa the infiltration had largely obliterated the carcinoma, the alveolar structure was indistinct and became visible only with high power. At the periphery towards healthy tissue was marked inflammation and superficial necrosis of the mucous membrane with vascular granulation tissue and giant cells. In these spots there was only little evidence of any carcinomatous changes, although these were somewhat more evident in the submucosa.

The etiological relation of the two diseases in this instance offers considerable difficulties. Tuberculosis existed in addition to the places mentioned only in the axillary glands. It appears possible that tubercle bacilli gained entrance through the carcinomatous ulcer of the stomach. A second possibility is the congenital presence of tuberculosis which has remained latent but is revived by the carcinomatous changes. In favor of this assumption is, according to Clement, the large solitary tubercle in the liver and the tuberculous axillary lymph gland.

A report of Metterhausen cited by Stetten, and which concerned an operative case only, describes an adenocarcinoma of the cecum combined with tuberculosis. It is mentioned that the tuberculous granulations extended into the carcinoma; the former was supposed to be old, but had been revived by the cancer growth. There existed catarrh of pulmonary apex.

An opportunity was recently offered in this institute to study a case of primary scirrhus cancer of the stomach with extensive metastases in which a combination of the cancerous growth with active tuberculosis existed in the lung. It concerned a woman who had been in the hospital for a considerable time with the clinical diagnosis of cancer of the stomach.

Autopsy protocol (abbreviated). — Body of a very emaciated woman, forty-four years old, one hundred and seventy-three centimeters tall, weight, eighty-one pounds. Skin with cachetic hue, diffuse petechial eruption over arms, legs, and central parts of thorax. Few superficial decubiti over sacral area. There is complete pleural synechia on the right side. Precordial area about normal. Apex of heart in fifth interspace about twelve centimeters to left of median line. Heart weight, two hundred and twenty grams; pericardium intact; valves show no abnormality, mitral slightly thickened. Coronaries intact, yellowish patches in first part of aorta. *Left lung :* Along its anterior border the left lung has a pale gray, slightly mottled appearance, showing marked emphysema. Posteriorly the lower part of the upper lobe and the upper part of the lower lobe show quite extensive consolidation. Over these areas there is no crepitation, but the lung has a distinctly bulky consistence. On section this consolidated area has a solid partly fibrous appearance with numerous small white tubercle-like nodules. Similar small nodules varying in size from a pin's head to a small pea are seen scattered over the adjacent pleura. *Right lung:* Extensive area of consolidation towards the apex. On section this resembles exactly the one described in the left lung, but there are also several smaller areas scattered throughout the middle and lower lobes. The remainder of the lung shows emphysema and edema.

The abdomen contains about one liter of straw-colored fluid. A large, hard, nodular growth occupies the gastric area. To this in places the lower surface of the liver is adherent. The great omentum shows small, hard, yellowish white nodules scattered throughout it and the mesentery of the small intestines. The glands along the attachment of the mesentery are slightly enlarged, hard and shotty, but the abdominal glands are very much enlarged. The peritoneum is smooth and glistening. The tumor itself involved practically the entire stomach, pancreas, great omentum, gastro-hepatic omentum, so as to make these structures almost one indistinguishable mass. On opening the stomach in situ, its walls are almost completely replaced by a thick, nodular, white, stone-hard tumor, which extends from the pyloric to the cardiac end. The surface mucous membrane of the stomach shows considerable ulceration. The stomach is small, its lumen very greatly diminished to almost that of the gut. The pancreas is adherent to the posterior part of the stomach and completely involved in the growth. The involvement of the glands along the aorta is very extensive, forming a tubular mass about the size of a man's arm from the diaphragm to the bifurcation. Scattered over the surface of the peritoneum posteriorly in the pelvis are small yellowish white growths about one centimeter in diameter.

Liver, one thousand three hundred grams, small and soft, shows no metastases, nutmeg and fatty appearance, gall bladder with one large stone.

Both ovaries are slightly enlarged and show a more or less uniform infiltration by a growth very similar to that of the stomach. Spleen, kidneys, and the gut are without important lesions. The uterus shows a hard intramuscular growth in its posterior wall. Anatomical diagnosis: Cancer of the stomach involving the pancreas, the lower surface of the liver and omentum. Metastases in abdominal and mesenteric glands, in the ovaries and lungs and peritoneum. Myoma of uterus.

The microscopic examination of the tumor from different parts showed what had been supposed on macroscopic grounds, a scirrhus cancer. The stroma was very abundant, sometimes thick and dense, sometimes more fibrillar and mucoid in character. Epithelial cells were seen partly in nests or they permeated the stroma in thin, compressed, frequently single columns. Thus they became much flattened, occasionally elongated by force of the narrow channels through which they passed. But generally the cancer cells were small cuboidal, with round, deeply chromatic nuclei. By compression they had become individually poorly outlined; in such places they appeared not infrequently atrophic, their protoplasm became pale, indistinct, and the nucleus showed similar evidences. Doubt was expressed at time of autopsy that the lesion in the lung represented cancer, or only cancer, but a tuberculosis possibly with cancer metastases.

The microscopic examination of the consolidated parts of the lung shows extensive typical cheesy tubercles, with central necrosis, at the periphery numerous giant cells, lymphoid and epithelial cells. Very little had remained of the normal lung structure. Between the typical tubercles are sometimes seen much distorted and compressed alveoli with thickened cellular walls or cellular fibroblastic tissue which has obliterated lung tissue. This has formed in places thick scars. The narrowed alveolar lumen shows in places a serous exudate frequently containing desquamated alveolar epithelium or other inflammatory cells (tuberculous pneumonia). Again the normal alveolar structure is seen to be entirely overgrown by tuberculous granulation tissue containing giant cells, lymphoid and fibroblastic cells, and tubercle bacilli, without cheesy degeneration, but leading to cicatrisation. Within these areas one sees at a glance groups of cells of entirely different character and apparently quite foreign to the tuberculous inflammation. They lie in nests usually somewhat retracted from the wall, in dilated perivascular and tissue lymphatics and occasionally possibly in remnants of alveolar spaces. These foreign cells can be easily recognized by their shape and the tinctorial character of protoplasm and nuclei as the cells previously described in the scirrhus carcinoma of the stomach.

Definite, but as a rule small cancer plugs lie in those parts of the tuberculous granulations where a complete overgrowth of inflammatory cells has not yet occurred; faint remnants of such nests may be seen within cellular tuberculous masses; typical better preserved and large cancer cells may be seen in the lymphatics of scar tissue immediately adjoining cellular portions. In this way has resulted a very close and thorough association of both lesions.

Now the relation of these metastatic cancer masses to the tuberculous inflammation is evident at once; cheesy areas

or those completely obliterated by cells of tuberculous
tissue show no or few indistinct cancer metastases, but they
occur abundantly only in lymphatics and possibly some
remnants of alveolar spaces at the periphery of such masses,
also in the fully formed scar tissue. Most instructive, however, are those portions where the tuberculous granulations
spread to parts containing cancer. Lymphoid and epithelioid cells invade the cancer tissue, become mixed with these
epithelial cells and overgrow it in the same fashion as they
overgrow lung tissue in non-cancerous portions. In such
places the cancer cells are seen to degenerate and disintegrate; their protoplasm becomes pale and vacuolated, and
similar are the nuclear changes. But even in places where
the cancer nests have been embedded in the scar tissue they
show little if any tendency to progress but are seen to rather
undergo atrophy. Nowhere is any evidence that the cancer
cells progress at the expense of the fully-developed scar
tissue or within the developing tuberculous granulations or
within the more exudative pneumonic areas.

It will therefore be seen that in this instance the active
tuberculosis had a strong tendency to eliminate the invading
carcinomatous metastases. The clinical history as well as
the anatomic histological evidence lend support to the view
that the cancer of the stomach was the older of the two
diseases. During the cachectic state an active tuberculosis
of the lung had developed. Two other questions cannot be
answered with equal positiveness, namely, whether there had
not been an old latent tuberculous focus which was set free
by the cancer invasion or whether the local conditions produced by the tuberculous infection created a vulnerable
ground which allowed the cancer metastases to settle? Of
these the last seems very improbable for the reason that the
antagonism which the tuberculous infection seems to exert
in this case upon the cancer tissue, as well as the mechanical
obstruction to cancer progress in the infiltrated lung, would
hardly be compatible with a marked cancer infiltration.
This, moreover, appeared strongest in those portions which
were unaffected by the tuberculous inflammation. But even

the supposition that there existed originally an old focus has not much in its favor, for the tuberculous lesions have generally the type of an active one (exudation, rapid spreading of tuberculous inflammatory cells). It is true that it showed the tendency to cicatrise and in some parts scar formation was complete, but in this connection it must be remembered that the tumor was a scirrhus, and much of the scar tissue belonged to the tumor. There is also a possibility that the same conditions which are responsible for the abundant stroma in the tumor might have exerted some influence on the development of the tuberculous inflammation. Furthermore, no other old tuberculous focus was discovered in the body. In all events this case may be grouped with the second and rare class of Lubarsch — in which an active tuberculosis is engrafted upon a cancerous disease.

A review of the findings of this and the other cited cases shows that the combination of tuberculous inflammation and cancer may exist in three forms: (1) The cancer may dominate, infiltrate the tuberculous granulation tissue and prevent a successful invasion of the tuberculous infection. To this group belong the primary cancer of the ilium of Naegeli, the primary cancer originating from the wall of a tuberculous cavity of Schwalbe, the primary cancer of the breast by Warthin, the metastatic cancers of Naegeli in a bronchial gland in cancer of the esophagus, of Borst's in tuberculous lung, of Nehrkorn's metastatic epithelioma in a tuberculous lymph gland and Moack's metastatic scirrhus in a lymph gland. (2) The tuberculous inflammation overgrows and destroys the cancer; the cases of Clement of cancer of the pylorus with extensive tuberculous involvement, the present case of scirrhus cancer metastases with tuberculosis in the lungs and possibly Metterhausen's case. (3) There exists a close association of the two diseases — "symbiosis" (Baumgarten) without any appreciable destructive influence of one upon the other; Baumgarten's case of cancer of larynx with recent tuberculous infection, Crone's case of epithelioma of the larynx on tuberculous basis and

probably Pepper's and Edsall's case of cancer and tuberculosis of the esophagus.

Other cases in which the combination has been described
probably belong to one of these groups, I should judge
mostly the last, but the meager attention which has been
paid in the descriptions to the points here discussed does
not allow an accurate classification in most of them.

Finally, all observations — excepting Moack's in a quiescent tuberculosis of a gland — agree that caseation puts a
stop to all cancerous invasion due, no doubt, as Borst and
others have pointed out, to unfavorable nutritive conditions.
It cannot be decided whether other more specific factors are
concerned in this.

It is plain from what has been presented, that neither
tuberculosis nor cancer seem to possess specific antagonistic
qualities, but that the results of such combinations depend
upon the conditions of individual cases. Thus it appears
probable that an actively malignant, rapidly growing cancer
may overcome the resistance imposed upon it by an
approaching or even freshly implanted tuberculous inflammation. On the other hand, an actively progressing, virulent
tuberculous infection may overgrow an approaching or
superadded cancer, particularly if the cancer, like the one
observed in this instance (scirrhus) or by Clement (cylindrical cell carcinoma), is not of the irregular cellular, rapidly-
proliferating type. Finally, both lesions may coexist and
intergrow without any marked influence upon one another,
they may be equally balanced in power or grow in symbiosis,
unless by caseation the tuberculous infection removes the
necessary conditions for any growth. In the light of present
knowledge it is not possible to determine whether specific
factors enter into this relation; one would have to consider
in this connection on the one hand the virulence of tubercle
bacilli and toxins with associated productive or exudative
inflammations, on the other hand phagocytic or other properties of the cancer cells.

The tuberculous inflammation does not differ in its effects

from that of other inflammations upon cancer. The types of combinations described above may be observed with other inflammatory changes. Any inflammation may either overgrow the tumor and thereby tend to check cancerous infiltration, or it may become invaded by cancer or frequently coexist in "symbiosis" or unknown relation. Any inflammation may even by the production of necrosis impede the cancer progress. Undoubtedly similar reasons for the various outcomes of these combinations prevail in simple or specific inflammations. They may explain therapeutic effect and defect of such means of treatment in malignant tumors. Finally I consider that these variable factors may be of considerable importance in the development and progress of cancerous disease which originates on the basis of inflammatory processes. For this reason the local relation of inflammatory processes and malignant growths deserves renewed observation and consideration along the lines which have been outlined above.

CITED BIBLIOGRAPHY.

Wolf. Der primäre Lungenkrebs. Fortschritte der Medizin, 1895, 18.

Friedlander. Cancroid in einer Lungencaverne. Fortschritte der Medizin, 1885, 10.

Lubarsch. Uber den primären Krebs des Illiums nebst Bemerkungen über das gleichzeitige Vorkommen von Krebs und Tuberkulose. Virch. Arch , cxi, 1888.

Schwalbe. Entwicklung eines primären Carcinoms in einer tuberculösen Caverne. Virch. Arch., cxlix, 1897.

Crone. Ein Beitrag zur Lehre vom Lupus Carcinom. Tuberkulo-Carcinom des Kehlkopfs. Arbeiten aus dem pathologischen Institut zu Tübingen, ii, 1894-99.

Naegeli. Die Combination von Tuberkulose und Carcinom. Virch. Arch., cxlviii, 1897.

Borst. Die Lehre von den Geschwülsten, Würzburg, ii.

Nehrkorn. Cited by Borst.

Moack. On the occurrence of carcinoma and tuberculosis in the same organ or tissue. Jour. of Med. Research, new series, iii, 1902.

Baumgarten. Ueber ein Kehlkopfcarcinom combinirt mit den histologischen Erscheinungen der Tuberkulose. Arb. aus den pathol. Inst. zu Tübingen, ii, 94-99.

Warthin. Coexistence of carcinoma and tuberculosis of the mammary gland. Amer. Jour. Med. Sci., new series, cxviii, 1899.

Clement. Ueber seltnere Arten der Combination von Krebs und Tuberkulose. Virch. Arch., cxxxix, 1895.

Pepper and Edsall. Tuberculous occlusion of esophagus with partial cancerous infiltration. Amer Jour. Med. Sc., cxiv, 1897.

OTHER LITERATURE.

Bastedo. Association of cancer and tuberculosis. N Y. Med. News, 1904 (Literature).

Stetten. Coexistence of tuberculosis and carcinoma in same portion of intestine. Festschrift zur vierzig jährigen Stiftungsfeier des deutschen Hosp., New York, 1909 (Literature).

Weyeneth. Uber einen Fall von Krebs und Tuberkulose des Oesophagus. Dissertation, Zürich, 1900.

Balduwein. Uber die Verbreitungsweise von Tuberkulose und Carcinom im menschlichen Organismus. Dissertation, Bonn, 1889.

Cohen. Carcinom und Phthise. Dissertation, Köln, 1885.

Zenker. Carcinom und Tuberkulose im selben Organ. Deutsches Arch. f. Kl. Med., 1890, xvii.

DESCRIPTION OF PLATE.

PLATE VI.

FIG. 1. — Zeiss Object., 16 mm., Oc. 4. Cheesy tuberculous masses, tuberculous granulation tissue and cancer nests, partly infiltrated by lymphocytes, or entirely overgrown by tuberculous granulation tissue.

FIG. 2. — Zeiss Object., 3 mm., Oc. 4. In the upper part remnants of cancer nests containing degenerated, vacuolized cancer cells overgrown by tuberculous tissue, at extreme right a somewhat better preserved cancer column but with disintegrating cells and lymphocytes. Below exudate.

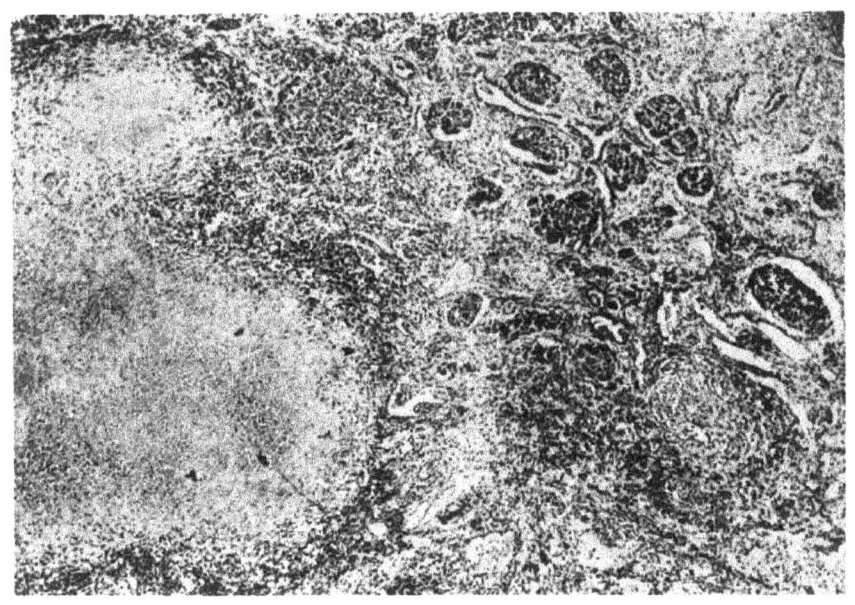

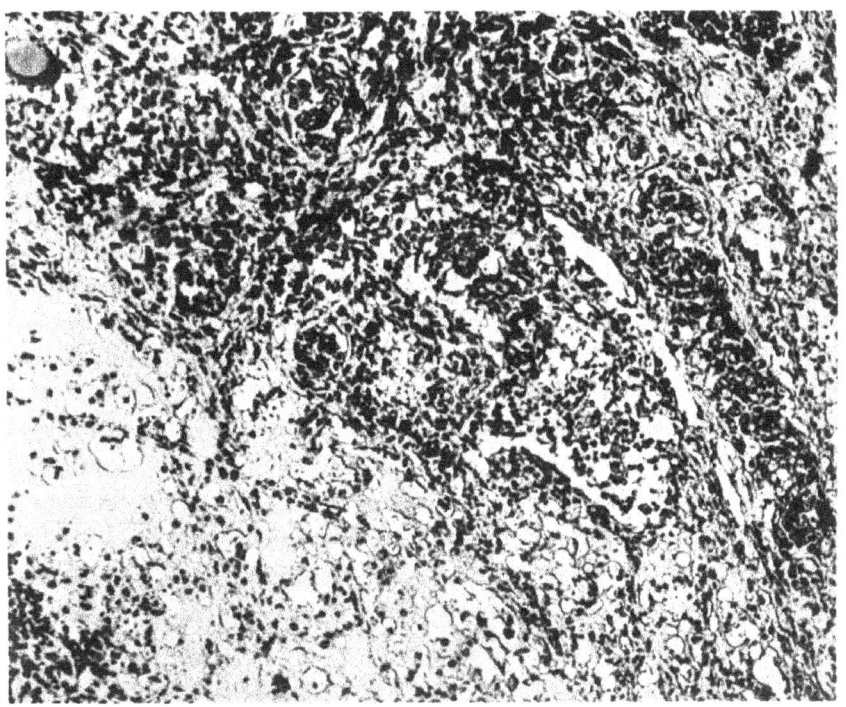

2

Lightning Source UK Ltd.
Milton Keynes UK
UKHW011555040119

334726UK00009B/490/P